Handbook of Data on Organic Compounds

2nd Edition

Supplement I
CAS Number Index

Editors

Robert C. Weast
Jeanette G. Grasselli

CRC Press, Inc.
Boca Raton, Florida

Library of Congress Cataloging-in-Publication Data
(Revised for volume.:Index Volume)

Handbook of data on organic compounds.

Rev. ed. of: CRC handbook of data on organic compounds. c1985.
Includes bibliographical references and indexes.
1. Organic compounds—Handbooks, manuals, etc.
I. Weast, Robert C., 1916- I. Grasselli, Jeanette G.
III. CRC handbook of data on organic compounds.
QD257.7.H36 1989 547 88-7485
ISBN0-8493-0420-2 (set)
ISBN 0-8493-0441-5 (Suppl. I)

Direct all inquiries to CRC Press, Inc., 2000 Corporate Blvd., N.W., Boca Raton, Florida, 33431.

International Standard Book Number 0-8493-0441-5 (Suppl. I)

Library of Congress Card Number 88-7485

Printed in the United States

PREFACE

Supplement I lists CAS numbers for all compounds found in the initial set of volumes of the *Handbook of Data on Organic Compounds,* 2nd Edition (HODOC II). Each CAS number is cross-referenced to the HODOC number of that compound. Consequently, when the CAS number is known, this index facilitates locating the basic entry for the compound.

THE EDITORS

Robert C. Weast retired as Editor-in-Chief of the *CRC Handbook of Chemistry and Physics* in 1989. He edited the handbook continuously since 1952. He received his doctorate degree from the university of Illinois, and in 1945 became a member of the faculty at Case Western Reserve University, Department of Chemistry and Chemical Engineering, Cleveland, Ohio. Within 2 years he was acting head of the department. In 1962, Dr. Weast resigned his professorship to accept a Vice Presidency in research with the Consolidated Natural Gas Service Company, and from 1972 to 1980 was the American Representative on the International Gas Union's Committee on Industrial Gas Utilization. Dr. Weast has since retired from the Consolidated Natural Gas Service Company and is currently Vice Chairman of the Board and the Director of Corporate Planning for Perfection Corporation. He is a Fellow in the American Association for the Advancement of Science, a member of Alpha Chi Sigma (the Chemical professional society), Sigma Xi (the honorary research society), the American Chemical Society, and CODATA.

Jeanette G. Grasselli, is a Distinguished Visiting Professor at Ohio University. She was formerly Director of Research and Analytical Sciences at the Research and Development Department of BP America, Cleveland, Ohio. Dr. Grasselli obtained her B.S. degree in chemistry at Ohio University, her M.S. at Case Western Reserve University, and has received D.Sc. (Hon.) degrees from Ohio University (1978) and Clarkson University (1986). She has 1 patent, 80 publications, and is co-editor of 8 books in the field of molecular spectroscopy. In 1970 she served as the National President of the Society for Applied Spectroscopy. She was active in organizing FACSS (The Federation of Analytical Chemistry and Spectroscopy Societies) and has been Chairman and Secretary of its Governing Board. She is on the Visiting Committee for the National Institute of Science and Technology (NBS), and serves on the Energy Research Advisory Board of the Department of Energy and the Chemical Sciences and Technology Board of the National Research Council. Dr. Grasselli is a Member of the U.S. National Committee for the International Union of Pure and Applied Chemistry (IUPAC), and is Secretary of the Spectroscopy Commission. She has received numerous awards and has been recognized for her many achievements in science, including the Anacham Award from the Detroit Association of Analytical Chemists, the Williams-Wright Award from the Coblentz Society, and the Garvan Medal of the American Chemical Society.

Handbook of Data on Organic Compounds, 2nd edition

Table of Contents — Volumes I to IX

Volume I — Explanations to the Database
Abbreviations
Chemical Abstracts Name Guide
Compounds A — Be
Volume II — Compounds Be — Bi
Volume III — Compounds Bi — D
Volume IV — Compounds E — L
Volume V — Compounds M — Po
Volume VI — Compounds Pr — Z
Volume VII — Synonym Index
Volume VIII — Molecular Formula Index
Molecular Weight Index
Melting Point Index
Boiling Point Index
Infrared Numerical Index of Source Curves
Infrared Spectral Data Index
Infrared Spectral Data Index to Hydrocarbons
Infrared Spectral Data Index to CHO Compounds
Infrared Spectral Data Index to CHN Compounds
Infrared Spectral Data Index to CHNO Compounds
Infrared Spectral Data Index to Halogens
Infrared Spectral Data Index to Boron Compounds
Infrared Spectral Data Index to Phosphorus Compounds
Infrared Spectral Data Index to Silicon Compounds
Infrared Spectral Data Index to Sulfur Compounds
Volume IX — Raman Numerical Index of Source Curves
Raman Spectral Data Index
Ultraviolet Numerical Index of Source Curves
Ultraviolet Spectral Data Index
^{1}H NMR Numerical Index of Source Curves
^{1}H NMR Spectral Data Index
^{1}H NMR Spectral Data Index to Hydrocarbons
^{1}H NMR Spectral Data Index to CHO Compounds
^{1}H NMR Spectral Data Index to CHN Compounds
^{1}H NMR Spectral Data Index to CHNO Compounds
^{1}H NMR Spectral Data Index to Halogen Compounds
^{1}H NMR Spectral Data Index to Boron Compounds
^{1}H NMR Spectral Data Index to Phosphorus Compounds
^{1}H NMR Spectral Data Index to Silicon Compounds
^{1}H NMR Spectral Data Index to Sulfur Compounds
^{13}C NMR Numerical Index of Source Curves
^{13}C NMR Spectral Data Index
Mass Spectra Numerical Index of Source Curves
Mass Spectra Spectral Data Index
Mass Spectrometry: Molecular Weight Index to Bromine Compounds
Mass Spectrometry: Molecular Weight Index to Chlorine Compounds
Mass Spectrometry: Molecular Weight Index to Nitrogen Compounds
Mass Spectrometry: Molecular Weight Index to Sulfur Compounds
Supplement I — CAS Number Index

50-00-0: 12801
50-01-1: 13373
50-02-2: 20314
50-03-3: 20346
50-04-4: 20356
50-06-6: 23206
50-11-3: 23192
50-13-5: 20249
50-14-6: 23938
50-22-6: 20347
50-23-7: 20355
50-27-1: 11673
50-28-2: 11658
50-29-3: 04012
50-30-6: 06150
50-31-7: 06604
50-32-8: 06941
50-34-0: 20488
50-35-1: 15379
50-36-2: 01483
50-43-1: 06606
50-44-2: 22216
50-45-3: 06146
50-50-0: 11661
50-53-3: 19699
50-55-5: 25573
50-63-5: 18286
50-66-8: 22159
50-69-1: 23920
50-70-4: 13182
50-71-5: 23161
50-73-7: 06603
50-74-8: 06587
50-75-9: 06602
50-78-2: 05855
50-79-3: 06149
50-81-7: 01453
50-82-8: 06605
50-84-0: 06147
50-85-1: 06375
50-86-2: 05850
50-89-5: 24911
50-91-9: 25504
51-00-3: 05846
51-06-9: 01808
51-12-7: 22876
51-16-1: 22142
51-17-2: 05606
51-18-3: 24993
51-19-4: 18986
51-20-7: 23128
51-21-8: 23137
51-26-3: 05231
51-28-5: 19332
51-30-9: 04660
51-35-4: 20373
51-36-5: 06155
51-39-8: 06607
51-41-2: 04645
51-43-4: 04658
51-44-5: 06151
51-45-6: 14821
51-48-9: 25255
51-49-0: 25242
51-52-5: 23255
51-55-8: 04232
51-56-9: 04198
51-59-2: 05852
51-63-8: 04806
51-64-9: 04805
51-66-1: 00301
51-67-2: 19019
51-79-6: 09081
51-80-9: 15730
52-31-3: 23188
52-39-1: 20344
52-43-7: 23195
52-49-3: 20274
52-51-7: 20925
52-52-8: 10593
52-67-5: 25531
52-86-8: 08607
52-90-4: 10864
53-03-2: 20316
53-05-4: 20331
53-06-5: 20357
53-16-7: 11676
53-41-8: 01061
53-42-9: 01062
53-46-3: 11739
53-70-3: 11092
53-89-4: 22483
53-96-3: 00247
54-04-6: 04814
54-06-8: 15219
54-11-5: 22701
54-12-6: 25223
54-16-0: 15182
54-25-1: 25006
54-36-4: 21509
54-71-7: 13099
54-85-3: 22872
54-88-6: 02524
54-95-5: 24435
54-96-6: 22881
55-10-7: 04166
55-16-3: 04226
55-18-5: 11721
55-21-0: 01800
55-22-1: 22829
55-43-6: 04894
55-63-0: 21045
55-73-2: 13364
55-81-2: 04801
55-91-4: 19973
55-99-2: 15759
56-03-1: 14945
56-04-2: 23254
56-05-3: 23023
56-08-6: 23169
56-09-7: 23243
56-12-2: 08305
56-18-8: 20753
56-23-5: 15694
56-25-7: 11609
56-29-1: 23186
56-33-7: 11384
56-35-9: 11390
56-36-0: 24213
56-38-2: 19977
56-40-6: 13272
56-41-7: 00996
56-45-1: 23957
56-49-5: 01560
56-53-1: 19250
56-54-2: 09437
56-55-3: 01975
56-57-5: 23690
56-75-7: 00200
56-81-5: 21033
56-82-6: 20388
56-84-8: 01461
56-85-9: 13258
56-86-0: 13246
56-87-1: 15539
56-89-3: 10879
56-92-8: 14822
57-00-1: 13282
57-06-7: 21719
57-08-9: 14202
57-10-3: 13875
57-11-4: 17393
57-13-6: 25355
57-14-7: 14611
57-15-8: 21427
57-24-9: 24258

57-27-2: 16064
57-37-4: 04236
57-41-0: 14855
57-43-2: 23201
57-44-3: 23191
57-47-6: 23437
57-50-1: 13212
57-53-4: 20945
57-55-6: 20889
57-56-7: 14682
57-57-8: 17979
57-62-5: 16145
57-63-6: 17349
57-66-9: 06254
57-68-1: 05319
57-71-6: 08242
57-85-2: 01092
57-87-4: 11623
57-88-5: 09378
57-91-0: 11666
57-97-6: 01983
58-00-4: 11179
58-05-9: 13253
58-08-2: 22207
58-14-0: 23112
58-15-1: 22486
58-18-4: 01091
58-22-0: 01090
58-25-3: 05697
58-27-5: 16815
58-32-2: 12152
58-33-3: 19697
58-39-9: 20120
58-54-8: 00582
58-55-9: 22199
58-56-0: 22908
58-61-7: 00985
58-63-9: 15285
58-72-0: 03368
58-73-1: 11709
58-74-2: 15447
58-85-5: 24612
58-86-6: 25570
58-89-9: 09879
58-90-2: 19609
58-93-5: 06981
58-94-6: 06978
58-95-7: 06813
58-96-8: 25502
58-97-9: 25507
59-00-7: 23747
59-02-9: 06812
59-23-4: 13165
59-26-7: 22807
59-30-3: 13252
59-31-4: 23826
59-40-5: 05337
59-46-1: 05878
59-47-2: 20911
59-48-3: 15273
59-49-4: 07183
59-50-7: 19177
59-51-8: 16049
59-52-9: 21321
59-66-5: 00087
59-67-6: 22828
59-87-0: 14694
59-92-7: 25254
60-00-4: 13309
60-01-5: 08516
60-09-3: 02459
60-10-6: 11079
60-11-7: 02222
60-12-8: 04818
60-15-1: 04804
60-18-4: 25238
60-23-1: 12064
60-24-2: 12187
60-27-5: 14908
60-29-7: 11858
60-32-2: 14203
60-33-3: 17359
60-34-4: 14631
60-35-5: 00053
60-54-8: 16147
60-56-0: 14832
60-57-1: 11251
60-70-8: 25541
60-80-0: 22473
60-82-2: 21502
60-87-7: 19696
60-91-3: 19694
61-00-7: 12374
61-12-1: 23736
61-19-8: 00991
61-25-6: 15448
61-28-9: 15278
61-50-7: 15223
61-54-1: 15222
61-78-9: 13278
61-80-3: 07174
61-82-5: 25033
61-90-5: 15525
61-94-9: 22860
62-23-7: 06523
62-38-4: 15590
62-44-2: 00242
62-47-5: 00072
62-48-6: 00667
62-50-0: 15754
62-53-3: 02014
62-55-5: 12049
62-56-6: 24848
62-57-7: 01011
62-59-9: 09296
62-75-9: 15615
62-90-8: 11681
63-00-3: 01080
63-05-8: 01078
63-25-2: 16937
63-37-6: 10884
63-68-3: 16050
63-74-1: 05303
63-75-2: 22858
63-91-2: 19714
63-98-9: 04106
63-99-0: 25469
64-04-0: 04787
64-10-8: 25483
64-17-5: 12088
64-18-6: 12840
64-19-7: 00444
64-47-1: 23438
64-65-3: 20258
64-67-5: 24290
64-69-7: 00651
64-75-5: 16148
64-77-7: 05345
64-85-7: 20353
64-86-8: 00377
65-19-0: 25575
65-29-2: 11737
65-45-2: 01900
65-46-3: 10881
65-49-6: 05921
65-61-2: 00971
65-64-5: 14651
65-71-4: 23143
65-82-7: 16054
65-85-0: 05838
65-86-1: 23102
66-02-4: 25239
66-22-8: 23115
66-25-1: 13975
66-27-3: 15755
66-28-4: 09269

66-32-0: 24261
66-56-8: 19331
66-71-7: 18956
66-75-1: 23127
66-76-2: 06863
66-77-3: 16589
66-81-9: 20257
66-99-9: 16590
67-20-9: 14867
67-43-6: 13287
67-47-0: 12948
67-51-6: 22404
67-52-7: 23173
67-56-1: 15805
67-63-0: 21293
67-64-1: 21433
67-66-3: 15705
67-68-5: 15689
67-71-0: 15690
67-72-1: 11843
67-97-0: 23936
68-11-1: 00654
68-12-2: 12819
68-19-9: 25543
68-34-8: 05384
68-35-9: 05336
68-41-7: 15500
68-88-2: 12129
68-94-0: 22222
68-96-2: 20352
69-09-0: 19700
69-24-9: 21476
69-65-8: 15570
69-72-7: 06297
69-89-6: 22197
69-91-0: 04122
69-93-2: 22172
70-07-5: 17963
70-11-1: 12305
70-18-8: 13313
70-22-4: 23429
70-23-5: 21087
70-25-7: 13370
70-26-8: 17870
70-29-1: 11874
70-30-4: 19460
70-34-8: 03459
70-47-3: 01455
70-53-1: 15542
70-54-2: 15538
70-55-3: 05378
70-69-9: 21436
70-70-2: 21497
70-78-0: 25258
71-00-1: 14568
71-23-8: 21292
71-30-7: 23227
71-36-3: 08523
71-41-0: 18476
71-43-2: 02529
71-44-3: 08093
71-55-6: 11892
71-63-6: 09263
72-14-0: 05339
72-18-4: 25528
72-19-5: 24907
72-23-1: 20358
72-40-2: 14808
72-43-5: 04013
72-44-6: 23508
72-48-0: 01252
72-54-8: 03078
72-55-9: 03071
72-63-9: 01047
73-05-2: 19254
73-22-3: 25224
73-24-5: 22139
73-31-4: 00293
73-32-5: 15420
73-40-5: 22220
73-49-4: 23478
73-70-1: 23035
73-78-9: 00207
74-11-3: 06058
74-31-7: 04493
74-39-5: 04722
74-69-1: 23037
74-79-3: 01379
74-82-8: 15621
74-83-9: 15624
74-84-0: 11745
74-85-1: 12541
74-86-2: 12636
74-87-3: 15640
74-88-4: 15679
74-89-5: 15599
74-90-8: 14721
74-93-1: 15765
74-95-3: 15654
74-96-4: 11753
74-97-5: 15625
74-98-6: 20489
74-99-7: 22100
75-00-3: 11772
75-01-4: 12549
75-02-5: 12580
75-03-6: 11845
75-04-7: 11682
75-05-8: 00869
75-07-0: 00031
75-08-1: 12063
75-09-2: 15662
75-10-5: 15670
75-11-6: 15672
75-12-7: 12806
75-15-0: 09182
75-17-2: 12805
75-18-3: 15700
75-19-4: 10768
75-21-8: 17987
75-22-9: 07752
75-25-2: 15702
75-26-3: 20500
75-27-4: 15629
75-28-5: 20647
75-29-6: 20517
75-30-9: 20625
75-31-0: 20440
75-33-2: 21013
75-34-3: 11810
75-35-4: 12564
75-36-5: 00928
75-37-6: 11830
75-38-7: 12573
75-39-8: 12090
75-43-4: 15666
75-44-5: 09211
75-45-6: 15641
75-46-7: 15709
75-47-8: 15713
75-50-3: 15605
75-52-5: 15683
75-54-7: 24023
75-55-8: 01532
75-56-9: 18019
75-60-5: 01412
75-61-6: 15658
75-62-7: 15637
75-63-8: 15638
75-64-9: 20480
75-65-0: 21409
75-66-1: 21020
75-69-4: 15706
75-71-8: 15664
75-72-9: 15651
75-73-0: 15695

75-75-2: 15753
75-76-3: 24091
75-77-4: 24010
75-78-5: 24018
75-79-6: 24107
75-80-9: 12243
75-81-0: 11802
75-82-1: 11805
75-83-2: 07983
75-84-3: 21322
75-85-4: 08588
75-86-5: 20985
75-87-6: 00052
75-88-7: 11795
75-89-8: 12249
75-91-2: 14724
75-93-4: 24294
75-94-5: 24103
75-95-6: 11865
75-96-7: 00757
75-97-8: 08645
75-98-9: 21165
75-99-0: 21134
76-00-6: 21426
76-01-7: 11867
76-02-8: 00941
76-03-9: 00760
76-04-0: 00524
76-05-1: 00792
76-06-2: 15708
76-08-4: 21425
76-09-5: 08226
76-11-9: 11882
76-12-0: 11883
76-13-1: 11898
76-14-2: 11823
76-15-3: 11793
76-16-4: 11844
76-17-5: 20726
76-19-7: 20673
76-20-0: 07906
76-24-4: 07700
76-28-8: 09266
76-30-2: 08200
76-36-8: 07824
76-37-9: 21349
76-39-1: 21338
76-43-7: 01087
76-44-8: 15790
76-45-9: 09305
76-51-7: 18965
76-57-3: 16073
76-59-5: 19061
76-60-8: 19063
76-61-9: 19065
76-62-0: 15330
76-65-3: 05823
76-67-5: 20835
76-68-6: 23189
76-74-4: 23200
76-75-5: 23152
76-76-6: 23203
76-80-2: 07711
76-83-5: 02881
76-84-6: 05055
76-86-8: 24011
76-89-1: 04239
76-93-7: 04235
76-94-8: 23218
76-99-3: 13708
77-02-1: 23216
77-03-2: 20254
77-04-3: 22911
77-06-5: 13179
77-09-8: 15334
77-10-1: 20201
77-16-7: 25184
77-25-8: 20817
77-26-9: 23219
77-28-1: 23184
77-30-5: 23198
77-40-7: 19525
77-41-8: 23390
77-42-9: 18725
77-46-3: 00373
77-47-4: 10484
77-48-5: 14849
77-49-6: 20942
77-50-9: 04841
77-52-1: 25509
77-53-2: 15800
77-55-4: 10605
77-59-8: 24199
77-60-1: 24202
77-63-4: 10851
77-65-6: 07830
77-67-8: 23393
77-71-4: 14852
77-74-7: 18530
77-75-8: 18797
77-76-9: 20582
77-77-0: 12590
77-78-1: 24291
77-79-2: 24709
77-83-8: 18033
77-84-9: 20936
77-85-0: 20939
77-86-1: 20920
77-89-4: 21025
77-90-7: 21024
77-92-9: 21026
77-93-0: 21028
77-94-1: 21027
77-99-6: 20935
78-04-6: 11305
78-07-9: 24122
78-08-0: 24057
78-09-1: 11849
78-10-4: 24170
78-11-5: 20924
78-13-7: 24171
78-26-2: 20944
78-27-3: 10144
78-30-8: 19937
78-32-0: 19939
78-33-1: 19292
78-39-7: 11899
78-40-0: 19928
78-42-2: 19935
78-43-3: 21316
78-45-5: 19807
78-50-2: 19763
78-54-6: 10143
78-59-1: 10433
78-62-6: 24033
78-67-1: 20965
78-74-0: 11891
78-75-1: 20543
78-77-3: 20512
78-78-4: 08017
78-79-5: 07799
78-80-8: 08984
78-81-9: 20459
78-82-0: 20991
78-83-1: 21334
78-84-2: 20396
78-85-3: 21629
78-87-5: 20558
78-88-6: 21699
78-89-7: 21303
78-90-0: 20746
78-92-2: 08524
78-93-3: 08603
78-94-4: 08946
78-95-5: 21537
78-97-7: 20982

78-98-8: 20399
78-99-9: 20556
79-00-5: 11893
79-01-6: 12599
79-02-7: 00035
79-03-8: 21595
79-04-9: 00930
79-05-0: 20403
79-06-1: 21638
79-07-2: 00122
79-08-3: 00482
79-09-4: 21058
79-10-7: 21772
79-14-1: 00637
79-15-2: 00101
79-16-3: 00303
79-17-4: 14703
79-19-6: 14662
79-20-9: 00678
79-21-0: 12034
79-22-1: 09228
79-24-3: 11857
79-27-6: 11879
79-28-7: 12591
79-29-8: 07984
79-30-1: 21607
79-31-2: 21223
79-33-4: 21180
79-34-5: 11881
79-35-6: 12567
79-36-7: 00933
79-37-8: 12026
79-38-9: 12558
79-39-0: 21651
79-40-3: 12029
79-41-4: 21886
79-42-5: 21211
79-43-6: 00570
79-44-7: 09125
79-46-9: 20671
79-49-2: 21575
79-54-9: 18914
79-55-0: 20196
79-57-2: 16146
79-58-3: 24180
79-63-0: 15521
79-69-6: 08972
79-70-9: 08971
79-72-1: 05329
79-74-3: 04743
79-77-6: 08974
79-83-4: 01001
79-89-0: 08966
79-94-7: 19467
79-96-9: 19468
79-97-0: 19469
80-00-2: 02932
80-05-7: 19466
80-06-8: 05007
80-07-9: 03923
80-08-0: 02492
80-09-1: 19602
80-10-4: 24019
80-11-5: 05363
80-13-7: 06156
80-14-8: 25206
80-15-9: 14727
80-18-2: 05456
80-26-2: 10395
80-28-4: 05380
80-32-0: 05312
80-33-1: 05424
80-35-3: 05323
80-39-7: 05368
80-40-0: 05463
80-41-1: 12121
80-42-2: 05482
80-44-4: 05422
80-46-6: 19328
80-48-8: 05464
80-49-9: 01543
80-55-7: 21204
80-58-0: 08328
80-59-1: 08901
80-62-6: 21905
80-63-7: 21798
80-68-2: 24906
80-69-3: 20843
80-70-6: 13380
80-71-7: 10752
80-72-8: 10749
80-75-1: 20351
80-82-0: 05475
80-89-7: 20325
80-90-0: 20326
80-91-1: 20323
80-92-2: 20324
80-93-3: 20342
80-97-7: 09359
81-04-9: 16823
81-05-0: 16870
81-07-2: 05655
81-08-3: 07160
81-10-7: 05320
81-13-0: 07848
81-14-1: 12375
81-15-2: 03224
81-16-3: 16867
81-20-9: 03254
81-23-2: 09347
81-24-3: 12042
81-25-4: 09345
81-30-1: 06805
81-38-9: 05678
81-45-8: 01838
81-46-9: 01810
81-49-2: 01205
81-50-5: 01200
81-51-6: 01206
81-54-9: 01315
81-55-0: 01263
81-60-7: 01313
81-61-8: 01309
81-62-9: 01198
81-64-1: 01254
81-68-5: 05316
81-77-6: 01360
81-81-2: 06852
81-83-4: 05654
81-84-5: 17041
81-86-7: 17043
81-90-3: 05984
81-92-5: 04987
82-02-0: 13126
82-07-5: 25546
82-09-7: 07712
82-12-2: 01292
82-27-9: 01203
82-28-0: 01214
82-29-1: 01317
82-33-7: 01241
82-34-8: 01302
82-35-9: 01284
82-38-2: 01299
82-39-3: 01296
82-40-6: 13152
82-43-9: 01248
82-44-0: 01224
82-45-1: 01193
82-46-2: 01245
82-48-4: 01325
82-49-5: 01332
82-50-8: 01334
82-54-2: 11340
82-57-5: 13127
82-58-6: 11614

82-62-2: 19654
82-64-4: 15931
82-66-6: 15061
82-68-8: 03861
82-69-9: 15934
82-71-3: 04740
82-75-7: 16872
82-86-0: 00024
83-01-2: 09126
83-05-6: 04154
83-07-8: 22469
83-12-5: 15065
83-13-6: 20872
83-14-7: 15480
83-15-8: 00212
83-25-0: 23402
83-27-2: 20859
83-30-7: 06614
83-32-9: 00014
83-33-0: 15083
83-34-1: 15155
83-38-5: 01629
83-40-9: 06373
83-41-0: 03252
83-42-1: 02892
83-44-3: 09337
83-46-5: 24255
83-47-6: 24256
83-48-7: 24249
83-49-8: 09338
83-53-4: 16324
83-54-5: 23848
83-55-6: 16904
83-56-7: 16732
83-60-3: 25576
83-64-7: 16874
83-66-9: 03232
83-67-0: 22201
83-68-1: 16707
83-72-7: 16807
83-73-8: 23813
83-74-9: 14741
83-79-4: 06806
83-81-8: 04522
83-88-5: 23919
83-95-4: 13155
84-11-7: 18925
84-15-1: 24312
84-16-2: 19247
84-17-3: 19252
84-26-4: 15283
84-31-1: 09439
84-45-7: 01249
84-46-8: 01204
84-47-9: 01282
84-48-0: 01336
84-49-1: 01328
84-50-4: 01327
84-51-5: 01288
84-52-6: 10883
84-54-8: 01298
84-55-9: 21475
84-58-2: 09690
84-60-6: 01260
84-61-7: 04554
84-62-8: 04565
84-65-1: 01191
84-66-2: 04557
84-67-3: 07607
84-68-4: 07603
84-69-5: 04540
84-74-2: 04549
84-75-3: 04558
84-76-4: 04562
84-77-5: 04555
84-79-7: 16811
84-80-0: 16816
84-83-3: 00036
84-85-5: 16927
84-86-6: 16868
84-87-7: 16877
84-88-8: 23768
84-89-9: 16869
84-95-7: 16161
84-99-1: 05672
85-01-8: 18828
85-02-9: 06959
85-06-3: 06963
85-18-7: 22195
85-19-8: 15883
85-23-4: 09719
85-28-9: 15895
85-29-0: 15889
85-34-7: 04296
85-36-9: 05854
85-38-1: 06391
85-41-6: 15363
85-42-7: 15317
85-44-9: 15307
85-45-0: 02327
85-46-1: 16889
85-47-2: 16865
85-52-9: 05962
85-55-2: 06482
85-56-3: 06073
85-57-4: 06336
85-63-2: 15450
85-64-3: 19437
85-66-5: 18975
85-73-4: 06591
85-74-5: 16831
85-75-6: 16830
85-81-4: 23670
85-82-5: 16982
85-83-6: 16992
85-84-7: 16232
85-86-9: 17002
85-90-5: 06933
85-91-6: 06477
85-95-0: 19253
85-97-2: 07669
85-98-3: 25418
86-00-0: 07534
86-21-5: 22931
86-26-0: 07519
86-28-2: 09166
86-29-3: 04346
86-30-6: 02441
86-34-0: 23399
86-38-4: 00963
86-39-5: 24901
86-47-5: 23751
86-48-6: 16676
86-51-1: 01652
86-52-2: 16287
86-53-3: 16581
86-55-5: 16611
86-56-6: 16165
86-57-7: 16465
86-59-9: 23744
86-60-2: 16871
86-65-7: 16826
86-68-0: 23755
86-73-7: 12718
86-74-8: 09159
86-75-9: 23807
86-76-0: 11129
86-78-2: 18290
86-80-6: 11688
86-81-7: 01794
86-84-0: 16422
86-86-2: 16568
86-87-3: 16570
86-88-4: 24886
86-89-5: 16478
86-93-1: 24433

86-95-3: 23842
86-96-4: 23468
86-97-5: 16961
86-98-6: 23617
86-99-7: 23784
87-00-3: 04197
87-05-8: 06838
87-10-5: 01858
87-13-8: 20827
87-17-2: 01914
87-19-4: 06312
87-20-7: 06309
87-22-9: 06320
87-24-1: 06460
87-25-2: 05866
87-26-3: 19448
87-28-5: 06306
87-31-0: 07159
87-32-1: 25225
87-39-8: 23162
87-40-1: 04018
87-41-2: 15328
87-42-3: 22153
87-44-5: 07424
87-47-8: 09067
87-48-9: 15206
87-51-4: 15180
87-52-5: 15228
87-58-1: 23317
87-59-2: 02181
87-60-5: 02095
87-61-6: 04002
87-62-7: 02185
87-64-9: 19171
87-65-0: 19225
87-66-1: 05581
87-68-3: 07796
87-69-4: 08134
87-73-0: 13181
87-79-6: 24184
87-81-0: 24303
87-82-1: 03501
87-83-2: 03857
87-85-4: 03513
87-86-5: 19549
87-87-6: 04768
87-88-7: 09715
87-90-1: 24998
87-99-0: 25567
88-02-8: 05566
88-03-9: 05591
88-04-0: 19154
88-05-1: 02520
88-06-2: 19642
88-09-5: 08406
88-11-9: 09141
88-12-0: 23422
88-13-1: 24801
88-14-2: 12959
88-15-3: 12519
88-16-4: 02956
88-17-5: 02512
88-18-6: 19289
88-19-7: 05376
88-20-0: 05457
88-21-1: 05396
88-22-2: 05403
88-24-4: 19458
88-27-7: 19281
88-32-4: 19301
88-44-8: 05414
88-58-4: 04741
88-60-8: 19303
88-65-3: 05986
88-66-4: 02790
88-67-5: 06413
88-68-6: 01802
88-69-7: 19462
88-72-2: 03710
88-73-3: 02909
88-74-4: 02422
88-75-5: 19535
88-82-4: 06623
88-87-9: 19164
88-88-0: 02961
88-89-1: 19672
88-95-9: 04512
88-96-0: 04519
88-97-1: 05892
88-98-2: 10381
88-99-3: 04526
89-00-9: 22887
89-01-0: 22366
89-02-1: 05441
89-05-4: 05531
89-20-3: 04548
89-21-4: 02864
89-24-7: 14868
89-25-8: 22481
89-31-6: 15475
89-32-7: 05688
89-36-1: 05438
89-37-2: 09135
89-39-4: 03185
89-40-7: 15396
89-41-8: 06455
89-50-9: 06270
89-51-0: 04143
89-52-1: 05843
89-54-3: 05899
89-55-4: 06015
89-56-5: 06377
89-57-6: 05926
89-58-7: 03256
89-59-8: 02902
89-60-1: 02898
89-61-2: 03107
89-62-3: 02394
89-63-4: 02109
89-64-5: 19189
89-68-9: 19179
89-69-0: 04031
89-71-4: 06461
89-72-5: 19521
89-73-6: 01870
89-74-7: 12378
89-75-8: 07205
89-77-0: 05896
89-79-2: 10166
89-82-7: 10273
89-83-8: 19481
89-84-9: 12360
89-86-1: 06173
89-87-2: 03257
89-92-9: 02684
89-93-0: 04931
89-95-2: 05109
89-96-3: 02827
89-97-4: 04891
89-98-5: 01600
89-99-6: 04918
90-00-6: 19356
90-01-7: 05079
90-02-8: 01693
90-04-0: 02312
90-05-1: 19402
90-11-9: 16250
90-12-0: 16435
90-13-1: 16281
90-14-2: 16420
90-15-3: 16901
90-19-7: 06909
90-20-0: 16832
90-24-4: 12402
90-26-6: 04109
90-27-7: 04177

90-30-2: 16196
90-33-5: 06851
90-39-1: 15779
90-41-5: 07557
90-42-6: 07410
90-43-7: 07662
90-44-8: 01347
90-45-9: 00959
90-46-0: 25548
90-47-1: 25551
90-52-8: 23542
90-53-9: 04167
90-59-5: 01625
90-60-8: 01636
90-65-3: 13960
90-68-6: 19076
90-69-7: 12436
90-74-4: 13237
90-81-3: 05115
90-84-6: 21467
90-86-8: 05129
90-87-9: 03178
90-89-1: 20103
90-90-4: 15878
90-93-7: 15852
90-94-8: 15859
90-97-1: 05005
90-98-2: 15851
90-99-3: 02871
91-00-9: 04946
91-01-0: 05142
91-02-1: 16034
91-04-3: 04638
91-10-1: 19259
91-13-4: 02548
91-14-5: 03129
91-15-6: 04507
91-16-7: 03166
91-17-8: 16308
91-18-9: 22128
91-19-0: 23858
91-20-3: 16238
91-21-4: 15465
91-22-5: 23546
91-23-6: 03604
91-40-7: 06549
91-44-1: 06827
91-49-6: 00120
91-52-1: 06189
91-53-2: 23644
91-55-4: 15137
91-56-5: 15204
91-57-6: 16436
91-58-7: 16282
91-59-8: 16153
91-60-1: 16897
91-61-2: 23707
91-62-3: 23676
91-63-4: 23672
91-64-5: 06817
91-65-6: 09770
91-66-7: 02153
91-68-9: 19244
91-73-6: 04951
91-76-9: 25020
91-77-0: 25027
91-78-1: 24984
91-80-5: 11948
91-81-6: 11946
91-84-9: 11953
91-88-3: 12172
91-96-3: 07853
91-99-6: 12208
92-03-5: 07678
92-05-7: 07636
92-06-8: 24313
92-07-9: 22638
92-13-7: 13097
92-24-0: 16139
92-29-5: 15726
92-35-3: 23277
92-36-4: 02355
92-39-7: 19678
92-40-0: 16886
92-41-1: 16825
92-43-3: 22460
92-44-4: 16736
92-46-6: 23591
92-48-8: 06860
92-50-2: 12174
92-52-4: 07445
92-53-5: 16116
92-54-6: 20087
92-55-7: 15738
92-59-1: 04916
92-61-5: 06850
92-62-6: 00970
92-66-0: 07451
92-67-1: 07559
92-68-2: 07411
92-69-3: 07664
92-70-6: 16678
92-71-7: 17936
92-75-1: 16605
92-81-9: 00964
92-82-0: 18968
92-83-1: 25544
92-84-2: 19676
92-85-3: 24474
92-86-4: 07466
92-87-5: 07602
92-88-6: 07637
92-90-0: 07852
92-91-1: 12281
92-92-2: 07592
92-93-3: 07536
92-94-4: 24314
92-99-9: 23537
93-01-6: 16885
93-02-7: 01654
93-03-8: 05036
93-04-9: 16430
93-05-0: 04479
93-06-1: 02883
93-07-2: 06194
93-08-3: 12485
93-09-4: 16612
93-10-7: 23738
93-11-8: 16890
93-12-9: 16888
93-14-1: 20909
93-15-2: 03193
93-16-3: 03191
93-17-4: 04317
93-18-5: 16396
93-22-1: 16479
93-25-4: 04245
93-26-5: 00299
93-27-6: 00295
93-28-7: 19433
93-35-6: 06848
93-36-7: 16963
93-37-8: 23632
93-39-0: 06841
93-40-3: 04169
93-42-5: 00287
93-44-7: 16965
93-45-8: 19532
93-46-9: 04491
93-48-1: 04907
93-50-5: 02090
93-51-6: 19410
93-52-7: 03013
93-53-8: 04090
93-54-9: 05062
93-55-0: 21521

93-56-1: 12008
93-58-3: 06457
93-59-4: 04382
93-60-7: 22852
93-61-8: 12829
93-67-4: 02115
93-68-5: 07861
93-69-6: 14947
93-70-9: 07843
93-71-0: 00150
93-76-5: 00783
93-78-7: 00787
93-79-8: 00785
93-88-9: 04797
93-89-0: 06266
93-90-3: 12205
93-91-4: 08231
93-92-5: 05108
93-97-0: 05955
93-98-1: 01952
93-99-2: 06548
94-01-9: 04692
94-02-0: 05246
94-05-3: 21816
94-06-4: 19449
94-07-5: 05089
94-08-6: 06471
94-09-7: 05880
94-11-1: 00591
94-12-2: 05885
94-13-3: 06334
94-14-4: 05883
94-18-8: 06333
94-20-2: 05352
94-25-7: 05877
94-26-8: 06329
94-28-0: 14227
94-30-4: 06446
94-32-6: 06031
94-33-7: 12006
94-36-0: 18813
94-37-1: 20160
94-41-7: 22008
94-44-0: 22855
94-45-1: 07039
94-46-2: 08555
94-48-4: 17547
94-49-5: 11995
94-52-0: 05627
94-53-1: 05747
94-58-6: 05739
94-59-7: 05738
94-60-0: 10046
94-62-2: 20134
94-65-5: 10281
94-67-7: 01695
94-68-8: 02274
94-69-9: 12826
94-70-2: 02257
94-71-3: 19353
94-74-6: 00536
94-75-7: 00585
94-80-4: 00588
94-81-5: 08361
94-93-9: 19345
94-96-2: 14162
94-97-3: 07148
94-98-4: 04906
94-99-5: 03049
95-00-1: 04899
95-01-2: 01644
95-03-4: 02091
95-05-6: 24652
95-08-9: 08407
95-11-4: 07373
95-12-5: 07383
95-13-6: 14990
95-14-7: 07146
95-15-8: 07091
95-16-9: 07046
95-20-5: 15154
95-21-6: 07179
95-24-9: 07036
95-26-1: 07055
95-44-3: 07827
95-45-4: 08240
95-46-5: 02667
95-47-6: 03196
95-48-7: 19440
95-49-8: 02866
95-50-1: 03037
95-51-2: 02065
95-52-3: 03470
95-53-4: 02347
95-54-5: 04420
95-55-6: 18985
95-56-7: 19091
95-57-8: 19137
95-58-9: 22319
95-59-0: 11274
95-63-6: 04055
95-64-7: 02186
95-65-8: 19270
95-66-9: 02793
95-68-1: 02182
95-69-2: 02098
95-70-5: 04496
95-71-6: 04759
95-72-7: 02796
95-73-8: 03093
95-74-9: 02096
95-75-0: 03089
95-76-1: 02143
95-77-2: 19226
95-78-3: 02184
95-79-4: 02101
95-80-7: 04466
95-81-8: 02093
95-82-9: 02141
95-83-0: 04424
95-84-1: 19022
95-85-2: 18992
95-86-3: 19201
95-87-4: 19265
95-88-5: 04688
95-89-6: 22318
95-92-1: 11982
95-93-2: 03970
95-94-3: 03940
95-95-4: 19641
95-99-8: 13308
96-00-4: 09601
96-01-5: 09597
96-02-6: 13025
96-04-8: 13592
96-05-9: 21916
96-06-0: 18004
96-09-3: 18023
96-11-7: 20713
96-12-8: 20548
96-14-0: 18258
96-18-4: 20720
96-19-5: 21742
96-21-9: 21377
96-22-0: 18535
96-23-1: 21380
96-24-2: 20895
96-26-4: 21544
96-27-5: 20907
96-29-7: 08668
96-31-1: 25422
96-32-2: 00487
96-33-3: 21885
96-35-5: 00640
96-37-7: 10546
96-38-8: 10485

96-40-2: 10701
96-41-3: 10644
96-43-5: 24690
96-45-7: 14879
96-47-9: 12929
96-48-0: 13070
96-49-1: 11331
96-50-4: 24483
96-53-7: 24555
96-54-8: 23306
96-68-4: 05452
96-70-8: 19299
96-71-9: 05471
96-73-1: 05427
96-76-4: 19070
96-77-5: 04780
96-80-0: 12104
96-82-2: 13190
96-86-6: 20795
96-88-8: 20236
96-91-3: 19013
96-94-6: 19084
96-96-8: 02334
96-97-9: 06393
96-98-0: 06499
96-99-1: 06112
97-00-7: 02801
97-02-9: 02230
97-05-2: 06408
97-08-5: 05501
97-13-2: 11970
97-16-5: 19223
97-17-6: 19975
97-18-7: 19626
97-23-4: 19457
97-24-5: 19625
97-25-6: 13229
97-30-3: 13217
97-35-8: 05315
97-39-2: 13360
97-41-6: 10805
97-50-7: 02074
97-51-8: 01727
97-52-9: 02328
97-53-0: 19432
97-54-1: 19427
97-56-3: 02380
97-59-6: 25434
97-60-9: 00269
97-62-1: 21233
97-63-2: 21897
97-65-4: 08184
97-67-6: 08167
97-72-3: 21224
97-73-4: 21181
97-74-5: 24653
97-77-8: 24664
97-78-9: 13329
97-85-8: 21243
97-86-9: 21910
97-87-0: 21225
97-88-1: 21889
97-90-5: 21894
97-95-0: 08544
97-96-1: 07816
97-97-2: 11776
97-99-4: 13051
98-00-0: 13044
98-01-1: 12943
98-02-2: 13043
98-03-3: 24791
98-05-5: 01440
98-06-6: 03222
98-07-7: 04019
98-08-8: 04040
98-09-9: 05493
98-10-2: 05300
98-11-3: 05395
98-13-5: 24116
98-14-6: 01433
98-16-8: 02513
98-17-9: 19658
98-18-0: 05302
98-19-1: 03223
98-27-1: 19306
98-28-2: 19158
98-29-3: 04657
98-31-7: 05503
98-37-3: 05412
98-46-4: 03822
98-47-5: 05476
98-50-0: 01425
98-51-1: 03235
98-52-2: 10132
98-53-3: 10241
98-54-4: 19291
98-56-6: 02957
98-58-8: 05499
98-59-9: 05515
98-60-2: 05500
98-61-3: 05511
98-62-4: 05525
98-64-6: 05349
98-66-8: 05423
98-67-9: 05451
98-68-0: 05513
98-71-5: 05449
98-72-6: 01439
98-73-7: 06228
98-74-8: 05518
98-78-2: 10603
98-79-3: 20367
98-80-6: 07767
98-81-7: 02627
98-82-8: 03669
98-83-9: 03664
98-85-1: 05104
98-86-2: 12501
98-87-3: 03087
98-88-4: 07192
98-89-5: 10008
98-91-9: 04388
98-92-0: 22796
98-94-2: 09772
98-95-3: 03795
98-96-4: 22360
98-97-5: 22364
98-98-6: 22827
99-02-5: 12326
99-03-6: 12266
99-04-7: 06463
99-05-8: 05873
99-06-9: 06324
99-07-0: 19277
99-08-1: 03711
99-09-2: 02423
99-10-5: 06178
99-11-6: 22867
99-12-7: 03255
99-14-9: 21023
99-15-0: 15530
99-17-2: 13221
99-18-3: 04325
99-20-7: 13215
99-24-1: 06620
99-27-4: 04589
99-28-5: 19219
99-30-9: 02149
99-31-0: 04588
99-32-1: 22260
99-33-2: 07211
99-34-3: 06235
99-35-4: 04075
99-36-5: 06467
99-40-1: 12314
99-45-6: 12368

99-48-9: 10401
99-50-3: 06177
99-51-4: 03253
99-52-5: 02388
99-53-6: 19496
99-54-7: 03104
99-55-8: 02389
99-56-9: 04446
99-57-0: 19038
99-59-2: 02329
99-60-5: 06104
99-61-6: 01761
99-62-7: 02569
99-63-8: 04513
99-64-9: 06218
99-65-0: 03275
99-71-8: 19522
99-73-0: 12292
99-75-2: 06473
99-76-3: 06331
99-77-4: 06527
99-82-1: 09910
99-84-3: 10330
99-85-4: 09687
99-86-5: 09681
99-87-6: 03696
99-88-7: 02367
99-89-8: 19464
99-90-1: 12304
99-91-2: 12327
99-92-3: 12267
99-93-4: 12433
99-94-5: 06469
99-96-7: 06328
99-97-8: 02523
99-98-9: 04486
99-99-0: 03712
100-00-5: 02911
100-01-6: 02424
100-02-7: 19537
100-03-8: 05291
100-06-1: 12452
100-07-2: 07223
100-09-4: 06443
100-10-7: 01665
100-11-8: 02694
100-12-9: 03428
100-13-0: 03364
100-14-1: 02897
100-15-2: 02399
100-16-3: 14649
100-17-4: 03606
100-18-5: 02570
100-19-6: 12491
100-20-9: 04514
100-21-0: 04528
100-22-1: 04505
100-23-2: 02216
100-25-4: 03276
100-26-5: 22889
100-27-6: 04862
100-28-7: 03565
100-29-8: 03388
100-32-3: 11405
100-36-7: 11935
100-37-8: 12140
100-39-0: 02666
100-40-3: 10315
100-41-4: 03394
100-42-5: 03336
100-43-6: 22644
100-44-7: 02865
100-45-8: 10360
100-46-9: 04879
100-47-0: 06659
100-48-1: 22771
100-49-2: 10083
100-50-5: 10363
100-51-6: 04965
100-52-7: 01567
100-53-8: 04960
100-54-9: 22770
100-55-0: 22925
100-60-7: 09788
100-61-8: 02353
100-63-0: 14650
100-64-1: 10277
100-65-2: 02291
100-66-3: 03573
100-68-5: 03783
100-69-6: 22642
100-70-9: 22769
100-71-0: 22648
100-72-1: 22264
100-74-3: 16101
100-75-4: 20193
100-76-5: 01476
100-79-8: 11325
100-80-1: 03360
100-81-2: 04932
100-82-3: 04919
100-83-4: 01696
100-84-5: 03595
100-86-7: 04839
100-88-9: 24273
100-97-0: 24326
101-01-9: 13382
101-02-0: 19993
101-14-4: 02360
101-17-7: 02116
101-18-8: 19558
101-21-3: 09056
101-31-5: 04219
101-38-2: 09760
101-39-3: 21630
101-40-6: 10077
101-41-7: 04258
101-43-9: 21890
101-48-4: 03169
101-53-1: 19577
101-54-2: 04502
101-55-3: 02712
101-57-5: 05478
101-58-6: 03852
101-60-0: 20307
101-61-1: 02361
101-63-3: 03850
101-67-7: 02443
101-70-2: 02319
101-72-4: 04498
101-75-7: 02475
101-76-8: 03649
101-77-9: 02359
101-80-4: 02444
101-81-5: 03646
101-82-6: 22724
101-83-7: 09768
101-84-8: 03834
101-90-6: 18024
101-91-7: 07857
101-93-9: 12082
101-94-0: 04271
101-97-3: 04180
101-98-4: 12209
101-99-5: 09104
102-01-2: 07863
102-02-3: 14948
102-04-5: 21548
102-05-6: 04942
102-06-7: 13365
102-07-8: 25435
102-08-9: 24875
102-09-0: 09202
102-13-6: 04273
102-14-7: 08501
102-15-8: 00650

102-20-5: 04289
102-24-9: 07770
102-25-0: 04037
102-27-2: 02275
102-28-3: 00077
102-29-4: 04717
102-32-9: 04165
102-39-6: 00729
102-45-4: 10633
102-47-6: 03046
102-50-1: 02323
102-51-2: 04441
102-52-3: 20704
102-53-4: 15728
102-54-5: 12647
102-56-7: 02176
102-60-3: 21395
102-69-2: 20451
102-70-5: 21662
102-71-6: 12218
102-76-1: 21043
102-79-4: 12114
102-80-7: 11963
102-81-8: 12134
102-82-9: 07871
102-84-1: 19994
102-85-2: 19990
102-86-3: 13984
102-87-4: 11482
102-92-1: 22037
102-93-2: 05172
102-94-3: 21957
102-97-6: 04936
103-01-5: 13336
103-02-6: 21580
103-03-7: 14696
103-04-8: 00736
103-05-9: 05266
103-08-2: 17268
103-09-3: 00621
103-11-7: 21850
103-14-0: 19578
103-16-2: 19575
103-19-5: 11401
103-23-1: 14117
103-24-2: 17221
103-25-3: 05241
103-28-6: 21247
103-29-7: 03300
103-30-0: 03320
103-32-2: 04947
103-34-4: 16098
103-37-7: 08512
103-38-8: 08464
103-40-2: 08192
103-43-5: 08114
103-44-6: 13519
103-45-7: 00730
103-46-8: 00735
103-48-0: 21246
103-49-1: 04949
103-50-4: 03846
103-52-6: 08511
103-54-8: 21991
103-55-9: 11945
103-59-3: 21248
103-61-7: 08515
103-63-9: 02645
103-65-1: 03906
103-67-3: 04934
103-69-5: 02269
103-70-8: 12834
103-71-9: 03556
103-72-0: 03567
103-73-1: 03369
103-74-2: 22914
103-75-3: 22237
103-76-4: 20119
103-79-7: 21579
103-80-0: 04349
103-81-1: 04104
103-82-2: 04118
103-83-3: 04908
103-84-4: 00359
103-85-5: 24889
103-86-6: 19049
103-88-8: 00110
103-89-9: 00319
103-90-2: 00276
103-95-7: 05170
103-96-8: 04474
103-99-1: 17371
104-01-8: 04247
104-03-0: 04277
104-04-1: 00349
104-06-3: 00090
104-07-4: 17401
104-09-6: 04093
104-10-9: 04822
104-12-1: 02853
104-13-2: 02061
104-15-4: 05458
104-19-8: 20118
104-21-2: 05096
104-22-3: 05390
104-24-5: 07235
104-27-8: 18739
104-29-0: 20899
104-36-9: 03036
104-42-7: 02248
104-45-0: 03624
104-46-1: 03619
104-47-2: 04333
104-49-4: 03163
104-50-7: 13067
104-51-8: 02757
104-52-9: 02949
104-53-0: 05167
104-57-4: 12855
104-62-1: 12854
104-63-2: 12237
104-65-4: 21993
104-66-5: 03310
104-67-6: 13100
104-68-7: 12231
104-72-3: 02991
104-75-6: 13986
104-76-7: 14276
104-78-9: 20755
104-80-3: 14532
104-81-4: 02686
104-82-5: 02886
104-83-6: 02782
104-84-7: 04933
104-85-8: 06721
104-86-9: 04893
104-87-0: 01756
104-88-1: 01603
104-90-5: 22660
104-91-6: 19544
104-92-7: 02665
104-93-8: 03596
104-94-9: 02314
104-96-1: 02418
104-97-2: 10639
104-98-3: 21871
105-03-3: 17220
105-05-5: 03142
105-06-6: 03131
105-07-7: 06709
105-08-8: 10050
105-10-2: 04485
105-11-3: 09726
105-13-5: 05095
105-14-6: 18548
105-21-5: 13091

105-30-6: 18493
105-31-7: 14556
105-34-0: 00553
105-36-2: 00486
105-37-3: 21175
105-38-4: 21173
105-39-5: 00510
105-40-8: 09087
105-41-9: 13994
105-42-0: 14346
105-44-2: 18566
105-45-3: 08490
105-46-4: 00692
105-48-6: 00515
105-50-0: 18318
105-53-3: 20818
105-54-4: 08405
105-55-5: 24868
105-56-6: 00551
105-57-7: 11826
105-58-8: 09198
105-59-9: 12201
105-60-2: 01520
105-65-7: 24662
105-66-8: 08518
105-67-9: 19262
105-68-0: 08560
105-70-4: 21037
105-72-6: 08401
105-74-8: 18811
105-75-9: 08800
105-76-0: 08810
105-83-9: 20754
105-85-1: 17834
105-86-2: 17548
105-96-4: 14120
105-99-7: 14125
106-00-3: 14118
106-01-4: 17258
106-02-5: 17891
106-06-9: 17247
106-07-0: 17416
106-14-9: 17413
106-18-3: 11516
106-19-4: 14140
106-20-7: 13987
106-24-1: 17543
106-25-2: 17544
106-26-3: 17520
106-27-4: 08471
106-28-5: 11552
106-30-9: 13623
106-31-0: 08323
106-32-1: 17676
106-33-2: 11520
106-35-4: 13696
106-36-5: 21280
106-37-6: 02998
106-38-7: 02669
106-39-8: 02592
106-40-1: 02025
106-41-2: 19093
106-42-3: 03198
106-43-4: 02868
106-44-5: 19442
106-45-6: 05556
106-46-7: 03039
106-47-8: 02068
106-48-9: 19139
106-49-0: 02351
106-50-3: 04422
106-51-4: 09694
106-52-5: 20283
106-53-6: 05543
106-54-7: 05546
106-56-9: 16091
106-57-0: 20111
106-58-1: 20056
106-61-6: 21035
106-62-7: 21329
106-63-8: 21935
106-65-0: 08151
106-67-2: 18491
106-68-3: 17753
106-69-4: 14196
106-70-7: 14244
106-73-0: 13632
106-74-1: 21847
106-79-6: 10940
106-85-4: 17882
106-87-6: 17879
106-88-7: 18011
106-89-8: 17995
106-90-1: 21952
106-91-2: 21912
106-92-3: 18025
106-93-4: 11800
106-94-5: 20499
106-95-6: 21674
106-96-7: 22101
106-97-8: 07903
106-98-9: 08705
106-99-0: 07774
107-00-6: 08988
107-02-8: 21613
107-03-9: 21012
107-04-0: 11755
107-05-1: 21680
107-06-2: 11811
107-07-3: 12116
107-08-4: 20624
107-10-8: 20439
107-11-9: 21657
107-12-0: 20962
107-13-1: 21754
107-14-2: 00874
107-15-3: 11917
107-16-4: 00886
107-18-6: 21978
107-19-7: 22122
107-20-0: 00032
107-21-1: 11989
107-22-2: 11910
107-25-5: 12582
107-29-9: 00047
107-30-2: 15648
107-31-3: 12848
107-34-6: 01443
107-35-7: 12038
107-37-9: 24118
107-39-1: 18649
107-40-4: 18684
107-41-5: 18343
107-43-7: 15619
107-45-9: 18175
107-46-0: 11381
107-47-1: 20709
107-48-2: 18006
107-49-3: 11351
107-50-6: 09655
107-51-7: 25205
107-52-8: 14372
107-53-9: 25318
107-58-4: 21646
107-59-5: 00508
107-68-6: 12040
107-70-0: 18562
107-72-2: 24115
107-74-4: 17648
107-75-5: 17559
107-80-2: 07946
107-81-3: 18182
107-82-4: 07921
107-83-5: 18257
107-84-6: 07939
107-85-7: 07879

107-86-8: 08696
107-87-9: 18534
107-88-0: 08202
107-89-1: 07817
107-91-5: 00185
107-92-6: 08294
107-93-7: 08852
107-94-8: 21102
107-95-9: 00993
107-96-0: 21213
107-97-1: 13327
107-98-2: 21406
108-00-9: 11940
108-01-0: 12144
108-03-2: 20670
108-05-4: 00850
108-08-7: 18230
108-10-1: 18564
108-11-2: 18515
108-12-3: 08687
108-13-4: 20737
108-16-7: 21391
108-18-9: 20482
108-19-0: 14941
108-20-3: 20675
108-21-4: 00680
108-22-5: 21980
108-23-6: 09231
108-24-7: 00473
108-26-9: 22477
108-27-0: 23426
108-28-1: 13109
108-29-2: 13085
108-30-5: 13028
108-31-6: 13022
108-32-7: 11333
108-33-8: 24448
108-36-1: 02997
108-37-2: 02591
108-38-3: 03197
108-39-4: 19441
108-40-7: 05555
108-41-8: 02867
108-42-9: 02066
108-43-0: 19138
108-44-1: 02349
108-45-2: 04421
108-46-3: 04641
108-47-4: 22626
108-48-5: 22628
108-50-9: 22325
108-52-1: 23028
108-54-3: 22274
108-55-4: 22246
108-56-5: 08194
108-57-6: 03130
108-58-7: 04691
108-59-8: 20823
108-60-1: 20678
108-61-2: 21344
108-64-5: 08456
108-67-8: 04056
108-68-9: 19273
108-69-0: 02187
108-70-3: 04004
108-72-5: 05565
108-73-6: 05583
108-75-8: 22757
108-77-0: 24989
108-78-1: 25026
108-82-7: 13683
108-83-8: 13718
108-84-9: 18516
108-85-0: 09800
108-86-1: 02582
108-87-2: 09893
108-88-3: 03626
108-89-4: 22680
108-90-7: 02773
108-91-8: 09766
108-93-0: 10105
108-94-1: 10216
108-95-2: 18983
108-97-4: 22271
108-98-5: 05538
108-99-6: 22678
109-00-2: 22966
109-01-3: 20076
109-02-4: 16105
109-04-6: 22584
109-06-8: 22677
109-07-9: 20077
109-08-0: 22343
109-09-1: 22595
109-12-6: 23018
109-15-9: 21244
109-16-0: 21895
109-17-1: 21913
109-19-3: 08453
109-21-7: 08344
109-30-8: 17445
109-31-9: 17224
109-43-3: 10936
109-44-4: 14116
109-46-6: 24866
109-49-9: 14510
109-52-4: 18395
109-53-5: 20596
109-55-7: 20758
109-56-8: 12200
109-57-9: 24893
109-59-1: 12199
109-60-4: 00740
109-61-5: 09238
109-63-7: 07753
109-64-8: 20544
109-65-9: 07909
109-66-0: 18177
109-67-1: 18619
109-69-3: 07925
109-70-6: 20503
109-73-9: 07868
109-74-0: 08246
109-75-1: 08847
109-76-2: 20747
109-77-3: 20771
109-78-4: 20983
109-79-5: 08266
109-80-8: 20961
109-81-9: 11954
109-82-0: 00890
109-83-1: 12196
109-84-2: 12182
109-85-3: 11727
109-86-4: 12190
109-87-5: 15673
109-89-7: 11715
109-90-0: 11846
109-92-2: 12578
109-93-3: 12587
109-94-4: 12845
109-95-5: 17145
109-96-6: 23288
109-97-7: 23279
109-98-8: 22399
109-99-9: 12925
110-00-9: 12862
110-01-0: 24769
110-02-1: 24671
110-03-2: 14168
110-04-3: 12028
110-05-4: 18807
110-06-5: 11392
110-12-3: 14347
110-13-4: 14173
110-14-5: 08089

110-15-6: 08105
110-16-7: 08808
110-17-8: 08795
110-18-9: 11962
110-19-0: 00695
110-20-3: 14690
110-21-4: 14712
110-22-5: 18812
110-27-0: 24378
110-33-8: 14132
110-36-1: 24370
110-38-3: 10963
110-39-4: 08482
110-40-7: 10938
110-41-8: 25267
110-42-9: 10967
110-43-0: 13695
110-44-1: 13955
110-45-2: 08558
110-46-3: 17149
110-47-4: 20993
110-49-6: 12191
110-51-0: 07755
110-52-1: 07947
110-53-2: 18181
110-54-3: 13995
110-56-5: 07966
110-57-6: 08772
110-58-7: 18161
110-59-8: 18364
110-60-1: 08091
110-61-2: 08102
110-62-3: 18137
110-63-4: 08203
110-65-6: 09008
110-66-7: 18371
110-67-8: 20990
110-68-9: 07880
110-69-0: 07821
110-70-3: 11939
110-71-4: 11835
110-72-5: 11951
110-73-6: 12170
110-74-7: 12857
110-75-8: 12552
110-76-9: 11711
110-77-0: 12176
110-78-1: 20629
110-80-5: 12164
110-81-6: 11415
110-82-7: 09797
110-83-8: 10297
110-85-0: 20026
110-86-1: 22578
110-87-2: 22235
110-88-3: 25185
110-89-4: 20132
110-91-8: 16083
110-93-0: 13817
110-94-1: 18294
110-95-2: 20766
110-96-3: 20462
110-97-4: 21403
110-99-6: 00719
111-01-3: 24334
111-02-4: 24328
111-03-5: 17487
111-06-8: 13881
111-11-5: 17682
111-12-6: 17858
111-13-7: 17752
111-14-8: 13610
111-15-9: 12165
111-16-0: 13577
111-17-1: 21285
111-18-2: 14107
111-19-3: 10949
111-20-6: 10933
111-21-7: 12157
111-24-0: 18209
111-25-1: 13997
111-26-2: 13983
111-27-3: 14270
111-28-4: 13963
111-29-5: 18326
111-30-8: 18281
111-31-9: 14189
111-34-2: 07986
111-37-5: 11952
111-40-0: 11920
111-41-1: 12094
111-42-2: 12184
111-43-3: 20674
111-44-4: 11861
111-45-5: 12238
111-46-6: 12223
111-47-7: 20705
111-48-8: 12242
111-49-9: 01511
111-50-2: 14185
111-51-3: 08099
111-55-7: 11994
111-56-8: 11513
111-60-4: 17417
111-61-5: 17409
111-62-6: 17488
111-63-7: 17408
111-64-8: 17768
111-65-9: 17576
111-66-0: 17794
111-68-2: 13465
111-69-3: 14110
111-70-6: 13638
111-71-7: 13459
111-74-0: 11934
111-75-1: 12113
111-76-2: 12109
111-77-3: 12193
111-78-4: 10449
111-80-8: 17327
111-81-9: 25333
111-82-0: 11528
111-83-1: 17577
111-84-2: 17197
111-85-3: 17582
111-86-4: 17565
111-87-5: 17701
111-88-6: 17662
111-90-0: 12167
111-92-2: 07869
111-94-4: 20987
111-95-5: 11730
111-96-6: 11863
111-97-7: 20997
112-04-9: 24113
112-05-0: 17240
112-06-1: 00634
112-10-7: 17436
112-12-9: 25310
112-13-0: 10992
112-14-1: 00710
112-15-2: 12168
112-16-3: 11546
112-17-4: 00561
112-20-9: 17195
112-21-0: 08789
112-23-2: 12846
112-24-3: 11926
112-25-4: 12181
112-26-5: 11746
112-27-6: 12156
112-28-7: 10918
112-29-8: 10909
112-30-1: 10981
112-31-2: 10904
112-32-3: 12852

112-34-5: 12110
112-35-6: 12194
112-36-7: 11862
112-37-8: 25285
112-38-9: 25331
112-39-0: 13897
112-40-3: 11485
112-41-4: 11555
112-42-5: 25299
112-43-6: 25337
112-44-7: 25266
112-45-8: 25320
112-46-9: 25339
112-47-0: 10944
112-49-2: 24399
112-50-5: 12169
112-51-6: 11419
112-52-7: 11488
112-53-8: 11535
112-54-9: 11472
112-55-0: 11507
112-57-2: 11921
112-58-3: 14074
112-60-7: 12227
112-61-8: 17420
112-62-9: 17489
112-63-0: 17363
112-64-1: 24389
112-66-3: 00614
112-67-4: 13915
112-70-9: 25162
112-71-0: 24353
112-72-1: 24382
112-73-2: 08037
112-76-5: 17459
112-79-8: 17499
112-80-1: 17483
112-82-3: 13859
112-85-6: 11455
112-86-7: 11465
112-88-9: 17480
112-89-0: 17375
112-91-4: 17482
112-92-5: 17454
112-95-8: 11576
112-99-2: 17373
113-00-8: 13359
113-15-5: 11638
113-38-2: 11664
113-52-0: 11101
113-92-8: 22930
114-03-4: 25226
114-07-8: 11648
114-25-0: 07427
114-33-0: 22812
114-38-5: 25400
114-80-7: 02527
114-83-0: 00732
114-90-9: 12035
114-91-0: 22674
115-07-1: 21669
115-10-6: 15684
115-11-7: 21723
115-17-3: 00051
115-18-4: 08942
115-19-5: 09017
115-20-8: 12244
115-21-9: 24105
115-22-0: 08659
115-24-2: 20492
115-25-3: 09537
115-32-2: 05008
115-37-7: 16062
115-38-8: 23204
115-39-9: 19062
115-40-2: 19060
115-43-5: 23223
115-53-7: 16079
115-56-0: 23154
115-58-2: 23205
115-61-7: 23937
115-63-9: 20125
115-69-5: 20921
115-70-8: 20919
115-71-9: 18721
115-76-4: 20930
115-77-5: 20923
115-80-0: 20727
115-82-2: 24172
115-84-4: 20926
115-85-5: 19917
115-86-6: 19931
115-87-7: 19916
115-88-8: 19927
115-96-8: 12122
115-99-1: 17542
116-09-6: 21559
116-11-0: 21720
116-14-3: 12593
116-15-4: 21715
116-16-5: 21555
116-17-6: 19997
116-26-7: 09689
116-29-0: 04006
116-43-8: 08503
116-44-9: 05334
116-52-9: 25375
116-54-1: 00577
116-58-5: 20359
116-63-2: 16873
116-69-8: 04543
116-71-2: 01118
116-77-8: 01216
116-85-8: 01210
117-01-1: 01199
117-06-6: 01811
117-08-8: 15324
117-10-2: 01258
117-11-3: 01202
117-12-4: 01255
117-18-0: 03952
117-21-5: 15310
117-22-6: 16880
117-34-0: 04284
117-37-3: 15062
117-39-5: 06911
117-59-9: 16878
117-61-3: 07654
117-62-4: 16828
117-73-7: 25578
117-78-2: 01174
117-79-3: 01194
117-80-6: 16783
117-81-7: 04536
117-82-8: 04537
117-83-9: 04533
117-84-0: 04563
117-93-1: 00267
117-99-7: 15992
118-00-3: 13384
118-02-5: 04699
118-08-1: 15341
118-10-5: 09421
118-12-7: 15132
118-29-6: 15384
118-31-0: 16843
118-32-1: 16827
118-34-3: 13223
118-41-2: 06628
118-44-5: 16171
118-45-6: 15311
118-46-7: 16964
118-48-9: 07170
118-55-8: 06319
118-58-1: 06321
118-60-5: 06304

118-61-6: 06303
118-69-4: 03090
118-71-8: 22293
118-74-1: 03502
118-75-2: 09738
118-76-3: 10396
118-78-5: 23175
118-79-6: 19633
118-82-1: 19456
118-83-2: 02918
118-88-7: 05416
118-90-1: 06458
118-91-2: 06048
118-92-3: 05864
118-93-4: 12431
118-96-7: 03792
118-97-8: 06075
119-13-1: 06811
119-15-3: 19339
119-18-6: 22436
119-20-0: 05908
119-26-6: 14620
119-27-7: 03582
119-28-8: 16882
119-30-2: 06354
119-32-4: 02395
119-33-5: 19501
119-34-6: 19043
119-36-8: 06308
119-41-5: 00716
119-42-6: 19196
119-44-8: 22132
119-45-9: 00739
119-48-2: 16129
119-51-7: 20954
119-52-8: 12399
119-54-0: 20102
119-56-2: 05019
119-58-4: 05048
119-59-5: 02490
119-60-8: 15923
119-61-9: 15961
119-64-2: 16504
119-65-3: 15436
119-67-5: 06282
119-68-6: 06475
119-75-5: 02430
119-80-2: 06256
119-81-3: 00297
119-84-6: 06828
119-90-4: 07606
119-91-5: 07701
119-93-7: 07608
120-07-0: 12234
120-12-7: 01123
120-14-9: 01655
120-15-0: 23705
120-18-3: 16866
120-20-7: 04794
120-21-8: 01641
120-23-0: 00700
120-25-2: 01678
120-29-6: 01493
120-32-1: 19193
120-33-2: 06466
120-36-5: 21151
120-43-4: 20106
120-46-7: 20957
120-47-8: 06330
120-48-9: 06525
120-50-3: 06505
120-51-4: 06557
120-53-6: 07073
120-55-8: 12225
120-57-0: 05743
120-61-6: 04626
120-66-1: 00317
120-67-2: 12137
120-70-7: 19585
120-71-8: 02321
120-72-9: 15113
120-73-0: 22147
120-75-2: 07059
120-80-9: 04640
120-82-1: 04003
120-83-2: 19221
120-85-4: 20094
120-86-5: 12030
120-87-6: 17402
120-89-8: 14882
120-92-3: 10660
120-93-4: 14883
120-94-5: 23358
121-03-9: 05473
121-06-2: 19269
121-14-2: 03640
121-17-5: 02916
121-21-1: 10806
121-29-9: 10812
121-30-2: 04777
121-32-4: 01674
121-33-5: 01715
121-34-6: 06362
121-39-1: 18034
121-43-7: 07742
121-44-8: 11699
121-45-9: 19992
121-46-0: 07257
121-47-1: 05397
121-50-6: 02124
121-51-7: 05517
121-52-8: 05386
121-53-9: 06581
121-57-3: 05398
121-59-5: 01421
121-60-8: 05495
121-61-9: 00084
121-66-4: 24496
121-69-7: 02191
121-71-1: 12432
121-72-2: 02522
121-73-3: 02910
121-75-5: 08150
121-79-9: 06622
121-81-3: 01877
121-82-4: 24983
121-86-8: 02900
121-87-9: 02104
121-88-0: 19039
121-89-1: 12490
121-90-4: 07228
121-91-5: 04527
121-92-6: 06519
121-95-9: 00316
121-97-1: 21506
121-98-2: 06447
122-00-9: 12473
122-01-0: 07200
122-03-2: 01759
122-04-3: 07229
122-05-4: 22367
122-07-6: 11703
122-09-8: 04795
122-11-2: 05317
122-20-3: 21415
122-25-8: 06485
122-28-1: 00348
122-32-7: 17493
122-35-0: 21610
122-37-2: 19559
122-39-4: 02455
122-40-7: 13463
122-42-9: 09106
122-46-3: 00690
122-47-4: 03189
122-48-5: 08658

122-51-0: 11854
122-52-1: 19991
122-57-6: 08969
122-59-8: 00724
122-60-1: 18022
122-62-3: 10934
122-63-4: 21278
122-66-7: 14624
122-67-8: 21963
122-69-0: 21967
122-70-3: 21277
122-73-6: 03632
122-78-1: 04083
122-79-2: 00728
122-80-5: 00079
122-84-9: 21566
122-85-0: 00254
122-87-2: 13321
122-88-3: 00543
122-94-1: 19128
122-95-2: 03134
122-97-4: 05262
122-98-5: 12233
122-99-6: 12230
123-00-2: 16134
123-01-3: 03296
123-02-4: 04034
123-04-6: 13494
123-05-7: 13977
123-06-8: 20776
123-07-9: 19359
123-08-0: 01699
123-11-5: 01743
123-15-9: 18142
123-17-1: 17275
123-18-2: 17286
123-19-3: 13697
123-20-6: 08402
123-22-8: 19986
123-24-0: 25168
123-25-1: 08132
123-28-4: 21286
123-29-5: 17249
123-30-8: 18988
123-31-9: 04642
123-32-0: 22324
123-33-1: 22527
123-35-3: 17528
123-38-6: 20377
123-39-7: 12825
123-42-2: 18559
123-43-3: 00749
123-44-4: 18499
123-45-5: 00751
123-48-8: 13765
123-51-3: 08553
123-54-6: 18346
123-56-8: 23382
123-57-9: 16093
123-59-1: 20603
123-62-6: 21073
123-63-7: 25188
123-66-0: 14225
123-68-2: 14266
123-70-6: 20759
123-72-8: 07808
123-73-9: 08689
123-75-1: 23346
123-76-2: 18463
123-77-3: 11082
123-79-5: 14137
123-80-8: 12003
123-81-9: 00662
123-82-0: 13466
123-84-2: 21357
123-85-3: 00257
123-86-4: 00500
123-90-0: 24661
123-91-1: 11267
123-92-2: 08554
123-93-3: 00756
123-95-5: 17399
123-98-8: 17236
123-99-9: 17219
124-02-7: 21667
124-04-9: 14112
124-06-1: 24374
124-07-2: 17666
124-09-4: 14100
124-10-7: 24377
124-11-8: 17295
124-12-9: 17661
124-13-0: 17557
124-17-4: 12111
124-18-5: 10908
124-19-6: 17191
124-22-1: 11481
124-25-4: 24346
124-26-5: 17370
124-30-1: 17372
124-38-9: 09181
124-40-3: 15612
124-42-5: 12085
124-47-0: 25475
124-48-1: 15655
124-63-0: 15760
124-68-5: 21300
124-70-9: 24021
124-73-2: 11808
124-80-1: 09295
124-83-4: 10619
124-97-0: 09307
125-15-5: 23939
125-20-2: 15332
125-24-6: 07428
125-28-0: 16076
125-29-1: 16081
125-33-7: 23156
125-55-3: 23180
125-56-4: 13712
125-58-6: 13709
125-64-4: 20255
126-00-1: 04369
126-07-8: 24187
126-11-4: 20940
126-14-7: 13219
126-17-0: 24200
126-18-1: 24204
126-19-2: 24205
126-29-4: 09276
126-30-7: 20931
126-33-0: 24770
126-39-6: 11318
126-54-5: 24404
126-68-1: 19981
126-71-6: 19940
126-72-7: 21313
126-73-8: 19941
126-81-8: 10073
126-84-1: 20572
126-90-9: 17540
126-98-7: 21764
126-99-8: 07784
127-00-4: 21367
127-06-0: 21574
127-07-1: 25449
127-17-3: 21263
127-18-4: 12592
127-19-5: 00219
127-25-3: 18913
127-27-5: 18917
127-29-7: 00950
127-40-2: 09279
127-43-5: 18742
127-48-0: 17953
127-54-8: 19470

127-63-9: 03921
127-66-2: 05073
127-69-5: 05318
127-71-9: 01816
127-79-7: 05330
128-08-5: 23384
128-09-6: 23386
128-13-2: 09340
128-20-1: 20339
128-21-2: 20340
128-23-4: 20333
128-27-8: 11621
128-33-6: 09351
128-37-0: 19079
128-39-2: 19071
128-50-7: 07382
128-53-0: 23340
128-56-3: 01333
128-62-1: 15339
128-67-6: 01175
128-68-7: 07395
128-70-1: 22303
128-83-6: 01201
128-93-8: 01222
128-94-9: 01239
128-95-0: 01231
128-97-2: 16894
129-00-0: 22497
129-15-7: 01301
129-35-1: 01226
129-39-5: 01287
129-40-8: 01227
129-42-0: 01235
129-43-1: 01293
129-44-2: 01232
129-51-1: 11611
129-64-6: 15803
129-66-8: 06641
129-73-7: 02474
129-79-3: 12800
130-15-4: 16749
130-16-5: 23809
130-26-7: 23810
130-61-0: 19683
130-73-4: 19249
130-80-3: 19251
130-85-8: 16689
130-86-9: 07709
130-89-2: 09434
131-01-1: 25580
131-07-7: 17921
131-09-9: 01225
131-11-3: 04561
131-14-6: 01237
131-16-8: 04567
131-17-9: 04566
131-18-0: 04564
131-28-2: 06221
131-53-3: 15975
131-54-4: 15862
131-55-5: 15853
131-56-6: 15936
131-57-7: 15976
131-58-8: 16019
131-70-4: 04572
131-73-7: 02526
131-89-5: 19200
131-91-9: 17001
132-18-3: 20156
132-29-6: 19915
132-32-1: 09157
132-53-6: 16946
132-54-7: 16677
132-57-0: 16879
132-60-5: 23758
132-64-9: 11128
132-65-0: 11207
132-66-1: 06511
132-69-4: 20449
132-75-2: 16575
132-86-5: 16730
133-06-2: 15364
133-08-4: 20803
133-11-9: 05925
133-13-1: 20830
133-26-6: 13132
133-32-4: 15187
133-37-9: 08138
133-38-0: 08830
133-43-7: 15571
133-49-3: 05559
133-53-9: 19230
133-58-4: 15598
133-59-5: 05514
133-64-2: 05472
133-67-5: 06980
133-74-4: 05401
133-90-4: 05905
133-91-5: 06338
134-11-2: 06259
134-20-3: 05869
134-31-6: 23825
134-32-7: 16152
134-49-6: 16107
134-54-3: 16881
134-55-4: 05859
134-58-7: 25099
134-62-3: 01865
134-65-6: 12435
134-67-8: 14229
134-71-4: 05117
134-80-5: 21468
134-81-6: 12019
134-83-8: 02788
134-84-9: 16021
134-85-0: 15902
134-96-3: 01707
135-00-2: 16040
135-01-3: 03140
135-02-4: 01740
135-09-1: 06987
135-13-7: 06045
135-19-3: 16902
135-20-6: 02297
135-22-8: 15160
135-48-8: 18043
135-58-0: 24475
135-68-2: 07574
135-70-6: 23449
135-73-9: 12282
135-88-6: 16231
135-97-7: 01494
136-25-4: 21143
136-35-6: 24959
136-36-7: 04718
136-47-0: 06030
136-60-7: 06029
136-72-1: 18110
136-77-6: 04702
136-80-1: 12206
136-84-5: 14884
136-95-8: 07033
137-00-8: 24534
137-05-3: 21813
137-06-4: 05554
137-07-5: 05539
137-09-7: 19202
137-16-6: 13330
137-17-7: 02519
137-18-8: 09723
137-19-9: 04693
137-26-8: 24665
137-32-6: 08549
137-43-9: 10501
137-44-0: 22472
137-58-6: 00206

137-97-3: 24853
137-99-5: 19340
138-12-5: 04222
138-15-8: 13255
138-36-3: 05419
138-41-0: 05953
138-42-1: 05477
138-52-3: 13220
138-53-4: 04324
138-65-8: 04644
138-86-3: 10333
138-87-4: 10162
138-89-6: 02220
138-91-0: 10366
139-13-9: 13288
139-25-3: 03652
139-45-7: 21046
139-59-3: 02454
139-65-1: 02504
139-66-2: 03974
139-85-5: 01647
139-87-7: 12171
140-07-8: 12161
140-08-9: 12123
140-10-3: 21956
140-11-4: 00734
140-18-1: 00520
140-22-7: 09214
140-25-0: 11532
140-26-1: 08463
140-27-2: 08465
140-28-3: 11931
140-29-4: 04300
140-31-8: 20117
140-38-5: 25401
140-39-6: 00691
140-40-9: 00351
140-50-1: 00362
140-53-4: 04313
140-55-6: 20281
140-57-8: 24296
140-65-8: 16085
140-66-9: 19619
140-67-0: 03621
140-75-0: 04920
140-76-1: 22645
140-77-2: 10637
140-79-4: 20064
140-80-7: 18287
140-82-9: 12142
140-87-4: 00552
140-88-5: 21848
140-95-4: 25366
141-03-7: 08129
141-04-8: 14122
141-05-9: 08813
141-06-0: 18472
141-08-2: 17505
141-10-6: 25319
141-12-8: 17546
141-18-4: 14114
141-20-8: 11524
141-22-0: 17503
141-27-5: 17519
141-28-6: 14127
141-30-0: 22523
141-31-1: 14098
141-32-2: 21795
141-34-4: 12158
141-40-2: 17880
141-43-5: 12091
141-46-8: 00041
141-57-1: 24119
141-62-8: 24408
141-63-9: 18604
141-73-1: 18511
141-75-3: 08677
141-76-4: 21208
141-78-6: 00851
141-79-7: 18749
141-82-2: 20788
141-83-3: 25357
141-84-4: 24589
141-85-5: 02067
141-86-6: 22880
141-90-2: 23256
141-91-3: 16092
141-93-5: 03141
141-97-9: 08489
142-04-1: 02290
142-08-5: 22977
142-09-6: 21901
142-26-7: 00258
142-28-9: 20560
142-29-0: 10699
142-30-3: 14552
142-45-0: 09004
142-48-3: 13331
142-50-7: 11551
142-60-9: 21262
142-61-0: 14366
142-62-1: 14200
142-63-2: 20070
142-64-3: 20055
142-65-4: 18166
142-68-7: 22239
142-73-4: 13293
142-77-8: 17486
142-78-9: 11477
142-79-0: 13601
142-82-5: 13478
142-83-6: 13929
142-84-7: 20468
142-88-1: 14124
142-90-5: 21893
142-91-6: 13898
142-92-7: 00635
142-94-9: 17398
142-95-0: 17568
142-96-1: 08035
143-07-7: 11508
143-08-8: 17260
143-10-2: 10951
143-13-5: 00708
143-15-7: 11486
143-16-8: 13988
143-22-6: 12112
143-24-8: 18598
143-27-1: 13857
143-28-2: 17510
143-37-3: 12081
143-57-7: 09308
143-62-4: 09264
143-70-4: 16071
143-74-8: 19058
144-11-6: 20273
144-19-4: 18333
144-48-9: 00283
144-49-0: 00623
144-62-7: 11972
144-68-3: 09277
144-79-6: 23997
144-80-9: 00080
144-83-2: 05335
145-13-1: 20362
145-39-1: 03244
145-94-8: 15078
146-06-5: 09432
146-48-5: 25574
146-56-5: 20124
146-77-0: 00986
147-24-0: 11710
147-47-7: 23626
147-58-0: 13322
147-71-7: 08140
147-73-9: 08139

147-82-0: 02505
147-85-3: 20366
147-93-3: 06433
148-03-8: 06814
148-24-3: 23781
148-25-4: 16834
148-53-8: 01709
148-72-1: 13098
148-75-4: 16835
148-86-7: 07675
149-29-1: 13148
149-30-4: 07068
149-32-6: 08261
149-57-5: 14226
149-73-5: 15714
149-74-6: 24026
149-91-7: 06617
150-05-0: 04659
150-13-0: 05876
150-19-6: 19407
150-25-4: 13289
150-30-1: 19713
150-46-9: 07740
150-60-7: 11407
150-61-8: 11949
150-68-5: 25402
150-69-6: 25440
150-76-5: 19408
150-77-6: 11961
150-78-7: 03168
150-84-5: 17833
150-86-7: 13924
151-01-9: 09244
151-06-4: 04790
151-10-0: 03167
151-13-3: 17504
151-18-8: 20964
151-19-9: 17733
151-56-4: 01529
152-16-9: 11349
152-58-9: 20348
152-93-2: 23251
152-95-4: 06921
153-78-6: 12711
153-94-6: 25222
154-17-6: 14533
154-41-6: 04975
154-42-7: 22214
154-92-7: 01384
154-93-8: 25360
156-06-9: 05244
156-08-1: 04798
156-10-5: 02440
156-38-7: 04205
156-41-2: 04789
156-43-4: 02259
156-59-2: 12566
156-60-5: 12565
156-83-2: 23110
156-87-6: 21299
157-40-4: 24198
176-25-0: 11299
176-63-6: 24191
177-10-6: 11298
177-58-2: 25220
180-43-8: 24212
180-97-2: 24414
181-15-7: 24194
184-97-4: 11389
188-52-3: 11172
189-55-9: 06743
190-47-6: 11252
190-70-5: 05690
191-07-1: 09471
191-13-9: 22302
191-24-2: 06744
191-26-4: 11115
191-28-6: 25561
191-68-4: 11104
192-47-2: 11171
192-65-4: 17045
192-94-9: 05664
192-97-2: 06942
194-33-2: 17040
194-58-1: 11123
194-65-0: 11255
195-19-7: 06745
195-21-1: 17089
197-61-5: 23921
197-69-3: 11170
198-55-0: 18816
202-94-8: 01558
202-98-2: 10476
203-64-5: 10687
203-80-5: 18820
204-02-4: 18805
204-03-5: 17046
205-25-4: 05687
205-43-6: 06649
205-82-3: 05765
205-83-4: 17050
205-99-2: 01561
206-44-0: 12704
206-75-7: 07083
207-08-9: 05766
207-83-0: 11120
207-94-3: 11254
208-96-8: 00008
211-02-9: 05673
212-54-4: 15108
212-74-8: 24406
213-46-7: 20021
214-17-5: 05689
215-58-7: 07157
216-58-0: 11228
217-54-9: 11103
217-59-4: 25196
218-01-9: 09397
218-08-6: 17090
220-06-4: 11229
220-97-3: 15107
222-24-2: 05667
222-93-5: 18599
224-41-9: 11093
224-42-0: 11091
225-11-6: 01562
225-51-4: 01564
225-54-7: 11155
225-61-6: 06753
225-74-1: 06773
225-83-2: 06771
226-06-2: 06772
226-86-8: 11149
226-88-0: 06644
227-09-8: 11169
229-87-8: 18953
230-07-9: 18958
230-27-3: 06960
230-46-6: 18957
235-06-3: 25100
238-84-6: 05768
239-01-0: 05685
239-60-1: 11122
239-64-5: 11102
239-85-0: 11121
241-34-9: 05665
241-37-2: 05666
242-47-7: 11119
242-51-3: 11253
243-17-4: 05769
243-28-7: 05686
243-59-4: 15284
244-42-8: 15110
244-99-5: 15109
246-02-6: 02013
247-16-5: 24611

248-70-4: 07084
250-84-0: 24616
253-52-1: 20002
253-66-7: 09456
253-72-5: 17128
253-82-7: 23452
254-03-5: 06778
254-04-6: 06777
254-60-4: 17129
254-79-5: 17127
254-87-5: 07145
255-50-5: 05762
256-96-2: 11095
257-89-6: 01563
258-04-8: 06774
258-31-1: 13842
260-94-6: 00960
261-31-4: 24898
262-12-4: 11118
262-30-6: 23943
267-61-8: 05675
267-65-2: 05674
271-44-3: 14965
271-58-9: 05662
271-63-6: 23441
271-89-6: 05774
271-95-4: 05661
272-16-2: 05657
272-30-0: 06974
273-09-6: 05832
273-13-2: 06988
273-15-4: 06968
273-21-2: 14911
273-53-0: 07175
274-07-7: 05711
274-09-9: 05727
274-40-8: 15238
275-51-4: 01549
278-06-8: 24343
278-30-8: 07415
279-23-2: 07265
279-49-2: 17876
279-59-4: 24439
279-91-4: 24442
280-33-1: 07413
280-41-1: 24441
280-57-9: 11030
280-65-9: 07412
281-09-4: 25182
281-23-2: 25111
285-67-6: 17883
285-69-8: 11259
286-08-8: 07266
286-18-0: 01473
286-20-4: 17877
286-28-2: 24440
286-62-4: 17885
287-23-0: 09520
287-27-4: 24619
287-92-3: 10500
288-13-1: 22378
288-14-2: 15493
288-32-4: 14754
288-36-8: 25040
288-37-9: 13117
288-39-1: 24452
288-42-6: 17929
288-47-1: 24502
288-92-6: 24451
288-94-8: 24425
288-99-3: 17894
289-06-5: 24453
289-80-5: 22521
289-95-2: 23043
290-37-9: 22305
290-38-0: 24981
290-67-5: 11310
290-87-9: 24982
290-96-0: 24419
291-21-4: 25215
291-64-5: 09627
292-46-6: 18606
292-64-8: 10452
293-55-0: 10440
293-96-9: 09595
294-62-2: 09606
294-94-0: 25195
294-95-1: 11265
295-48-7: 10477
295-65-8: 09672
298-06-6: 19971
298-12-4: 00711
298-18-0: 07443
298-45-3: 05680
298-46-4: 11099
298-81-7: 13131
298-96-4: 24434
299-42-3: 05114
299-84-3: 19978
300-08-3: 22859
300-37-8: 22766
300-39-0: 25250
300-57-2: 03899
300-62-9: 04803
300-87-8: 15494
301-00-8: 17473
301-02-0: 17478
301-21-3: 19438
302-01-2: 14583
302-17-0: 11991
302-27-2: 00948
302-72-7: 00995
302-84-1: 23956
303-04-8: 08776
303-07-1: 06176
303-21-9: 19483
303-38-8: 06172
303-43-5: 09385
304-06-3: 07595
304-19-8: 15391
304-21-2: 23001
304-84-7: 01862
304-91-6: 06424
305-01-1: 06830
305-03-3: 04363
305-15-7: 14606
305-33-9: 22874
305-67-9: 04224
305-84-0: 14571
305-85-1: 19257
306-03-6: 04220
306-08-1: 04209
306-37-6: 14613
306-44-5: 20400
306-67-2: 08095
306-83-2: 11825
306-92-3: 16239
306-94-5: 16471
306-98-9: 09821
307-11-9: 25201
307-20-0: 07960
307-25-5: 14058
307-26-6: 14075
307-30-2: 17717
307-34-6: 17629
307-59-5: 11496
309-00-2: 11250
309-18-2: 00201
309-23-9: 22058
311-89-7: 07884
312-30-1: 03927
312-40-3: 24042
312-73-2: 05635
312-84-5: 23955
313-72-4: 16472
313-74-6: 01981

314-45-4: 16284
315-30-0: 22493
315-80-0: 11116
316-49-4: 01992
317-52-2: 14109
319-03-9: 15316
319-78-8: 15418
319-85-7: 09877
319-86-8: 09878
319-89-1: 09740
319-94-8: 10351
320-51-4: 02125
320-60-5: 03123
320-72-9: 06157
320-77-4: 18600
320-96-7: 04568
321-14-2: 06089
321-28-8: 03466
321-38-0: 16411
321-60-8: 07510
321-98-2: 05116
323-09-1: 16412
324-74-3: 07512
326-06-7: 08235
326-56-7: 05749
326-59-0: 05741
326-62-5: 04320
326-91-0: 08236
327-19-5: 05628
327-54-8: 03961
327-57-1: 17343
327-69-5: 00818
327-70-8: 00831
327-92-4: 03154
328-13-2: 00403
328-38-1: 15523
328-39-2: 15524
328-42-7: 08193
328-50-7: 18316
328-74-5: 02022
328-96-1: 02228
329-71-5: 19334
329-89-5: 22802
329-98-6: 04959
330-54-1: 25414
331-39-5: 21834
332-43-4: 02832
332-77-4: 12883
333-18-6: 11937
333-41-5: 19976
333-88-0: 00798
334-25-8: 14206
334-28-1: 13628
334-41-8: 00796
334-42-9: 13484
334-43-0: 13491
334-48-5: 10953
334-56-5: 10921
334-59-8: 10964
334-61-2: 10910
334-62-3: 10912
334-64-5: 10984
334-88-3: 15653
334-99-6: 15712
335-02-4: 15711
335-05-7: 15764
335-44-4: 08072
335-57-9: 13532
335-60-4: 17561
335-67-1: 17693
335-68-2: 17619
335-74-0: 10979
335-76-2: 10975
336-08-3: 14148
336-19-6: 10303
336-30-1: 09020
336-50-5: 09540
336-59-4: 08417
338-69-2: 00994
338-75-0: 20570
338-83-0: 20455
339-43-5: 05307
341-27-5: 06278
341-70-8: 19695
342-16-5: 22075
343-21-5: 19688
343-65-7: 04364
344-01-4: 14380
344-03-6: 03032
344-04-7: 02711
344-07-0: 02919
344-25-2: 20365
344-62-7: 00400
345-16-4: 06279
345-17-5: 02841
345-18-6: 02843
345-35-7: 02879
345-69-7: 15968
345-83-5: 15969
345-92-6: 15861
346-18-9: 06984
347-93-3: 21453
348-51-6: 02834
348-52-7: 03461
348-54-9: 02280
348-67-4: 16048
348-70-9: 00531
349-01-9: 03473
349-03-1: 02710
349-04-2: 05526
349-46-2: 10877
349-75-7: 05151
349-78-0: 05528
349-88-2: 05509
350-03-8: 12515
350-28-7: 06280
350-30-1: 02842
350-31-2: 02840
350-46-9: 03483
350-50-5: 03469
350-51-6: 03352
350-90-3: 21852
351-36-0: 00401
351-50-8: 14566
351-54-2: 01686
352-11-4: 02880
352-28-3: 08799
352-32-9: 03472
352-33-0: 02836
352-34-1: 03462
352-70-5: 03471
352-91-0: 20511
352-93-2: 11887
352-97-6: 13281
353-25-3: 17677
353-29-7: 11487
353-36-6: 11841
353-42-4: 07754
353-49-1: 09210
353-50-4: 09212
353-55-9: 15656
353-58-2: 15630
353-59-3: 15626
353-61-7: 20612
353-66-2: 24041
353-82-2: 00881
353-83-3: 11902
353-85-5: 00899
353-89-9: 24065
353-91-3: 01450
354-04-1: 11809
354-06-3: 11758
354-14-3: 11884
354-19-8: 00580
354-21-2: 11895
354-23-4: 11824

354-25-6: 11794
354-26-7: 11796
354-28-9: 00144
354-32-5: 00942
354-33-6: 11869
354-38-1: 00395
354-47-2: 11866
354-55-2: 11770
354-56-3: 11868
354-58-5: 11897
354-72-3: 11871
354-80-3: 11722
354-81-4: 11870
355-02-2: 09984
355-04-4: 18280
355-19-1: 08041
355-20-4: 07977
355-25-9: 07943
355-42-0: 14076
355-63-5: 13743
355-68-0: 09848
355-75-9: 10302
355-80-6: 18498
356-16-1: 11886
356-18-3: 09526
356-24-1: 08420
356-27-4: 08419
356-28-5: 08418
357-20-0: 08742
357-26-6: 08739
357-39-1: 15749
357-49-3: 22059
357-57-3: 24259
357-67-5: 23193
357-70-0: 13172
358-06-5: 24040
358-43-0: 24131
358-60-1: 24123
358-81-6: 24032
358-99-6: 12548
359-01-3: 07999
359-02-4: 12570
359-10-4: 12550
359-11-5: 12601
359-13-7: 12143
359-18-2: 07727
359-25-1: 00494
359-28-4: 11896
359-29-5: 12600
359-32-0: 00931
359-37-5: 12602
359-54-6: 01418
359-55-7: 01420
359-69-3: 01408
359-70-6: 11732
359-72-8: 18645
359-76-2: 01419
360-56-5: 11028
360-68-9: 09364
360-89-4: 08785
363-03-1: 09737
363-04-2: 22046
363-72-4: 03865
363-73-5: 02495
363-81-5: 02511
364-32-9: 19374
364-76-1: 02288
366-18-7: 07692
366-29-0: 07616
366-77-8: 04368
367-11-3: 03151
367-12-4: 19371
367-21-5: 02084
367-23-7: 04038
367-25-9: 02165
367-29-3: 02287
367-31-7: 04440
367-57-7: 18362
367-64-6: 00795
368-34-3: 00528
368-43-4: 05523
368-47-8: 24134
368-54-7: 23258
368-77-4: 06737
368-88-7: 05448
368-97-8: 01393
368-98-9: 07730
370-14-9: 19447
370-32-1: 12250
370-35-4: 20887
371-40-4: 02282
371-41-5: 19373
371-62-0: 12177
371-67-5: 11798
371-78-8: 15701
372-06-5: 25209
372-09-8: 00549
372-18-9: 03152
372-19-0: 02281
372-20-3: 19372
372-31-6: 08521
372-38-3: 04039
372-42-9: 18688
372-46-3: 09871
372-47-4: 22664
372-48-5: 22663
372-54-3: 14057
372-63-4: 11048
372-64-5: 11409
372-75-8: 17871
372-93-0: 08545
373-05-7: 14234
373-14-8: 14056
373-28-4: 14000
373-32-0: 14280
373-44-4: 17634
373-49-9: 13922
373-52-4: 15633
373-53-5: 15678
373-74-0: 24133
373-80-8: 24287
373-88-6: 11736
374-01-6: 21431
374-07-2: 11822
374-10-7: 24288
374-25-4: 08711
374-77-6: 09822
375-01-9: 08546
375-07-5: 08378
375-14-4: 18509
375-16-6: 08686
375-22-4: 08416
375-34-8: 08056
375-43-9: 08001
375-45-1: 08055
375-83-7: 13552
376-27-2: 17695
376-65-8: 10563
376-77-2: 10508
376-79-4: 14407
376-87-4: 18630
377-70-8: 09685
377-93-5: 09586
378-22-3: 18782
378-68-7: 21152
378-72-3: 18589
379-30-6: 00843
379-50-0: 24072
379-79-3: 11639
380-78-9: 11764
381-61-3: 07795
381-73-7: 00599
381-88-4: 08673
382-21-8: 21733
382-44-5: 01079
382-45-6: 01081

383-29-9: 03926
383-50-6: 12178
383-63-1: 00810
383-66-4: 00838
383-67-5: 00837
383-90-4: 07736
384-04-3: 07801
384-22-5: 03821
384-46-3: 05626
385-00-2: 06165
386-17-4: 15331
388-11-4: 00436
388-14-7: 07643
388-69-2: 07639
388-82-9: 07476
389-40-2: 03867
390-79-4: 20894
392-44-9: 07788
392-52-9: 14382
392-56-3: 03511
392-69-8: 03491
392-83-6: 02728
392-85-8: 03488
393-38-4: 03158
393-52-2: 07212
393-56-6: 02283
393-75-9: 02805
394-31-0: 05917
394-33-2: 19376
394-46-7: 03351
394-47-8: 06706
396-01-0: 22134
396-64-5: 07477
396-86-1: 07646
398-21-0: 07452
398-23-2: 07478
399-92-8: 21284
400-39-5: 00819
400-52-2: 00801
400-61-3: 00815
401-14-9: 19627
401-54-7: 12553
401-55-8: 00495
401-56-9: 00529
401-78-5: 02729
401-80-9: 03489
402-31-3: 02580
402-42-6: 03487
402-43-7: 02730
402-44-8: 03490
402-46-0: 05369
402-67-5: 03482
403-21-4: 06281
403-41-8: 05077
403-42-9: 12392
403-43-0: 07214
404-24-0: 00402
404-72-8: 03464
404-86-4: 17294
405-42-5: 03155
405-50-5: 04184
405-79-8: 00627
405-84-5: 00828
405-99-2: 03353
406-46-2: 21686
406-74-6: 00624
406-76-8: 20484
407-25-0: 00793
407-38-5: 00841
407-75-0: 18425
407-97-6: 18185
407-98-7: 18199
408-16-2: 13645
408-27-5: 17711
408-38-8: 13867
420-04-2: 09497
420-05-3: 09514
420-12-2: 24623
420-24-6: 01448
420-26-8: 20611
420-42-8: 01405
420-45-1: 20576
420-46-2: 11900
420-52-0: 19751
420-90-6: 21676
420-97-3: 20565
420-99-5: 20519
421-05-6: 00875
421-06-7: 11771
421-13-6: 15643
421-17-0: 15745
421-30-7: 01444
421-32-9: 01447
421-50-1: 21588
421-56-7: 15642
421-58-9: 19883
421-59-0: 19884
421-69-2: 11804
421-83-0: 15762
421-85-2: 15751
422-94-6: 19771
423-01-8: 19766
423-26-7: 20617
423-33-6: 20616
425-23-0: 08421
425-75-2: 15758
425-87-6: 11797
426-92-6: 09495
428-59-1: 18029
430-40-0: 01397
430-51-3: 21552
430-57-9: 11818
430-58-0: 12571
430-66-0: 11901
430-71-7: 21611
430-85-3: 12562
430-91-1: 00104
430-95-5: 21701
431-03-8: 08230
431-06-1: 11815
431-27-6: 21670
431-47-0: 00817
431-52-7: 21746
431-53-8: 21704
431-76-5: 01400
431-97-0: 19793
432-02-0: 01409
432-03-1: 15618
432-04-2: 19758
432-05-3: 24247
432-24-6: 10369
432-25-7: 10368
433-19-2: 02581
433-52-3: 00530
433-53-4: 00601
433-98-7: 05527
434-03-7: 20364
434-13-9: 09343
434-16-2: 09350
434-45-7: 12526
434-64-0: 03870
434-75-3: 06079
434-84-4: 07256
434-87-7: 08237
434-90-2: 07463
436-05-5: 25235
437-09-2: 23945
437-38-7: 20435
437-50-3: 25560
437-86-5: 03476
438-08-4: 01051
438-32-4: 16172
439-14-5: 05707
440-17-5: 19682
440-60-8: 05140
441-30-5: 00419

442-16-0: 00972
442-51-3: 23002
443-09-4: 21479
443-48-1: 14829
443-79-8: 15419
443-83-4: 02838
444-15-5: 23211
444-27-9: 24541
444-30-4: 19657
445-01-2: 02607
445-28-3: 01895
445-29-4: 06273
445-69-2: 04515
446-10-6: 03479
446-33-3: 03480
446-34-4: 03478
446-35-5: 03157
446-51-5: 05074
446-52-6: 01683
446-72-0: 06900
447-31-4: 12316
447-53-0: 16343
447-60-9: 06736
448-71-5: 20309
450-95-3: 12393
451-13-8: 04164
451-40-1: 12386
451-46-7: 06276
451-69-4: 21853
451-80-9: 03377
452-06-2: 22138
452-09-5: 02837
452-35-7: 07067
452-58-4: 22877
452-71-1: 02286
452-72-2: 19375
452-73-3: 02839
452-77-7: 02285
452-84-6: 02284
452-86-8: 04665
453-13-4: 21388
453-17-8: 20389
453-18-9: 00626
453-20-3: 13057
454-14-8: 21532
454-29-5: 14574
454-31-9: 00600
454-39-7: 17346
454-92-2: 06608
455-14-1: 02514
455-18-5: 06738
455-19-6: 01788
455-31-2: 03156
455-32-3: 07240
455-37-8: 01896
455-38-9: 06274
455-88-9: 03474
456-05-3: 04980
456-07-5: 01898
456-22-4: 06275
456-41-7: 02681
456-47-3: 05075
456-48-4: 01684
456-49-5: 03467
457-87-4: 04799
458-03-7: 03378
458-09-3: 04185
458-35-5: 19390
458-36-6: 21628
458-37-7: 13448
458-87-7: 03460
458-88-8: 20216
459-22-3: 04321
459-26-7: 03379
459-32-5: 21854
459-47-2: 03406
459-56-3: 05076
459-57-4: 01685
459-60-9: 03468
459-67-6: 10747
459-72-3: 00625
459-73-4: 13310
459-80-3: 17538
460-00-4: 02653
460-11-7: 12632
460-12-8: 07806
460-13-9: 20610
460-16-2: 12555
460-19-5: 11971
460-35-5: 20541
460-55-9: 24062
460-96-8: 19734
461-08-5: 09255
461-35-8: 20729
461-58-5: 13362
461-68-7: 20375
461-72-3: 14844
461-88-1: 22878
461-89-2: 25003
461-98-3: 23036
462-02-2: 25193
462-06-6: 03458
462-08-8: 22532
462-18-0: 25172
462-23-7: 09032
462-38-4: 20524
462-39-5: 20575
462-43-1: 21328
462-60-2: 13279
462-72-6: 07917
462-73-7: 07935
462-94-2: 18285
462-95-3: 11852
463-04-7: 17154
463-11-6: 17618
463-16-1: 17250
463-17-2: 25292
463-23-0: 17205
463-24-1: 17263
463-33-2: 25272
463-40-1: 17468
463-49-0: 20374
463-51-4: 12618
463-52-5: 15771
463-58-1: 09184
463-71-8: 09245
463-72-9: 09124
463-82-1: 20583
464-06-2: 08077
464-07-3: 08582
464-10-8: 15704
464-15-3: 07278
464-17-5: 07367
464-41-5: 07271
464-43-7: 07322
464-45-9: 07323
464-48-2: 07352
464-49-3: 07351
464-72-2: 12010
464-85-7: 13145
464-86-8: 13144
464-88-0: 10606
465-24-7: 15799
465-42-9: 09289
465-74-7: 25508
465-76-9: 09298
465-77-0: 09299
465-78-1: 09304
465-99-6: 17864
466-14-8: 07837
466-37-5: 12518
466-43-3: 01472
466-49-9: 01471
466-72-8: 11645
466-77-3: 11642
466-81-9: 17904

466-96-6: 16077
466-97-7: 16063
467-14-1: 16075
467-55-0: 24207
467-60-7: 20267
467-75-4: 25179
467-85-6: 14354
467-96-9: 24437
467-98-1: 16080
468-28-0: 09751
468-45-1: 24181
468-46-2: 13225
469-22-7: 23435
469-91-0: 15796
469-92-1: 15795
470-03-1: 24203
470-35-9: 18027
470-38-2: 09281
470-40-6: 10767
470-44-0: 14843
470-49-5: 18420
470-53-1: 13017
470-67-7: 17878
470-68-8: 20246
470-82-6: 17886
470-99-5: 10406
471-03-4: 11876
471-04-5: 13620
471-12-5: 07399
471-16-9: 07401
471-25-0: 22115
471-35-2: 11029
471-43-2: 11813
471-44-3: 00527
471-46-5: 11912
471-47-6: 00463
471-54-5: 12617
471-83-0: 17347
471-87-4: 23414
471-95-4: 07771
472-15-1: 15536
472-20-8: 10390
472-70-8: 09284
472-93-5: 09272
473-02-9: 25133
473-06-3: 07391
473-13-2: 16241
473-19-8: 07279
473-29-0: 05354
473-30-3: 24484
473-34-7: 05356
473-55-2: 07280
473-68-7: 09551
473-72-3: 09541
473-81-4: 21157
473-91-6: 10722
473-98-3: 15535
474-25-9: 09339
474-40-8: 24251
474-62-4: 11626
474-69-1: 11622
474-77-1: 09387
474-86-2: 11654
475-00-3: 13226
475-03-6: 16535
475-20-7: 15797
475-25-2: 05652
475-31-0: 13349
475-38-7: 16788
475-67-2: 11185
475-81-0: 11178
476-28-8: 13173
476-32-4: 09329
476-45-9: 16791
476-60-8: 01338
476-70-0: 11181
476-73-3: 05529
477-27-0: 00375
477-32-7: 08448
477-33-8: 08908
477-47-4: 13146
477-60-1: 25234
477-75-8: 05600
477-89-4: 11343
477-90-7: 22267
478-29-5: 01323
478-36-4: 17081
478-40-0: 16792
478-43-3: 01172
478-61-5: 07250
478-71-7: 18922
478-94-4: 11612
478-95-5: 11613
479-00-5: 11610
479-13-0: 05836
479-18-5: 22209
479-21-0: 15929
479-22-1: 09717
479-23-2: 01559
479-27-6: 16702
479-33-4: 10493
479-41-4: 15277
479-45-8: 02417
479-47-0: 05530
479-58-3: 18821
479-59-4: 06964
479-61-8: 15557
480-11-5: 06903
480-15-9: 06936
480-16-0: 06910
480-18-2: 06906
480-38-6: 06883
480-40-0: 06904
480-44-4: 06901
480-63-7: 06635
480-64-8: 06180
480-66-0: 12529
480-67-1: 06182
480-68-2: 23220
480-70-6: 05827
480-72-8: 00010
480-91-1: 15417
480-93-3: 15264
480-96-6: 05833
481-06-1: 17051
481-20-9: 09355
481-21-0: 09354
481-26-5: 20319
481-29-8: 01060
481-37-8: 01486
481-39-0: 16808
481-49-2: 18041
481-72-1: 01264
481-74-3: 01266
482-05-3: 07619
482-27-9: 13128
482-28-0: 15227
482-44-0: 13133
482-48-4: 13125
482-66-6: 10688
482-68-8: 23935
482-74-6: 05757
482-89-3: 15281
483-04-5: 17920
483-07-8: 25579
483-17-0: 11607
483-18-1: 11605
483-35-2: 01259
483-55-6: 16810
483-64-7: 06452
483-65-8: 18883
483-76-1: 16242
483-78-3: 16369
483-84-1: 16887
483-87-4: 18847
484-11-7: 18959

484-17-3: 18946
484-28-6: 05745
484-29-7: 13154
484-38-8: 04131
484-47-9: 14797
484-65-1: 02904
484-67-3: 19611
484-73-1: 11353
484-89-9: 09731
484-93-5: 01497
485-34-7: 23757
485-38-1: 04597
485-47-2: 15060
485-64-3: 09427
485-65-4: 09428
485-71-2: 09419
485-72-3: 06925
485-89-2: 23754
486-25-9: 12783
486-34-0: 16550
486-35-1: 06831
486-39-5: 15477
486-63-5: 06931
486-66-8: 06922
486-70-4: 23850
486-71-5: 23849
486-74-8: 23740
486-84-0: 23003
486-86-2: 15794
486-89-5: 15782
487-06-9: 06835
487-10-5: 11054
487-11-6: 04052
487-19-4: 22703
487-21-8: 22130
487-26-3: 06895
487-27-4: 15788
487-49-0: 12367
487-59-2: 25227
487-68-3: 01798
487-70-7: 01790
487-79-6: 23372
487-89-8: 15189
487-93-4: 15269
488-10-8: 10758
488-17-5: 04664
488-18-6: 22288
488-23-3: 03968
488-43-7: 13184
488-44-8: 01029
488-45-9: 14743
488-47-1: 04678
488-59-5: 15290
488-70-0: 19616
488-71-1: 02498
488-73-3: 15292
488-81-3: 23918
488-82-4: 01363
488-86-8: 10746
488-87-9: 04695
488-92-6: 23303
488-93-7: 12960
488-97-1: 25134
488-98-2: 03486
489-58-7: 06912
489-72-5: 15781
489-75-8: 23756
489-84-9: 01550
489-86-1: 01554
489-98-5: 02525
490-02-8: 22373
490-03-9: 10415
490-06-2: 04669
490-10-8: 10690
490-11-9: 22891
490-28-8: 22955
490-46-0: 06940
490-53-9: 15467
490-55-1: 24533
490-59-5: 06775
490-64-2: 06627
490-65-3: 16446
490-78-8: 12361
490-79-9: 06175
490-91-5: 09736
491-04-3: 10402
491-07-6: 10266
491-09-8: 10429
491-11-2: 19187
491-14-5: 18762
491-18-9: 23302
491-30-5: 15482
491-31-6: 06816
491-34-9: 23706
491-35-0: 23674
491-36-1: 23484
491-37-2: 06880
491-38-3: 06818
491-39-4: 07139
491-67-8: 06938
491-70-3: 06908
491-72-5: 06183
491-80-5: 06902
491-88-3: 13079
492-17-1: 07599
492-21-7: 25547
492-22-8: 24900
492-27-3: 23748
492-37-5: 04255
492-38-6: 04265
492-39-7: 04973
492-62-6: 13193
492-73-9: 12024
492-80-8: 02064
492-85-3: 16119
492-86-4: 04157
492-88-6: 01673
492-89-7: 04674
492-93-3: 15575
492-94-4: 12017
492-97-7: 07716
492-99-9: 10069
493-01-6: 16309
493-02-7: 16310
493-05-0: 06779
493-08-3: 06780
493-09-4: 05717
493-33-4: 12419
493-52-7: 06224
493-57-2: 07130
493-74-3: 09762
493-77-6: 24991
494-03-1: 16200
494-12-2: 06786
494-19-9: 11097
494-40-6: 04053
494-47-3: 15725
494-52-0: 22731
494-55-3: 11337
494-90-6: 05803
494-97-3: 22741
494-99-5: 03171
495-08-9: 04756
495-18-1: 01903
495-20-5: 20269
495-40-9: 08623
495-41-0: 08951
495-45-4: 08950
495-54-5: 04470
495-61-4: 10350
495-69-2: 13285
495-76-1: 05754
495-85-2: 22272
495-86-3: 21639
495-91-0: 20135
496-10-6: 15053

496-11-7: 15008
496-15-1: 15127
496-16-2: 05783
496-41-3: 05810
496-42-4: 15221
496-46-8: 14751
496-49-1: 21570
496-64-0: 22282
496-67-3: 07831
496-72-0: 04443
496-73-1: 04709
496-74-2: 04785
496-76-4: 23174
496-77-5: 17766
496-78-6: 19665
496-93-5: 14577
497-03-0: 08695
497-04-1: 20927
497-06-3: 08839
497-09-6: 20390
497-18-7: 09213
497-20-1: 10486
497-25-6: 17955
497-26-7: 11320
497-30-3: 14823
497-32-5: 07275
497-38-1: 07332
497-39-2: 19077
497-56-3: 19451
497-59-6: 22259
497-62-1: 07362
497-76-7: 13222
497-95-0: 10800
498-00-0: 05086
498-02-2: 12420
498-12-4: 07379
498-14-6: 07380
498-15-7: 07372
498-16-8: 14496
498-23-7: 08834
498-24-8: 08833
498-59-9: 10875
498-60-2: 12944
498-62-4: 24792
498-66-8: 07364
498-67-9: 03079
498-89-5: 13009
498-94-2: 20241
498-95-3: 20240
498-96-4: 22856
498-98-6: 22836
499-02-5: 10827
499-04-7: 22857
499-06-9: 06213
499-44-5: 09665
499-49-0: 04606
499-50-3: 22897
499-70-7: 10263
499-75-2: 19476
499-76-3: 06384
499-80-9: 22888
499-81-0: 22892
499-83-2: 22890
499-89-8: 09658
499-99-0: 10338
500-05-0: 22257
500-22-1: 22786
500-38-9: 04656
500-44-7: 22933
500-49-2: 04733
500-55-0: 04266
500-66-3: 04727
500-72-1: 00713
500-85-6: 09764
500-99-2: 19260
501-15-5: 04666
501-16-6: 21835
501-19-9: 19434
501-24-6: 19550
501-30-4: 22292
501-36-0: 04703
501-52-0: 05199
501-53-1: 09237
501-65-5: 03455
501-89-3: 04145
501-91-7: 24798
501-92-8: 19590
501-94-0: 04849
501-97-3: 05229
501-98-4: 21870
502-41-0: 09642
502-42-1: 09647
502-44-3: 17973
502-47-6: 17818
502-49-8: 10459
502-50-1: 13584
502-52-3: 13896
502-56-7: 17282
502-59-0: 13476
502-61-4: 11547
502-65-8: 09273
502-71-6: 17444
502-72-7: 10479
502-73-8: 13396
502-75-0: 13891
502-97-6: 11280
502-99-8: 17779
503-01-5: 13734
503-09-3: 18015
503-11-7: 18032
503-17-3: 08989
503-28-6: 11052
503-29-7: 01524
503-30-0: 17974
503-38-8: 09240
503-40-2: 15743
503-46-8: 10356
503-60-6: 08763
503-64-0: 08853
503-66-2: 21196
503-74-2: 08450
503-76-4: 20539
503-80-0: 01415
503-87-7: 14902
503-93-5: 09625
504-01-8: 10053
504-03-0: 20149
504-07-4: 23132
504-08-5: 25014
504-15-4: 04710
504-17-6: 23151
504-20-1: 13455
504-24-5: 22533
504-29-0: 22531
504-31-4: 22270
504-40-5: 17419
504-53-0: 18609
504-57-4: 17175
504-61-0: 08930
504-63-2: 20890
504-64-3: 20376
504-70-1: 22451
504-78-9: 24539
504-79-0: 24508
504-88-1: 21261
504-99-4: 17688
505-02-2: 10894
505-20-4: 11431
505-22-6: 11266
505-23-7: 11432
505-29-3: 11433
505-32-8: 13923
505-34-0: 20640
505-47-5: 00998
505-48-6: 17636
505-52-2: 25154

505-54-4: 13871
505-60-2: 11889
505-65-7: 11308
505-66-8: 11085
505-84-0: 20649
505-92-0: 11559
505-94-2: 25139
505-95-3: 11522
506-03-6: 20906
506-12-7: 13415
506-13-8: 13892
506-14-9: 13921
506-23-0: 17465
506-24-1: 17516
506-30-9: 11586
506-31-0: 11602
506-32-1: 11599
506-33-2: 11464
506-35-4: 11468
506-36-5: 11462
506-38-7: 18050
506-42-3: 17509
506-46-7: 13847
506-48-9: 17356
506-50-3: 24952
506-51-4: 24340
506-52-5: 13849
506-59-2: 15613
506-68-3: 09516
506-77-4: 09517
506-78-5: 15297
506-80-9: 09185
506-93-4: 13374
506-96-7: 00924
507-02-8: 00944
507-09-5: 12052
507-19-7: 20513
507-20-0: 20533
507-25-5: 15696
507-32-4: 01431
507-36-8: 07922
507-40-4: 14736
507-45-9: 07976
507-47-1: 12099
507-60-8: 07772
508-02-1: 17866
508-32-7: 25135
508-44-1: 11264
508-54-3: 16078
508-65-6: 09297
508-66-7: 09301
508-67-8: 09303
509-14-8: 15699
509-15-9: 13174
509-20-6: 00947
509-64-8: 16074
509-87-5: 23209
510-18-9: 13059
510-20-3: 20816
511-07-9: 11635
511-08-0: 11634
511-09-1: 11632
511-10-4: 11633
511-18-2: 17338
511-20-6: 11617
511-21-7: 11618
511-34-2: 24208
511-67-1: 23190
511-70-6: 07836
511-96-6: 24201
511-97-7: 24209
512-04-9: 24210
512-06-1: 24211
512-12-9: 17951
512-13-0: 07314
512-16-3: 09989
512-22-1: 25218
512-48-1: 18617
512-56-1: 19929
512-60-7: 25136
512-63-0: 10854
512-69-6: 13214
512-85-6: 11261
513-02-0: 19936
513-08-6: 19933
513-12-2: 12175
513-23-5: 07402
513-31-5: 21692
513-35-9: 08784
513-36-0: 20532
513-37-1: 21683
513-38-2: 20627
513-42-8: 21986
513-44-0: 21017
513-49-5: 07888
513-81-5: 07792
513-86-0: 08652
513-88-2: 21540
513-92-8: 12594
513-96-2: 20999
514-10-3: 18912
514-12-5: 07339
514-42-1: 11608
514-92-1: 09282
515-12-8: 09869
515-25-3: 23415
515-30-0: 04211
515-40-2: 02799
515-46-8: 05445
515-49-1: 05306
515-59-3: 05331
515-81-1: 09577
515-82-2: 12838
515-83-3: 12246
515-84-4: 00767
515-88-8: 07331
515-90-2: 07361
515-94-6: 01008
515-96-8: 00465
516-05-2: 20850
516-06-3: 25527
516-12-1: 23396
516-53-0: 20320
516-54-1: 20338
516-55-2: 20337
516-58-5: 20341
516-78-9: 11627
516-79-0: 11615
516-84-7: 11624
516-85-8: 11620
516-91-6: 09372
516-92-7: 09362
516-95-0: 09361
517-04-4: 11667
517-06-6: 11677
517-07-7: 11670
517-09-9: 11650
517-10-2: 09386
517-22-6: 23301
517-23-7: 13060
517-25-9: 15715
517-51-1: 16144
517-60-2: 04874
517-66-8: 05677
517-82-8: 16805
517-89-5: 16789
517-92-0: 01269
517-97-5: 11183
518-05-8: 16716
518-17-2: 15282
518-28-5: 13147
518-44-5: 06185
518-55-8: 17054
518-61-6: 19679
518-63-8: 19692
518-75-2: 06796

518-77-4: 11189
518-82-1: 01324
518-83-2: 01253
518-90-1: 04559
519-02-8: 15586
519-05-1: 06286
519-09-5: 01480
519-34-6: 15938
519-37-9: 22202
519-44-8: 04698
519-62-0: 15556
519-72-2: 04902
519-73-3: 03680
519-74-4: 20157
519-87-9: 00234
520-03-6: 15398
520-18-3: 06937
520-26-3: 06877
520-29-6: 06887
520-31-0: 06914
520-33-2: 06881
520-36-5: 06899
520-45-6: 22245
520-69-4: 23304
520-72-9: 07133
520-79-6: 06352
520-85-4: 20354
521-04-0: 24250
521-11-9: 01068
521-12-0: 01069
521-18-6: 01067
521-31-3: 20004
521-35-7: 11173
521-62-0: 01228
521-74-4: 23811
521-85-7: 07708
522-09-8: 15081
522-12-3: 06879
522-27-0: 12018
522-34-9: 12020
522-57-6: 15939
522-66-7: 09430
522-75-8: 07086
522-80-5: 25577
523-27-3: 01132
523-31-9: 04541
523-68-2: 00265
523-80-8: 05732
524-15-2: 13153
524-30-1: 06843
524-34-5: 23760
524-38-9: 15382
524-40-3: 22776
524-42-5: 16748
524-63-0: 09417
524-80-1: 09178
524-96-9: 04881
525-03-1: 12713
525-06-4: 12619
525-14-4: 13228
525-15-5: 23004
525-18-8: 15072
525-27-9: 16973
525-37-1: 16824
525-48-4: 23753
525-52-0: 05585
525-64-4: 12770
525-68-8: 23628
525-75-7: 22702
525-82-6: 06934
526-08-9: 05333
526-31-8: 25233
526-35-2: 17952
526-37-4: 01650
526-55-6: 15226
526-73-8: 04054
526-75-0: 19261
526-84-1: 08831
526-85-2: 19662
526-86-3: 09722
526-95-4: 13188
526-99-8: 13156
527-00-4: 01028
527-17-3: 09741
527-18-4: 04770
527-20-8: 02446
527-21-9: 09739
527-35-5: 19618
527-52-6: 14535
527-53-7: 03969
527-54-8: 19668
527-55-9: 04696
527-60-6: 19667
527-61-7: 09724
527-62-8: 19004
527-69-5: 12942
527-72-0: 24800
527-73-1: 14786
527-84-4: 03694
527-85-5: 01925
527-89-9: 04380
527-93-5: 11117
528-15-4: 06339
528-21-2: 12527
528-23-4: 10600
528-29-0: 03274
528-44-9: 05572
528-45-0: 06234
528-48-3: 06907
528-53-0: 06957
528-71-2: 18976
528-75-6: 01668
528-76-7: 05284
528-79-0: 19482
528-81-4: 01454
528-90-5: 06634
529-00-0: 10260
529-08-8: 01551
529-16-8: 07365
529-17-9: 01477
529-19-1: 06719
529-20-4: 01752
529-21-5: 22657
529-23-7: 01571
529-27-1: 14636
529-28-2: 03538
529-33-9: 16952
529-34-0: 17009
529-35-1: 16953
529-36-2: 16896
529-38-4: 01479
529-61-3: 25555
529-64-6: 04215
529-65-7: 00246
529-68-0: 05857
529-84-0: 06834
529-86-2: 01346
529-92-0: 23558
530-14-3: 12397
530-22-3: 05831
530-40-5: 22819
530-43-8: 13882
530-44-9: 15946
530-48-3: 03354
530-50-7: 14623
530-55-2: 09721
530-57-4: 06346
530-62-1: 14759
530-64-3: 23657
530-78-9: 06609
530-85-8: 22119
530-91-6: 17003
530-93-8: 17025
531-02-2: 07685
531-05-5: 25219
531-18-0: 15811

531-29-3: 13224
531-37-3: 19404
531-39-5: 18436
531-40-8: 21960
531-59-9: 06855
531-67-9: 07687
531-75-9: 06842
531-81-7: 06794
531-84-0: 16680
531-85-1: 07605
531-88-4: 15335
531-91-9: 07610
532-09-2: 00113
532-18-3: 16221
532-24-1: 01496
532-27-4: 12329
532-48-9: 12422
532-54-7: 04101
532-96-7: 06556
533-17-5: 00173
533-18-6: 00689
533-24-4: 04726
533-30-2: 07034
533-31-3: 05758
533-32-4: 07846
533-45-9: 24507
533-48-2: 14877
533-49-3: 07828
533-50-6: 08674
533-58-4: 19393
533-60-8: 10246
533-63-1: 07832
533-68-6: 08330
533-70-0: 02168
533-73-3: 05582
533-75-5: 09662
533-86-8: 17498
533-98-2: 07945
534-00-9: 07920
534-07-6: 21541
534-13-4: 24870
534-15-6: 11834
534-22-5: 12908
534-26-9: 14763
534-52-1: 19452
534-59-8: 20802
534-68-9: 13163
534-84-9: 10280
534-85-0: 04447
534-97-4: 14537
535-11-5: 21075
535-13-7: 21094
535-15-9: 00574
535-17-1: 21060
535-46-6: 24248
535-75-1: 20239
535-77-3: 03695
535-80-8: 06054
535-83-1: 22961
535-86-4: 10424
535-87-5: 06134
535-89-7: 23033
536-08-3: 06184
536-17-4: 24567
536-20-9: 22956
536-25-4: 05920
536-32-3: 01724
536-33-4: 22783
536-38-9: 12294
536-40-3: 06063
536-50-5: 05042
536-57-2: 05294
536-59-4: 10388
536-60-7: 05121
536-66-3: 06490
536-69-6: 22831
536-74-3: 03456
536-75-4: 22650
536-78-7: 22649
536-80-1: 03554
536-88-9: 22659
536-89-0: 14638
536-90-3: 02313
536-95-8: 06562
537-09-7: 06351
537-17-7: 25021
537-26-8: 01492
537-29-1: 04217
537-39-3: 17508
537-45-1: 09758
537-46-2: 04796
537-47-3: 14697
537-49-5: 25240
537-64-4: 15592
537-65-5: 04472
537-67-7: 02456
537-91-7: 11404
537-92-8: 00318
537-98-4: 21863
538-08-9: 09507
538-23-8: 17697
538-24-9: 11533
538-32-9: 25489
538-39-6: 03307
538-41-0: 02018
538-43-2: 20917
538-44-3: 18738
538-49-8: 22718
538-51-2: 02472
538-56-7: 21958
538-58-9: 18129
538-62-5: 11081
538-64-7: 08798
538-68-1: 03878
538-74-9: 03981
538-75-0: 09781
538-81-8: 02735
538-86-3: 03593
538-93-2: 03772
539-00-4: 21581
539-03-7: 00175
539-08-2: 20421
539-15-1: 19280
539-23-1: 22560
539-30-0: 03382
539-32-2: 22591
539-47-9: 21855
539-52-6: 12913
539-74-2: 21080
539-80-0: 09659
539-82-2: 18424
539-83-3: 20406
539-86-6: 21767
539-88-8: 18465
539-89-9: 09096
539-90-2: 08479
539-92-4: 09194
539-93-5: 11527
540-07-8: 14262
540-08-9: 13426
540-09-0: 25105
540-10-3: 13890
540-12-5: 17502
540-18-1: 08504
540-23-8: 02352
540-36-3: 03153
540-37-4: 02300
540-38-5: 19395
540-42-1: 21255
540-43-2: 13475
540-47-6: 10779
540-51-2: 12106
540-54-5: 20516
540-61-4: 00870
540-63-6: 12033
540-67-0: 11850

540-69-2: 12841
540-73-8: 14612
540-80-7: 17143
540-82-9: 24293
540-84-1: 18277
540-88-5: 00606
540-97-6: 10295
541-01-5: 13727
541-02-6: 10693
541-05-9: 10853
541-14-0: 20486
541-15-1: 20485
541-16-2: 20796
541-28-6: 08006
541-31-1: 08271
541-33-3: 07963
541-35-5: 07829
541-41-3: 09226
541-42-4: 17150
541-46-8: 07859
541-47-9: 08907
541-50-4: 08485
541-53-7: 24660
541-58-2: 24514
541-59-3: 23337
541-69-5: 04455
541-73-1: 03038
541-88-8: 00503
541-89-9: 08868
541-91-3: 10481
542-05-2: 18317
542-08-5: 08491
542-10-9: 11990
542-11-0: 02289
542-15-4: 02421
542-16-5: 02488
542-18-7: 09815
542-28-9: 22285
542-37-0: 08454
542-46-1: 09623
542-50-7: 13400
542-52-9: 09197
542-54-1: 18369
542-55-2: 12851
542-56-3: 17152
542-58-5: 12117
542-59-6: 12005
542-69-8: 08002
542-76-7: 20971
542-81-4: 11790
542-85-8: 11848
542-88-1: 15686
542-90-5: 24636
542-91-6: 24039
542-92-7: 10482
543-20-4: 08243
543-21-5: 09002
543-24-8: 13273
543-27-1: 09232
543-28-2: 09093
543-29-3: 17138
543-38-4: 14578
543-50-0: 24641
543-54-4: 22742
543-59-9: 18193
543-67-9: 17156
543-75-9: 11311
543-82-8: 13473
543-83-9: 13368
543-86-2: 08556
543-87-3: 08559
544-00-3: 07883
544-01-4: 08038
544-02-5: 08060
544-10-5: 14002
544-13-8: 18292
544-16-1: 17141
544-25-2: 09657
544-35-4: 17362
544-48-9: 10895
544-57-0: 24338
544-62-7: 20913
544-63-8: 24367
544-65-0: 24393
544-66-1: 24394
544-72-9: 17466
544-73-0: 17464
544-76-3: 13858
544-77-4: 13869
544-84-3: 13850
544-85-4: 11569
545-06-2: 00898
545-26-6: 09268
545-47-1: 15537
545-68-6: 11643
545-93-7: 23179
546-06-5: 09463
546-33-8: 10857
546-43-0: 17053
546-45-2: 10858
546-56-5: 10845
546-97-4: 12616
547-25-1: 12859
547-52-4: 05305
547-63-7: 21237
547-65-9: 13088
547-81-9: 11674
547-91-1: 23769
547-94-4: 03238
548-39-0: 18822
548-40-3: 18042
548-51-6: 06388
548-59-4: 09718
548-61-8: 02371
548-66-3: 04162
548-93-6: 05915
548-98-1: 09334
550-10-7: 11335
550-23-2: 14794
550-24-3: 09720
550-44-7: 15389
550-60-7: 16993
550-74-3: 22478
550-82-3: 19711
550-90-3: 15784
550-97-0: 06313
550-99-2: 14764
551-01-9: 20756
551-06-4: 16427
551-09-7: 11958
551-35-9: 01911
551-62-2: 03959
551-72-4: 15289
551-76-8: 19652
551-77-9: 19651
551-78-0: 19649
551-88-2: 18263
551-92-8: 14773
552-16-9: 06515
552-30-7: 15306
552-32-9: 00347
552-41-0: 12416
552-45-4: 02884
552-46-5: 16173
552-58-9: 06905
552-59-0: 06923
552-62-5: 22204
552-63-6: 04216
552-70-5: 01474
552-72-7: 20252
552-79-4: 05049
552-82-9: 02404
552-86-3: 12351
552-89-6: 01760
552-94-3: 06299
553-03-7: 23833

553-17-3: 19405
553-19-5: 05671
553-20-8: 00350
553-26-4: 07696
553-39-9: 16857
553-53-7: 22849
553-60-6: 22853
553-68-4: 12213
553-69-5: 05150
553-82-2: 03086
553-86-6: 05820
553-90-2: 11985
553-94-6: 02619
553-97-9: 09735
554-00-7: 02140
554-01-8: 23230
554-12-1: 21222
554-14-3: 24749
554-21-2: 11334
554-35-8: 20981
554-52-9: 19020
554-55-2: 02017
554-68-7: 11700
554-70-1: 19750
554-84-7: 19536
554-95-0: 05573
555-03-3: 03605
555-10-2: 10328
555-16-8: 01762
555-21-5: 04344
555-28-2: 20288
555-30-6: 25256
555-43-1: 17449
555-44-2: 13900
555-45-3: 24380
555-48-6: 00075
555-57-7: 04943
555-59-9: 08915
555-77-1: 11690
555-89-5: 03657
555-96-4: 14652
556-02-5: 25236
556-03-6: 25237
556-08-1: 05845
556-10-5: 25479
556-18-3: 01575
556-24-1: 08457
556-27-4: 01026
556-33-2: 13315
556-48-9: 10054
556-50-3: 13314
556-52-5: 18035
556-53-6: 20456
556-56-9: 21716
556-61-6: 15682
556-64-9: 24640
556-67-2: 10844
556-68-3: 10460
556-69-4: 17769
556-71-8: 10442
556-82-1: 08935
556-88-7: 13375
556-89-8: 25478
556-90-1: 24603
556-96-7: 02617
556-97-8: 02794
557-00-6: 08468
557-11-9: 25493
557-17-5: 20642
557-22-2: 11437
557-24-4: 08857
557-25-5: 08382
557-30-2: 11911
557-31-3: 21710
557-40-4: 21730
557-44-8: 13902
557-59-5: 24337
557-61-9: 17357
557-66-4: 11724
557-68-6: 15634
557-89-1: 01416
557-91-5: 11799
557-93-7: 21673
557-94-8: 11766
557-95-9: 15632
557-98-2: 21679
557-99-3: 00943
558-13-4: 15693
558-17-8: 20628
558-20-3: 15720
558-25-8: 15763
558-30-5: 18000
558-37-2: 08733
558-38-3: 21311
558-42-9: 21369
558-43-0: 20910
558-53-2: 11909
558-59-8: 11908
559-70-6: 17869
559-82-0: 09309
560-05-4: 10618
560-07-6: 10621
560-08-7: 09038
560-21-4: 18278
560-22-5: 18650
560-23-6: 18647
560-27-0: 20820
560-95-2: 15639
561-05-7: 00954
561-07-9: 00953
561-10-4: 14355
561-20-6: 23940
561-25-1: 16061
561-27-3: 16066
561-83-1: 23194
561-94-4: 11636
562-23-2: 17008
562-48-1: 16051
562-49-2: 18231
562-74-3: 10408
562-90-3: 00752
563-03-1: 19406
563-04-2: 19938
563-16-6: 14037
563-41-7: 14693
563-45-1: 08738
563-46-2: 08737
563-47-3: 21684
563-52-0: 08717
563-57-5: 21700
563-58-6: 21694
563-60-0: 12607
563-70-2: 15636
563-76-8: 21592
563-78-0: 08732
563-79-1: 08781
563-80-4: 08663
563-83-7: 20426
563-84-8: 08573
563-93-9: 12608
564-00-1: 07442
564-02-3: 18276
564-03-4: 18651
564-04-5: 18580
564-36-3: 11630
564-37-4: 11631
564-94-3: 07376
565-00-4: 07272
565-53-7: 00567
565-59-3: 18229
565-60-6: 18514
565-62-8: 18748
565-63-9: 08902
565-64-0: 21144
565-67-3: 18529
565-68-4: 18798

565-69-5: 18587
565-70-8: 08424
565-75-3: 18279
565-76-4: 18648
565-77-5: 18683
565-80-0: 18581
565-81-1: 08306
566-02-9: 20330
566-27-8: 09376
566-56-3: 20322
566-57-4: 20321
566-65-4: 20332
566-93-8: 09353
567-18-0: 16884
567-61-3: 06379
567-72-6: 09352
568-02-5: 17091
568-21-8: 11182
568-93-4: 01267
568-98-9: 01209
568-99-0: 01212
569-06-2: 01290
569-08-4: 01257
569-10-8: 01256
569-15-3: 18924
569-31-3: 15336
569-34-6: 22203
569-41-5: 16361
569-42-6: 16735
569-51-7: 05571
569-57-3: 02815
569-65-3: 20045
569-77-7: 05692
570-08-1: 20789
570-10-5: 06369
570-22-9: 14820
570-23-0: 05918
570-24-1: 02390
570-30-9: 11672
570-74-1: 09371
571-31-3: 01063
571-57-3: 16906
571-58-4: 16357
571-60-8: 16731
571-61-9: 16358
571-98-2: 05963
572-45-2: 12021
572-51-0: 18939
572-59-8: 09436
572-67-8: 17919
572-83-8: 01220
572-93-0: 01261
573-11-5: 06626
573-12-6: 18923
573-17-1: 18832
573-20-6: 16744
573-22-8: 18947
573-26-2: 00308
573-27-3: 00278
573-33-1: 14766
573-34-2: 17941
573-54-6: 06018
573-56-8: 19335
573-69-3: 04615
573-79-5: 05959
573-90-0: 00314
573-97-7: 16966
573-98-8: 16355
574-00-5: 16729
574-09-4: 12387
574-12-9: 06935
574-17-4: 15205
574-19-6: 12428
574-25-4: 15286
574-39-0: 09161
574-45-8: 02241
574-61-8: 15964
574-66-3: 15963
574-84-5: 06832
574-93-6: 20020
574-96-9: 16600
574-98-1: 15369
575-36-0: 00330
575-37-1: 16360
575-38-2: 16734
575-41-7: 16356
575-43-9: 16359
575-44-0: 16733
575-61-1: 15353
575-74-6: 01828
575-82-6: 06307
575-89-3: 00790
576-15-8: 15118
576-22-7: 02618
576-23-8: 02615
576-24-9: 19220
576-26-1: 19268
576-54-5: 09876
576-55-6: 19605
576-83-0: 02733
576-92-1: 05418
577-16-2: 12471
577-19-5: 02704
577-27-5: 10836
577-33-3: 01340
577-37-7: 15780
577-47-9: 14890
577-55-9: 02568
577-56-0: 05839
577-66-2: 22210
577-68-4: 04560
577-71-9: 19336
577-72-0: 02335
577-85-5: 06927
577-91-3: 05230
577-95-7: 01184
578-06-3: 00955
578-07-4: 00958
578-11-0: 05849
578-22-3: 06385
578-39-2: 06383
578-46-1: 02396
578-51-8: 02594
578-54-1: 02266
578-57-4: 02663
578-58-5: 03594
578-66-5: 23518
578-67-6: 23778
578-68-7: 23514
578-86-9: 06886
578-95-0: 00977
579-04-4: 14536
579-07-7: 20951
579-10-2: 00320
579-11-3: 00176
579-18-0: 05966
579-21-5: 12475
579-43-1: 12000
579-44-2: 12409
579-52-2: 06404
579-58-8: 00279
579-60-2: 19436
579-66-8: 02152
579-72-6: 01663
579-75-9: 06436
579-92-0: 06412
579-93-1: 05970
579-94-2: 00618
579-98-6: 13275
580-02-9: 05858
580-13-2: 16251
580-15-4: 23516
580-16-5: 23779
580-17-6: 23513
580-18-7: 23776
580-19-8: 23517

580-20-1: 23780
580-22-3: 23512
580-48-3: 25019
580-51-8: 07663
581-08-8: 00240
581-28-2: 00956
581-29-3: 00957
581-30-6: 19706
581-40-8: 16362
581-42-0: 16363
581-43-1: 16737
581-45-3: 22732
581-46-4: 07694
581-47-5: 07693
581-49-7: 07698
581-55-5: 15742
581-64-6: 19705
581-89-5: 16466
581-96-4: 16571
581-97-5: 00331
582-02-5: 11057
582-08-1: 11056
582-16-1: 16364
582-17-2: 16738
582-22-9: 04807
582-24-1: 12434
582-33-2: 05874
582-60-5: 05616
582-61-6: 07187
582-62-7: 08620
582-69-4: 05961
582-77-4: 01934
582-78-5: 01935
582-80-9: 05974
582-85-4: 19512
583-03-9: 05000
583-04-0: 06569
583-05-1: 18347
583-06-2: 08913
583-17-5: 21865
583-19-7: 02635
583-33-5: 12962
583-39-1: 05645
583-46-0: 14900
583-48-2: 14038
583-50-6: 07826
583-53-9: 02996
583-55-1: 02655
583-58-4: 22630
583-61-9: 22625
583-63-1: 09695
583-68-6: 02039
583-69-7: 04744
583-70-0: 02616
583-71-1: 02620
583-75-5: 02044
583-78-8: 19224
583-80-2: 09664
583-86-8: 17453
584-02-1: 18478
584-04-3: 21389
584-12-3: 12864
584-13-4: 25034
584-26-9: 14889
584-27-0: 08187
584-44-1: 06389
584-45-2: 20878
584-48-5: 02623
584-49-6: 05421
584-70-3: 01933
584-84-9: 03164
584-87-2: 06288
584-90-7: 11040
584-92-9: 17656
584-94-1: 14035
584-98-5: 08115
584-99-6: 08820
585-05-7: 00458
585-25-1: 17650
585-32-0: 04905
585-34-2: 19290
585-47-7: 04781
585-48-8: 22582
585-68-2: 12997
585-70-6: 12961
585-71-7: 02640
585-74-0: 12472
585-76-2: 05991
585-79-5: 02705
585-81-9: 06382
585-84-2: 21770
586-10-7: 02333
586-11-8: 19337
586-27-6: 10403
586-30-1: 06381
586-35-6: 04618
586-37-8: 12451
586-38-9: 06440
586-39-0: 03363
586-42-5: 05840
586-61-8: 02679
586-62-9: 10348
586-63-0: 10349
586-75-4: 07194
586-76-5: 05995
586-77-6: 02035
586-78-7: 02706
586-84-5: 12907
586-89-0: 05841
586-91-4: 05958
586-95-8: 22926
586-96-9: 03820
586-98-1: 22924
587-02-0: 02267
587-03-1: 05111
587-04-2: 01602
587-48-4: 05844
587-54-2: 05973
587-55-3: 05960
587-61-1: 22767
587-65-5: 00172
587-85-9: 15596
588-01-2: 25187
588-07-8: 00174
588-16-9: 00300
588-22-7: 00594
588-23-8: 00311
588-30-7: 21867
588-32-9: 00542
588-46-5: 00364
588-47-6: 02416
588-52-3: 11035
588-53-4: 19580
588-63-6: 02723
588-64-7: 01771
588-67-0: 02752
588-68-1: 01774
588-72-7: 02631
588-73-8: 02632
588-88-5: 20104
588-96-5: 02637
589-08-2: 04809
589-09-3: 02484
589-10-6: 02638
589-14-0: 13046
589-15-1: 02585
589-16-2: 02268
589-17-3: 02598
589-18-4: 05112
589-21-9: 14597
589-29-7: 04635
589-33-3: 23284
589-38-8: 14336
589-40-2: 12850
589-41-3: 09084
589-44-6: 08307

589-53-7: 13543
589-55-9: 13639
589-59-3: 08462
589-63-9: 17754
589-68-4: 24372
589-75-3: 17674
589-87-7: 02657
589-92-4: 10255
589-93-5: 22627
589-96-8: 21315
589-98-0: 17702
590-01-2: 21090
590-02-3: 00504
590-11-4: 12561
590-12-5: 12560
590-13-6: 21672
590-14-7: 21671
590-16-9: 11767
590-17-0: 00872
590-18-1: 08746
590-19-2: 07773
590-26-1: 12576
590-34-1: 01442
590-35-2: 18228
590-36-3: 18512
590-42-1: 20639
590-50-1: 18550
590-60-3: 08589
590-66-9: 09838
590-67-0: 10148
590-72-7: 01426
590-73-8: 14034
590-86-3: 07819
590-87-4: 18248
590-90-9: 08655
590-92-1: 21079
590-93-2: 09012
591-07-1: 00065
591-08-2: 00089
591-09-3: 00474
591-11-7: 13108
591-12-8: 13102
591-17-3: 02668
591-18-4: 02656
591-19-5: 02024
591-20-8: 19092
591-22-0: 22631
591-27-5: 18987
591-28-6: 23234
591-31-1: 01742
591-33-3: 00241
591-34-4: 21254
591-35-5: 19227
591-38-8: 00066
591-47-9: 10327
591-49-1: 10324
591-50-4: 03526
591-54-8: 23019
591-55-9: 23020
591-60-6: 08486
591-62-8: 09049
591-68-4: 18408
591-73-1: 13909
591-76-4: 14063
591-78-6: 14335
591-80-0: 18694
591-81-1: 08432
591-82-2: 20638
591-84-4: 24644
591-86-6: 14688
591-87-7: 00738
591-93-5: 18089
591-95-7: 18088
591-96-8: 18090
592-13-2: 14036
592-17-6: 25473
592-20-1: 21530
592-22-3: 11411
592-27-8: 13540
592-31-4: 25379
592-34-7: 09217
592-35-8: 09051
592-41-6: 14389
592-42-7: 13933
592-44-9: 13931
592-45-0: 13932
592-46-1: 13935
592-49-4: 13934
592-50-7: 18247
592-51-8: 18693
592-55-2: 11763
592-56-3: 00040
592-57-4: 09673
592-65-4: 20708
592-70-1: 09085
592-72-3: 08029
592-76-7: 13735
592-80-3: 18492
592-82-5: 08009
592-84-7: 12842
592-88-1: 21739
592-90-5: 17972
593-03-3: 13906
593-08-8: 25169
593-11-3: 11521
593-12-4: 17581
593-14-6: 17587
593-45-3: 17374
593-49-7: 13397
593-50-0: 24953
593-51-1: 15610
593-52-2: 01399
593-53-3: 15676
593-54-4: 19743
593-55-5: 11723
593-56-6: 14732
593-57-7: 01394
593-58-8: 01401
593-59-9: 01396
593-60-2: 12544
593-61-3: 12637
593-63-5: 12638
593-66-8: 12581
593-67-9: 12540
593-68-0: 19740
593-70-4: 15646
593-71-5: 15647
593-79-3: 15688
593-80-6: 15692
593-81-7: 15607
593-84-0: 24630
593-87-3: 13372
593-88-4: 01406
593-92-0: 12559
593-94-2: 15661
593-96-4: 11754
593-98-6: 15628
594-00-3: 11784
594-02-5: 11832
594-03-6: 12032
594-04-7: 15667
594-07-0: 09131
594-08-1: 09254
594-09-2: 19752
594-10-5: 24242
594-11-6: 10780
594-15-0: 15703
594-16-1: 20545
594-18-3: 15657
594-20-7: 20561
594-21-8: 11904
594-22-9: 15707
594-27-4: 24229
594-34-3: 20554
594-36-5: 07940
594-37-6: 20567

594-38-7: 08007
594-39-8: 07891
594-42-3: 15744
594-44-5: 12043
594-45-6: 12037
594-48-9: 21126
594-51-4: 07959
594-52-5: 07915
594-56-9: 08744
594-57-0: 07932
594-58-1: 21113
594-60-5: 08580
594-61-6: 21202
594-65-0: 00380
594-67-2: 11872
594-68-3: 00847
594-70-7: 20663
594-71-8: 20540
594-72-9: 11821
594-73-0: 11842
594-81-0: 07955
594-82-1: 08057
594-83-2: 08600
594-84-3: 07970
594-85-4: 07971
594-89-8: 20613
594-90-1: 20672
594-91-2: 08033
595-30-2: 17887
595-31-3: 17888
595-32-4: 10620
595-37-9: 08384
595-39-1: 15489
595-40-4: 15490
595-41-5: 18525
595-44-8: 20569
595-49-3: 20588
595-50-6: 20687
595-84-6: 20833
595-89-1: 20305
595-90-4: 24230
595-91-5: 04175
595-92-6: 11388
596-24-7: 24195
596-27-0: 15333
596-29-2: 15343
596-41-8: 11145
596-43-0: 02682
596-48-5: 05137
596-51-0: 23416
596-75-8: 20814
596-85-0: 16859
596-88-3: 11637
597-04-6: 08394
597-05-7: 18528
597-09-1: 20937
597-12-6: 13216
597-32-0: 21383
597-33-1: 20897
597-35-3: 11875
597-43-3: 08152
597-44-4: 08171
597-49-9: 18527
597-50-2: 19762
597-51-3: 19768
597-52-4: 24165
597-55-7: 20798
597-59-1: 21022
597-67-1: 24061
597-71-7: 20922
597-72-8: 20641
597-76-2: 14318
597-77-3: 14320
597-81-9: 19982
597-84-2: 19983
597-93-3: 17276
597-96-6: 14323
598-01-6: 13692
598-02-7: 19918
598-03-8: 20686
598-04-9: 08045
598-05-0: 24292
598-10-7: 10818
598-16-3: 12598
598-17-4: 20542
598-19-6: 21982
598-21-0: 00925
598-22-1: 21591
598-23-2: 08996
598-25-4: 07779
598-26-5: 21995
598-29-8: 20577
598-31-2: 21533
598-32-3: 08929
598-37-8: 11819
598-38-9: 12135
598-40-3: 21612
598-41-4: 00063
598-42-5: 00255
598-45-8: 20634
598-50-5: 25459
598-52-7: 24881
598-53-8: 20643
598-55-0: 09088
598-56-1: 11704
598-57-2: 15616
598-58-3: 17135
598-60-7: 01536
598-61-8: 09535
598-64-1: 09134
598-65-2: 13363
598-74-3: 07892
598-76-5: 20566
598-77-6: 20717
598-78-7: 21091
598-79-8: 21797
598-80-1: 21207
598-81-2: 20423
598-88-9: 12568
598-89-0: 00603
598-92-5: 11791
598-94-7: 25423
598-96-9: 18685
598-98-1: 21167
598-99-2: 00771
599-01-9: 21290
599-04-2: 13077
599-07-5: 17863
599-50-8: 09287
599-61-1: 02491
599-62-2: 05364
599-64-4: 19513
599-66-6: 03930
599-67-7: 05126
599-69-9: 05393
599-70-2: 03444
599-71-3: 05372
599-75-7: 05466
599-79-1: 06407
599-86-0: 05381
599-89-3: 05460
599-91-7: 05469
599-94-0: 03313
599-97-3: 11894
599-99-5: 00758
600-00-0: 21086
600-05-5: 21128
600-06-6: 07930
600-07-7: 08445
600-11-3: 18218
600-13-5: 09492
600-14-6: 18345
600-15-7: 08425
600-17-9: 08923
600-18-0: 08483
600-20-4: 08665

600-21-5: 01016
600-22-6: 21265
600-25-9: 20536
600-26-0: 00046
600-28-2: 23290
600-30-6: 08367
600-31-7: 20800
600-32-8: 08373
600-33-9: 20804
600-36-2: 18526
600-37-3: 08436
600-40-8: 11836
600-76-0: 05653
601-34-3: 09383
601-54-7: 09389
601-57-0: 09388
601-75-2: 20829
601-76-3: 20586
601-77-4: 20483
601-79-6: 20855
601-84-3: 06138
601-87-6: 02391
601-88-7: 03105
601-89-8: 04719
601-97-8: 04569
601-99-0: 06394
602-00-6: 06395
602-01-7: 03639
602-02-8: 02800
602-03-9: 02229
602-09-5: 07438
602-13-1: 04070
602-15-3: 18860
602-21-1: 04582
602-23-3: 04071
602-25-5: 01244
602-27-7: 03272
602-29-9: 03790
602-37-9: 16297
602-38-0: 16378
602-55-1: 01161
602-56-2: 00969
602-60-8: 01158
602-63-1: 01265
602-64-2: 01314
602-65-3: 01318
602-69-7: 01170
602-73-3: 01243
602-82-4: 04531
602-87-9: 00018
602-92-6: 06145
602-94-8: 06541
602-96-0: 04072
602-99-3: 19527
603-02-1: 03209
603-06-5: 03206
603-10-1: 19670
603-11-2: 04576
603-12-3: 06233
603-13-4: 04073
603-26-9: 15675
603-28-1: 02020
603-32-7: 01407
603-33-8: 07715
603-34-9: 02237
603-35-0: 19753
603-36-1: 24243
603-38-3: 02240
603-39-4: 03882
603-40-7: 02473
603-44-1: 19472
603-45-2: 09756
603-48-5: 02372
603-49-6: 03682
603-50-9: 19597
603-51-0: 14687
603-52-1: 09076
603-53-2: 13381
603-54-3: 25436
603-62-3: 15395
603-69-0: 08297
603-71-4: 04064
603-76-9: 15153
603-77-0: 14616
603-78-1: 06135
603-79-2: 06201
603-80-5: 06380
603-81-6: 06130
603-83-8: 02387
603-84-9: 19183
603-85-0: 19037
603-86-1: 19186
603-87-2: 19040
604-32-0: 09380
604-35-3: 09379
604-44-4: 16913
604-53-5: 07430
604-59-1: 17088
604-61-5: 05964
604-68-2: 13199
604-69-3: 13207
604-83-1: 18859
604-88-6: 03510
604-94-4: 18930
604-95-5: 18934
605-01-6: 03864
605-02-7: 16484
605-03-8: 23697
605-04-9: 23700
605-22-1: 01236
605-27-6: 01303
605-32-3: 01294
605-36-7: 16501
605-37-8: 16787
605-39-0: 07483
605-40-3: 01250
605-42-5: 01242
605-43-6: 01251
605-44-7: 01238
605-45-8: 04539
605-48-1: 01137
605-50-5: 04538
605-54-9: 04534
605-55-0: 18943
605-61-8: 16293
605-62-9: 16942
605-63-0: 16294
605-65-2: 16891
605-69-6: 16925
605-70-9: 16712
605-71-0: 16375
605-78-7: 15959
605-82-3: 01151
605-83-4: 01150
605-84-5: 11055
605-85-6: 12486
605-87-8: 18944
605-89-0: 16746
605-90-3: 01893
605-99-2: 22173
606-04-2: 22193
606-10-0: 15842
606-12-2: 15988
606-17-7: 06249
606-18-8: 05936
606-19-9: 04602
606-20-2: 03642
606-21-3: 02803
606-22-4: 02232
606-23-5: 15056
606-26-8: 06517
606-27-9: 06518
606-28-0: 05965
606-31-5: 01669
606-33-7: 23471
606-34-8: 01799

606-35-9: 03625
606-37-1: 16373
606-38-2: 00224
606-40-6: 16911
606-42-8: 23619
606-43-9: 23844
606-45-1: 06438
606-46-2: 02159
606-57-5: 16223
606-75-7: 07620
606-79-1: 16985
606-80-4: 07621
606-81-5: 07498
606-83-7: 05250
606-84-8: 21843
606-86-0: 21529
606-87-1: 04911
606-88-2: 14661
606-91-7: 05834
606-93-9: 23774
606-97-3: 07541
606-99-5: 02279
607-00-1: 12822
607-01-2: 19741
607-04-5: 07460
607-07-8: 07502
607-09-0: 16762
607-12-5: 07670
607-20-5: 16761
607-22-7: 16291
607-23-8: 16188
607-24-9: 16940
607-28-3: 15217
607-31-8: 23663
607-32-9: 15459
607-33-0: 16860
607-34-1: 23691
607-35-2: 23694
607-37-4: 16304
607-38-5: 16229
607-39-6: 16998
607-42-1: 01167
607-44-3: 16226
607-45-4: 16995
607-46-5: 16376
607-50-1: 16439
607-52-3: 16475
607-53-4: 16544
607-54-5: 04398
607-55-6: 16905
607-56-7: 25430
607-57-8: 12747
607-59-0: 02202
607-60-3: 16198
607-61-4: 01186
607-62-5: 01185
607-66-9: 23846
607-67-0: 23788
607-71-6: 06858
607-81-8: 20876
607-85-2: 06310
607-86-3: 06459
607-90-9: 06323
607-91-0: 05734
607-97-6: 08412
607-99-8: 03992
608-05-9: 15214
608-07-1: 15224
608-08-2: 15267
608-16-2: 24613
608-21-9: 03985
608-22-0: 02128
608-23-1: 02792
608-25-3: 04708
608-26-4: 19168
608-27-5: 02139
608-28-6: 03532
608-29-7: 04041
608-30-0: 02131
608-31-1: 02142
608-32-2: 05563
608-33-3: 19212
608-34-4: 23142
608-36-6: 08128
608-37-7: 08827
608-42-4: 08829
608-43-5: 04752
608-44-6: 04748
608-45-7: 01368
608-50-4: 04057
608-53-7: 01367
608-54-8: 01362
608-55-9: 18602
608-66-2: 13157
608-68-4: 08136
608-69-5: 08145
608-71-9: 19548
608-72-0: 04001
608-74-2: 03512
608-80-0: 04878
608-90-2: 03856
608-91-3: 19650
608-92-4: 19653
608-93-5: 03858
608-96-8: 03871
608-97-9: 05159
608-98-0: 04738
609-02-9: 20852
609-08-5: 20851
609-09-6: 20867
609-11-0: 08368
609-12-1: 08338
609-13-2: 08341
609-14-3: 08475
609-16-5: 02506
609-19-8: 19643
609-20-1: 04478
609-21-2: 19002
609-22-3: 19215
609-23-4: 19660
609-24-5: 19629
609-25-6: 05584
609-26-7: 18243
609-27-8: 18507
609-29-0: 18352
609-30-3: 08434
609-31-4: 08563
609-39-2: 12915
609-40-5: 24759
609-41-6: 23283
609-46-1: 05450
609-63-2: 05324
609-65-4: 07198
609-66-5: 01829
609-67-6: 07219
609-69-8: 10016
609-70-1: 22851
609-71-2: 22846
609-72-3: 02521
609-73-4: 03549
609-86-9: 05909
609-87-0: 06595
609-88-1: 04063
609-92-7: 03215
609-93-8: 19453
609-98-3: 05578
609-99-4: 06349
610-02-6: 06610
610-03-7: 03214
610-04-8: 06287
610-09-3: 10032
610-10-6: 10038
610-11-7: 09691
610-14-0: 07227
610-15-1: 01941
610-16-2: 06217

610-17-3: 02214
610-22-0: 04579
610-23-1: 03208
610-25-3: 03791
610-26-4: 19671
610-27-5: 04578
610-28-6: 06232
610-29-7: 04628
610-30-0: 06230
610-31-1: 04074
610-33-3: 04278
610-34-4: 06516
610-35-5: 04570
610-36-6: 05946
610-37-7: 06399
610-38-8: 02626
610-39-9: 03644
610-40-2: 02804
610-46-8: 01139
610-48-0: 01154
610-49-1: 01119
610-50-4: 01344
610-53-7: 00230
610-54-8: 03370
610-66-2: 04342
610-67-3: 03386
610-69-5: 00705
610-71-9: 06137
610-72-0: 06206
610-74-2: 06132
610-78-6: 19190
610-81-1: 19044
610-89-9: 18710
610-90-2: 06613
610-91-3: 04062
610-92-4: 04625
610-93-5: 15351
610-94-6: 05990
610-96-8: 06051
610-97-9: 06414
610-99-1: 21496
611-00-7: 06136
611-01-8: 06202
611-03-0: 06131
611-05-2: 02392
611-06-3: 03108
611-07-4: 19191
611-08-5: 23146
611-09-6: 15216
611-10-9: 10602
611-12-1: 21319
611-13-2: 12974
611-14-3: 03421
611-15-4: 03359
611-17-6: 02670
611-19-8: 02780
611-20-1: 06710
611-21-2: 02188
611-22-3: 02292
611-23-4: 03720
611-24-5: 19443
611-32-5: 23678
611-33-6: 23580
611-34-7: 23515
611-35-8: 23576
611-36-9: 23777
611-41-6: 23391
611-43-8: 07484
611-44-9: 24942
611-45-0: 16487
611-49-4: 16464
611-61-0: 07486
611-62-1: 07632
611-69-8: 05119
611-70-1: 21516
611-71-2: 04190
611-72-3: 04189
611-73-4: 04281
611-74-5: 01874
611-76-7: 06761
611-79-0: 15843
611-80-3: 15864
611-91-6: 21462
611-92-7: 25424
611-94-9: 16004
611-95-0: 05968
611-97-2: 15868
611-98-3: 15844
611-99-4: 15865
612-00-0: 03407
612-01-1: 25460
612-02-2: 23952
612-03-3: 00309
612-08-8: 11027
612-11-3: 22622
612-12-4: 02552
612-13-5: 06683
612-14-6: 04633
612-15-7: 03356
612-16-8: 05092
612-17-9: 16344
612-18-0: 23625
612-19-1: 06267
612-20-4: 06371
612-22-6: 03426
612-23-7: 02891
612-24-8: 06724
612-25-9: 05133
612-27-1: 06534
612-28-2: 02397
612-29-3: 03806
612-35-1: 06558
612-36-2: 02427
612-37-3: 22175
612-40-8: 06038
612-42-0: 06043
612-45-3: 00315
612-52-2: 16210
612-53-3: 15236
612-54-4: 15280
612-55-5: 16421
612-57-7: 23578
612-58-8: 23673
612-59-9: 23575
612-60-2: 23677
612-61-3: 23579
612-62-4: 23574
612-64-6: 02278
612-69-1: 07649
612-71-5: 24318
612-75-9: 07490
612-76-0: 07634
612-78-2: 07432
612-79-3: 07706
612-80-6: 07704
612-81-7: 07702
612-84-0: 07647
612-87-3: 07622
612-94-2: 16485
612-95-3: 23699
612-96-4: 23695
612-97-5: 23454
612-98-6: 04945
613-01-4: 11647
613-03-6: 05588
613-06-9: 01140
613-07-0: 01136
613-08-1: 01168
613-11-6: 19693
613-12-7: 01155
613-13-8: 01120
613-14-9: 01345
613-15-0: 00968
613-17-2: 16684
613-18-3: 23610
613-19-4: 23782

613-20-7: 16750
613-21-8: 23801
613-29-6: 02138
613-30-9: 23680
613-31-0: 01138
613-32-1: 18969
613-33-2: 07494
613-35-4: 00093
613-36-5: 02978
613-37-6: 07521
613-39-8: 20065
613-40-1: 07505
613-42-3: 07542
613-43-4: 11034
613-45-6: 01653
613-46-7: 16582
613-48-9: 02160
613-49-0: 16861
613-50-3: 23692
613-51-4: 23693
613-53-6: 23711
613-54-7: 12302
613-56-9: 15960
613-59-2: 16488
613-62-7: 16486
613-64-9: 14619
613-65-0: 02203
613-69-4: 01670
613-70-7: 19403
613-73-0: 04417
613-75-2: 15737
613-76-3: 16705
613-77-4: 23613
613-78-5: 06314
613-79-6: 16440
613-80-9: 16476
613-81-0: 16545
613-84-3: 01720
613-85-4: 14615
613-88-7: 04326
613-89-8: 12268
613-90-1: 04345
613-91-2: 12502
613-92-3: 04396
613-93-4: 01928
613-94-5: 06293
613-97-8: 02277
614-00-6: 02402
614-16-4: 05193
614-17-5: 01894
614-18-6: 22848
614-20-0: 05245
614-21-1: 12492
614-22-2: 01805
614-23-3: 01817
614-27-7: 05247
614-28-8: 01819
614-29-9: 04863
614-30-2: 04940
614-32-4: 06503
614-33-5: 21044
614-34-6: 06504
614-39-1: 01809
614-45-9: 04384
614-47-1: 22009
614-48-2: 22032
614-54-0: 05078
614-60-8: 21866
614-61-9: 00541
614-63-1: 24955
614-66-4: 18119
614-68-6: 03560
614-70-0: 19360
614-71-1: 03383
614-72-2: 02817
614-73-3: 03380
614-75-5: 04203
614-76-6: 00108
614-77-7: 25468
614-78-8: 24882
614-80-2: 00274
614-96-0: 15157
614-97-1: 05622
614-99-3: 12968
615-00-9: 14614
615-01-0: 19505
615-05-4: 04463
615-09-8: 13111
615-10-1: 12993
615-11-2: 12969
615-13-4: 15103
615-15-6: 05620
615-16-7: 05648
615-18-9: 07177
615-20-3: 07052
615-21-4: 07082
615-22-5: 07062
615-30-5: 18140
615-35-0: 11915
615-36-1: 02023
615-37-2: 03542
615-38-3: 10149
615-39-4: 10150
615-41-8: 02844
615-42-9: 03160
615-43-0: 02298
615-48-5: 05542
615-52-1: 11988
615-54-3: 03986
615-55-4: 02132
615-56-5: 19213
615-57-6: 02129
615-58-7: 19209
615-59-8: 03020
615-60-1: 02797
615-62-3: 19174
615-65-6: 02092
615-67-8: 04745
615-68-9: 04042
615-71-4: 05564
615-72-5: 19205
615-74-7: 19170
615-76-9: 24966
615-77-0: 23141
615-79-2: 18423
615-80-5: 17905
615-81-6: 11975
615-82-7: 13323
615-83-8: 18403
615-84-9: 11914
615-85-0: 11906
615-86-1: 01101
615-87-2: 03007
615-89-4: 04697
615-90-7: 04753
615-93-0: 09710
615-94-1: 09716
615-98-5: 11987
615-99-6: 11986
616-02-4: 13034
616-04-6: 14862
616-06-8: 17344
616-10-4: 07420
616-12-6: 18679
616-16-0: 08550
616-19-3: 20568
616-20-6: 18195
616-21-7: 07964
616-23-9: 21314
616-24-0: 18163
616-25-1: 18715
616-27-3: 08631
616-28-4: 09009
616-29-5: 21376
616-30-8: 20892
616-31-9: 18373

616-34-2: 13326
616-38-6: 09200
616-39-7: 11719
616-40-0: 14609
616-42-2: 24299
616-43-3: 23308
616-44-4: 24750
616-45-5: 23417
616-46-6: 24666
616-47-7: 14777
616-55-7: 19078
616-62-6: 20884
616-69-3: 03207
616-72-8: 03213
616-75-1: 20875
616-76-2: 06289
616-79-5: 05939
616-82-0: 06398
616-85-3: 05453
616-86-4: 02264
616-91-1: 10869
617-00-5: 02377
617-02-7: 06502
617-05-0: 06364
617-07-2: 25369
617-09-4: 09191
617-20-9: 02251
617-22-1: 14588
617-23-2: 14640
617-30-1: 14157
617-31-2: 18428
617-32-3: 08241
617-33-4: 00565
617-35-6: 21264
617-36-7: 00464
617-41-4: 08120
617-42-5: 08823
617-43-6: 08822
617-45-8: 01460
617-48-1: 08165
617-50-5: 21241
617-51-6: 21192
617-52-7: 08186
617-53-8: 08837
617-54-9: 08838
617-60-7: 11514
617-65-2: 13245
617-68-5: 15541
617-72-1: 13550
617-73-2: 17680
617-75-4: 01404
617-78-7: 18239
617-80-1: 08251
617-83-4: 09501
617-84-5: 12818
617-85-6: 24241
617-86-7: 24130
617-88-9: 12872
617-89-0: 13037
617-90-3: 12941
617-92-5: 23298
617-94-7: 05039
617-99-2: 05596
618-01-9: 05439
618-03-1: 05415
618-07-5: 01438
618-13-3: 05446
618-27-9: 20372
618-30-4: 12964
618-31-5: 03016
618-32-6: 07191
618-34-8: 02808
618-36-0: 04928
618-39-3: 04391
618-40-6: 14635
618-41-7: 05287
618-45-1: 19463
618-46-2: 07199
618-47-3: 01926
618-48-4: 01830
618-49-5: 01901
618-51-9: 06415
618-58-6: 06140
618-61-1: 02393
618-62-2: 03106
618-63-3: 19188
618-64-4: 19042
618-65-5: 01687
618-68-8: 05251
618-71-3: 06238
618-73-5: 01962
618-74-6: 06601
618-76-8: 06340
618-77-9: 06596
618-80-4: 19240
618-83-7: 04604
618-84-8: 05944
618-85-9: 03641
618-86-0: 02802
618-87-1: 02233
618-88-2: 04609
618-89-3: 05994
618-91-7: 06417
618-94-0: 06521
618-95-1: 06522
618-98-4: 06520
619-02-3: 10259
619-03-4: 06139
619-04-5: 06211
619-05-6: 06133
619-08-9: 19184
619-10-3: 19185
619-12-5: 20000
619-14-7: 06396
619-15-8: 03643
619-17-0: 05937
619-18-1: 02231
619-19-2: 06392
619-20-5: 06268
619-21-6: 06283
619-22-7: 01664
619-23-8: 02894
619-24-9: 06725
619-25-0: 05134
619-26-1: 02398
619-27-2: 14648
619-28-3: 03100
619-31-8: 02215
619-33-0: 11812
619-37-4: 25412
619-39-6: 10976
619-41-0: 12301
619-42-1: 05997
619-44-3: 06419
619-45-4: 05882
619-50-1: 06529
619-52-3: 10347
619-55-6: 01927
619-56-7: 01831
619-57-8: 01902
619-58-9: 06418
619-60-3: 19278
619-64-7: 06269
619-65-8: 06127
619-66-9: 06284
619-67-0: 06296
619-68-1: 06536
619-72-7: 06726
619-73-8: 05135
619-75-0: 03022
619-78-3: 03102
619-80-7: 01943
619-81-8: 10044
619-82-9: 10048
619-84-1: 06219
619-86-3: 06263

619-93-2: 03330
619-99-8: 14042
620-02-0: 12949
620-03-1: 12952
620-05-3: 03541
620-08-6: 22672
620-11-1: 18522
620-12-2: 21041
620-13-3: 02685
620-14-4: 03422
620-16-6: 02828
620-17-7: 19357
620-18-8: 19350
620-19-9: 02885
620-20-2: 02781
620-22-4: 06720
620-23-5: 01754
620-24-6: 05080
620-25-7: 02293
620-26-8: 03721
620-27-9: 01441
620-32-6: 03931
620-40-6: 04882
620-42-8: 19999
620-43-9: 07881
620-47-3: 03738
620-49-5: 04449
620-52-0: 09192
620-54-2: 06559
620-59-7: 25244
620-63-3: 08466
620-71-3: 20434
620-72-4: 00489
620-73-5: 00519
620-79-1: 05200
620-80-4: 08502
620-81-5: 11916
620-82-6: 11981
620-83-7: 03739
620-85-9: 03432
620-86-0: 06560
620-88-2: 03810
620-92-8: 19455
620-93-9: 02379
620-94-0: 03980
620-95-1: 22725
621-01-2: 24854
621-02-3: 09193
621-03-4: 00189
621-04-5: 25444
621-06-7: 04116
621-07-8: 04923
621-08-9: 03920
621-09-0: 12084
621-10-3: 20740
621-12-5: 14686
621-13-6: 08811
621-14-7: 08155
621-15-8: 13982
621-18-1: 05957
621-19-2: 21945
621-21-6: 18124
621-23-8: 04050
621-25-0: 14717
621-26-1: 14589
621-27-2: 19595
621-29-4: 03561
621-31-8: 19361
621-32-9: 03384
621-33-0: 02258
621-34-1: 19354
621-36-3: 04261
621-37-4: 04204
621-38-5: 00109
621-40-9: 24883
621-42-1: 00275
621-44-3: 25260
621-50-1: 04343
621-51-2: 06261
621-52-3: 03387
621-56-7: 20903
621-58-9: 19352
621-59-0: 01712
621-62-5: 11775
621-63-6: 12139
621-64-7: 20466
621-66-9: 19507
621-70-5: 14265
621-72-7: 05631
621-76-1: 20653
621-77-2: 18165
621-83-0: 24890
621-84-1: 09117
621-87-4: 21578
621-88-5: 00358
621-91-0: 02578
621-94-3: 14641
621-95-4: 02249
622-04-8: 21393
622-06-0: 08809
622-08-2: 12235
622-09-3: 16100
622-15-1: 15774
622-16-2: 02309
622-20-8: 03314
622-21-9: 17291
622-22-0: 03312
622-24-2: 02830
622-26-4: 20261
622-29-7: 15617
622-30-0: 04922
622-31-1: 01766
622-32-2: 01767
622-37-7: 02532
622-38-8: 03445
622-39-9: 22737
622-40-2: 16131
622-42-4: 03799
622-45-7: 00557
622-46-8: 09111
622-47-9: 04263
622-50-4: 00284
622-51-5: 25470
622-52-6: 24884
622-56-0: 24350
622-57-1: 02276
622-58-2: 03562
622-60-6: 03385
622-61-7: 02819
622-62-8: 19355
622-63-9: 03448
622-68-4: 01430
622-74-2: 24962
622-75-3: 04419
622-76-4: 02771
622-77-5: 09512
622-78-6: 03569
622-79-7: 02537
622-80-0: 02487
622-82-2: 14629
622-85-5: 03903
622-86-6: 02820
622-87-7: 03838
622-93-5: 21320
622-95-7: 02672
622-96-8: 03423
622-97-9: 03361
622-98-0: 02829
623-00-7: 06673
623-03-0: 06679
623-04-1: 04968
623-05-2: 05081
623-08-5: 02190
623-09-6: 04497
623-10-9: 02294
623-11-0: 03722

623-12-1: 02859
623-15-4: 08958
623-17-6: 13045
623-18-7: 21858
623-19-8: 13050
623-20-1: 21857
623-21-2: 08415
623-24-5: 02550
623-25-6: 02554
623-26-7: 04509
623-27-8: 04518
623-30-3: 21625
623-32-5: 10743
623-33-6: 13311
623-34-7: 07975
623-36-9: 18615
623-39-2: 20908
623-40-5: 18572
623-42-7: 08444
623-43-8: 08899
623-46-1: 11816
623-47-2: 22117
623-48-3: 00652
623-49-4: 09242
623-50-7: 00639
623-51-8: 00663
623-53-0: 09205
623-56-3: 14363
623-57-4: 20904
623-59-6: 00304
623-60-9: 21368
623-65-4: 13877
623-66-5: 17670
623-69-8: 21390
623-70-1: 08893
623-71-2: 21104
623-72-3: 21197
623-73-4: 00563
623-76-7: 25416
623-78-9: 09080
623-80-3: 09243
623-81-4: 24298
623-82-5: 14144
623-84-7: 20901
623-85-8: 09120
623-86-9: 00449
623-87-0: 21038
623-91-6: 08801
623-93-8: 17261
623-95-0: 25438
623-96-1: 09203
623-97-2: 12119
623-99-4: 17467
624-03-3: 13888
624-04-4: 11519
624-08-8: 13424
624-09-9: 13629
624-13-5: 17698
624-16-8: 10989
624-17-9: 17223
624-18-0: 04481
624-20-4: 14012
624-24-8: 18437
624-29-3: 09842
624-31-7: 03544
624-38-4: 03162
624-39-5: 04784
624-40-8: 24420
624-41-9: 08552
624-42-0: 13717
624-43-1: 21040
624-44-2: 14348
624-45-3: 18466
624-48-6: 08814
624-49-7: 08803
624-54-4: 21269
624-57-7: 18016
624-58-8: 18030
624-61-3: 12639
624-64-6: 08745
624-65-7: 22103
624-66-8: 22108
624-67-9: 22092
624-70-4: 11785
624-72-6: 11831
624-73-7: 11833
624-74-8: 12641
624-75-9: 00888
624-76-0: 12185
624-78-2: 11731
624-79-3: 11847
624-80-6: 14627
624-81-7: 11725
624-82-8: 15775
624-83-9: 15681
624-84-0: 14680
624-85-1: 14737
624-86-2: 14731
624-89-5: 11856
624-91-9: 17148
624-92-0: 11416
624-95-3: 08543
624-96-4: 07974
624-97-5: 18735
625-01-4: 09331
625-04-7: 18543
625-05-8: 08312
625-06-9: 18506
625-08-1: 08438
625-15-0: 25427
625-23-0: 14298
625-25-2: 13660
625-27-4: 18678
625-28-5: 08254
625-30-9: 18162
625-31-0: 18718
625-36-5: 21600
625-37-6: 08699
625-38-7: 08851
625-44-5: 20644
625-45-6: 00673
625-46-7: 21729
625-47-8: 11792
625-48-9: 12220
625-49-0: 12804
625-50-3: 00243
625-52-5: 25441
625-54-7: 20599
625-55-8: 12849
625-56-9: 15806
625-57-0: 09139
625-58-1: 17134
625-59-2: 24642
625-60-5: 12060
625-65-0: 18667
625-66-1: 07972
625-67-2: 18219
625-68-3: 08350
625-69-4: 18328
625-71-8: 08428
625-74-1: 20662
625-75-2: 00701
625-76-3: 15674
625-77-4: 00054
625-80-9: 20706
625-82-1: 23291
625-84-3: 23292
625-86-5: 12886
625-88-7: 24710
625-92-3: 22615
625-95-6: 03543
625-96-7: 10252
625-98-9: 02835
625-99-0: 02845
626-00-6: 03161
626-01-7: 02299

626-02-8: 19394
626-03-9: 22988
626-04-0: 04783
626-05-1: 22614
626-10-8: 08824
626-13-1: 02378
626-15-3: 02549
626-16-4: 02553
626-17-5: 04508
626-18-6: 04634
626-19-7: 04517
626-20-0: 03357
626-21-1: 22620
626-22-2: 04418
626-23-3: 07894
626-25-5: 00638
626-26-6: 08058
626-27-7: 13614
626-29-9: 24368
626-31-3: 08111
626-33-5: 13724
626-35-7: 00702
626-36-8: 09039
626-37-9: 09098
626-39-1: 03987
626-40-4: 02133
626-41-5: 19214
626-43-7: 02144
626-44-8: 04043
626-46-0: 19206
626-47-1: 22983
626-48-2: 23144
626-53-9: 13609
626-55-1: 22585
626-58-4: 20182
626-60-8: 22596
626-61-9: 22597
626-62-0: 09889
626-64-2: 22967
626-67-5: 20176
626-70-0: 14142
626-71-1: 14113
626-72-2: 08398
626-77-7: 14267
626-82-4: 14218
626-83-5: 21307
626-84-6: 12192
626-85-7: 24297
626-86-8: 14146
626-87-9: 18208
626-88-0: 18188
626-89-1: 18495
626-91-5: 08015
626-92-6: 18215
626-94-8: 14480
626-95-9: 18325
626-96-0: 18146
626-97-1: 18149
626-99-3: 18108
627-00-9: 08353
627-01-0: 13312
627-02-1: 20601
627-03-2: 00615
627-04-3: 00622
627-05-4: 08032
627-06-5: 25494
627-07-6: 20465
627-08-7: 20651
627-09-8: 22123
627-11-2: 09218
627-12-3: 09121
627-13-4: 17140
627-16-7: 22104
627-18-9: 21302
627-19-0: 18768
627-20-3: 18653
627-21-4: 18769
627-22-5: 07781
627-24-7: 08783
627-26-9: 08848
627-27-0: 08928
627-30-5: 21305
627-31-6: 20578
627-32-7: 21331
627-35-0: 20460
627-36-1: 20633
627-37-2: 21663
627-39-4: 20398
627-40-7: 21721
627-41-8: 22110
627-42-9: 11789
627-45-2: 12823
627-48-5: 09515
627-49-6: 19737
627-51-0: 12596
627-52-1: 24289
627-53-2: 11873
627-54-3: 11877
627-58-7: 13942
627-59-8: 14300
627-61-2: 18397
627-63-4: 08844
627-67-8: 08027
627-70-3: 21567
627-73-6: 08161
627-82-7: 20916
627-83-8: 17407
627-84-9: 24373
627-88-3: 17435
627-89-4: 17490
627-90-7: 25291
627-91-8: 14147
627-93-0: 14135
627-96-3: 14014
627-97-4: 13748
627-98-5: 14286
628-02-4: 13980
628-03-5: 08010
628-04-6: 07992
628-05-7: 18261
628-08-0: 08932
628-09-1: 21306
628-11-5: 09222
628-12-6: 09227
628-13-7: 22667
628-16-0: 13973
628-17-1: 18249
628-20-6: 08248
628-21-7: 07982
628-28-4: 08011
628-29-5: 08031
628-30-8: 20636
628-32-0: 20598
628-33-1: 22106
628-34-2: 11779
628-36-4: 14711
628-37-5: 18814
628-41-1: 09674
628-44-4: 17727
628-46-6: 14254
628-50-2: 09229
628-61-5: 17584
628-62-6: 13464
628-63-7: 00723
628-64-8: 09225
628-66-0: 20928
628-67-1: 08215
628-68-2: 12224
628-71-7: 13825
628-72-8: 13438
628-73-9: 14187
628-76-2: 18216
628-77-3: 18227
628-79-5: 18144
628-80-8: 18256
628-81-9: 07990

628-83-1: 24627
628-87-5: 00887
628-89-7: 12124
628-92-2: 09668
628-94-4: 14099
628-96-6: 11997
628-97-7: 13889
629-01-6: 17562
629-03-8: 14015
629-04-9: 13479
629-05-0: 17843
629-06-1: 13486
629-08-3: 13602
629-09-4: 14032
629-11-8: 14156
629-12-9: 18255
629-14-1: 11827
629-15-2: 11996
629-16-3: 17144
629-17-4: 24635
629-19-6: 11422
629-20-9: 10461
629-22-1: 17414
629-23-2: 24388
629-27-6: 17620
629-30-1: 13589
629-31-2: 13462
629-32-3: 13538
629-33-4: 12847
629-35-6: 15595
629-36-7: 20677
629-37-8: 17628
629-39-0: 17139
629-40-3: 17635
629-41-4: 17645
629-43-6: 17146
629-45-8: 11413
629-46-9: 17153
629-49-2: 11566
629-50-5: 25142
629-54-9: 13855
629-56-1: 13918
629-57-2: 13927
629-59-4: 24352
629-60-7: 25156
629-62-9: 18059
629-63-0: 24364
629-64-1: 13551
629-65-2: 13556
629-66-3: 17176
629-70-9: 13910
629-72-1: 18060
629-73-2: 13917
629-74-3: 13925
629-75-4: 13926
629-76-5: 18070
629-78-7: 13403
629-79-8: 13872
629-80-1: 13853
629-81-2: 18057
629-82-3: 17630
629-89-0: 17513
629-90-3: 13401
629-91-4: 13854
629-92-5: 17164
629-93-6: 17382
629-94-7: 13387
629-96-9: 11594
629-97-0: 11445
629-98-1: 11467
629-99-2: 18046
630-01-3: 13844
630-02-4: 17355
630-03-5: 17157
630-06-8: 14373
630-07-9: 18607
630-08-0: 09183
630-10-4: 23946
630-13-7: 20579
630-16-0: 11878
630-17-1: 20509
630-18-2: 20976
630-19-3: 20391
630-20-6: 11880
630-25-1: 11807
630-51-3: 08201
630-60-4: 09259
630-72-8: 00900
630-76-2: 03571
631-01-6: 17865
631-22-1: 20813
631-25-4: 21031
631-28-7: 20715
631-31-2: 08162
631-36-7: 24090
631-57-2: 20994
631-64-1: 00564
631-65-2: 07941
631-67-4: 12050
632-02-0: 21308
632-05-3: 08063
632-07-5: 21120
632-14-4: 25500
632-15-5: 24714
632-16-6: 24711
632-20-2: 24905
632-21-3: 21585
632-22-4: 25497
632-24-6: 05951
632-25-7: 06579
632-32-6: 11628
632-38-2: 11651
632-46-2: 06210
632-47-3: 05681
632-50-8: 03315
632-51-9: 03331
632-52-0: 14656
632-56-4: 12047
632-58-6: 04581
632-77-9: 01308
632-79-1: 15323
632-80-4: 15327
632-82-6: 01312
632-83-7: 01219
632-89-3: 25499
632-92-8: 03273
632-93-9: 22905
632-94-0: 22957
632-95-1: 22954
632-99-5: 02015
633-12-5: 06600
633-35-2: 01070
633-99-8: 16969
634-19-5: 12798
634-20-8: 15731
634-36-6: 04049
634-41-3: 16181
634-42-4: 01939
634-43-5: 16182
634-44-6: 14618
634-47-9: 23585
634-65-1: 04694
634-66-2: 03938
634-67-3: 02507
634-68-4: 03963
634-83-3: 02493
634-85-5: 09743
634-87-7: 05567
634-89-9: 03933
634-90-2: 03939
634-91-3: 02510
634-92-4: 03964
634-93-5: 02509
634-94-6: 05535
634-95-7: 25417
634-97-9: 23326

635-03-0: 21009
635-10-9: 05534
635-11-0: 03966
635-12-1: 01190
635-21-2: 05897
635-22-3: 02110
635-26-7: 14637
635-27-8: 23577
635-46-1: 23703
635-51-8: 08195
635-53-0: 06041
635-54-1: 05956
635-65-4: 07426
635-67-6: 04649
635-78-9: 19710
635-81-4: 03958
635-84-7: 12283
635-88-1: 16438
635-90-5: 23312
635-93-8: 01613
635-96-1: 19903
636-04-4: 04387
636-09-9: 04622
636-10-2: 14632
636-21-5: 02348
636-24-8: 04444
636-25-9: 19203
636-28-2: 03934
636-30-6: 02508
636-31-7: 03965
636-32-8: 05536
636-36-2: 14239
636-38-4: 21621
636-41-9: 23307
636-44-2: 13000
636-46-4: 04603
636-53-3: 04596
636-60-2: 08181
636-61-3: 08166
636-72-6: 24817
636-73-7: 22937
636-76-0: 05952
636-78-2: 06582
636-79-3: 22850
636-82-8: 10370
636-84-0: 24073
636-86-2: 14897
636-93-1: 19414
636-97-5: 06528
636-98-6: 03551
637-01-4: 04506
637-03-6: 01402
637-04-7: 14639
637-07-0: 21118
637-19-4: 24857
637-21-8: 04199
637-27-4: 21276
637-31-0: 04487
637-32-1: 14946
637-39-8: 12219
637-44-5: 22120
637-47-8: 14590
637-50-3: 03896
637-53-6: 12051
637-55-8: 17447
637-56-9: 02260
637-57-0: 21949
637-58-1: 16086
637-59-2: 02725
637-61-6: 09757
637-64-9: 13052
637-65-0: 13054
637-69-4: 03358
637-70-7: 09693
637-78-5: 21252
637-87-6: 02846
637-88-7: 10067
637-89-8: 19401
637-92-3: 20602
637-95-6: 12039
637-98-9: 09140
638-00-6: 24712
638-02-8: 24713
638-03-9: 02350
638-04-0: 09841
638-07-3: 08363
638-08-4: 17395
638-10-8: 08909
638-11-9: 08473
638-13-1: 23233
638-16-4: 25002
638-17-5: 11435
638-20-0: 00121
638-21-1: 19747
638-23-3: 10871
638-25-5: 17696
638-26-6: 17412
638-28-8: 14003
638-29-9: 18592
638-37-9: 08088
638-41-5: 09235
638-45-9: 14059
638-46-0: 07995
638-49-3: 12853
638-51-7: 17147
638-53-9: 25158
638-54-0: 17633
638-58-4: 24348
638-60-8: 24396
638-65-3: 17391
638-66-4: 17369
638-67-5: 25101
638-68-6: 24950
638-73-3: 15644
638-95-9: 25511
639-11-2: 09300
639-46-3: 16069
639-58-7: 24218
639-81-6: 11640
639-99-6: 10084
640-06-2: 17890
640-19-7: 00250
640-57-3: 03743
640-59-5: 05355
640-61-9: 05358
640-68-6: 25526
641-06-5: 00194
641-16-7: 19621
641-29-2: 02016
641-36-1: 11184
641-70-3: 15321
641-74-7: 15576
641-85-0: 20318
641-91-8: 16548
641-96-3: 23447
642-04-6: 22357
642-18-2: 17922
642-28-4: 16852
642-29-5: 16024
642-31-9: 01165
642-32-0: 03957
642-36-4: 14905
642-44-4: 23123
642-71-7: 19661
642-84-2: 01837
642-91-1: 05663
643-01-6: 15572
643-03-8: 01034
643-12-9: 15288
643-14-1: 09488
643-28-7: 02366
643-43-6: 04171
643-58-3: 07522
643-60-7: 15478
643-62-9: 00976
643-65-2: 16020

643-75-4: 21047
643-79-8: 04516
643-84-5: 06956
643-93-6: 07523
643-94-7: 04650
644-08-6: 07524
644-13-3: 16025
644-15-5: 16219
644-16-6: 16220
644-21-3: 14628
644-31-5: 18806
644-32-6: 11412
644-33-7: 01879
644-34-8: 21998
644-35-9: 19594
644-36-0: 04259
644-49-5: 21251
644-71-3: 19557
644-80-4: 08821
644-87-1: 08175
644-90-6: 17667
644-97-3: 19882
644-98-4: 22683
645-00-1: 03550
645-08-9: 06361
645-09-0: 01942
645-12-5: 12981
645-13-6: 12463
645-35-2: 14572
645-36-3: 11696
645-41-0: 07882
645-45-4: 05278
645-48-7: 14676
645-49-8: 03321
645-55-6: 02425
645-56-7: 19596
645-59-0: 05189
645-62-5: 14385
645-66-9: 11511
645-67-0: 18470
645-69-2: 08154
645-72-7: 13911
645-92-1: 24972
645-93-2: 25008
645-96-5: 05281
646-04-8: 18652
646-05-9: 18751
646-06-0: 11312
646-07-1: 18448
646-13-9: 17437
646-14-0: 14072
646-19-5: 13574
646-20-8: 13576
646-24-2: 17217
646-25-3: 10931
646-30-0: 17169
646-31-1: 24329
650-52-2: 19791
650-53-3: 24236
650-69-1: 01361
651-55-8: 11653
651-77-4: 22777
652-45-9: 02239
652-46-0: 19069
653-37-2: 01769
653-49-6: 14660
654-42-2: 04754
655-04-9: 01189
655-32-3: 12524
655-48-1: 11998
655-86-7: 18973
656-73-5: 17406
657-26-1: 15547
658-78-6: 00826
659-04-1: 21946
659-34-7: 19362
659-49-4: 02439
659-70-1: 08460
659-86-9: 22109
660-68-4: 11716
660-88-8: 18399
661-11-0: 13530
661-19-8: 11458
661-54-1: 22113
661-68-7: 11358
661-71-2: 09522
661-95-0: 20552
661-96-1: 20621
662-50-0: 07855
663-93-4: 09306
665-66-7: 25108
666-52-4: 21589
666-99-9: 17167
667-27-6: 00493
668-45-1: 06681
668-94-0: 14774
669-49-8: 11740
670-54-2: 12613
670-95-1: 14790
670-96-2: 14789
671-35-2: 23262
671-56-7: 08940
672-13-9: 01711
672-15-1: 14575
672-21-9: 04223
672-45-7: 23148
672-76-4: 09666
672-89-9: 22284
673-06-3: 19712
673-22-3: 01710
673-31-4: 05265
673-32-5: 03912
673-66-5: 01542
674-76-0: 18681
674-82-8: 17986
675-09-2: 22278
675-10-5: 22283
675-20-7: 20289
676-49-3: 15717
676-55-1: 15718
676-59-5: 19738
676-80-2: 15719
676-92-6: 11760
677-21-4: 21747
677-43-0: 19892
677-56-5: 20622
677-68-9: 20703
677-69-0: 20615
678-26-2: 18236
679-37-8: 14738
679-77-6: 11406
679-86-7: 20684
679-89-0: 11757
680-31-9: 19942
681-04-9: 19893
681-06-1: 19868
681-57-2: 18299
681-84-5: 24174
682-01-9: 24176
682-30-4: 19823
683-08-9: 19833
683-50-1: 20380
683-51-2: 21617
683-57-8: 00100
683-68-1: 11803
683-72-7: 00197
683-73-8: 12606
684-16-2: 21556
684-84-4: 08212
684-93-5: 25464
685-73-4: 13171
685-87-0: 20801
685-91-6: 00205
686-50-0: 13324
687-10-5: 00814
687-26-3: 18596

687-28-5: 08160
687-30-9: 01464
687-47-8: 21186
687-57-0: 24268
687-69-4: 13277
687-72-9: 22045
688-14-2: 15531
688-52-8: 14530
688-71-1: 07745
688-74-4: 07738
688-82-4: 13508
688-84-6: 21898
688-99-3: 14481
689-00-9: 14563
689-31-6: 14150
689-42-9: 22062
689-85-0: 00655
689-89-4: 13959
689-97-4: 08981
690-02-8: 18815
690-08-4: 18671
690-37-9: 18521
690-39-1: 20623
690-92-6: 14438
690-93-7: 14437
691-37-2: 18643
691-38-3: 18682
691-83-8: 08836
691-93-0: 21107
692-04-6: 15545
692-24-0: 14444
692-35-3: 20522
692-42-2: 01392
692-45-5: 12844
692-47-7: 14448
692-50-2: 09000
692-86-4: 25332
692-93-3: 00237
692-96-6: 13760
693-02-7: 14540
693-05-0: 20992
693-07-2: 11781
693-13-0: 20478
693-19-6: 17689
693-23-2: 11503
693-39-0: 13549
693-54-9: 10987
693-57-2: 11510
693-58-3: 17198
693-61-8: 25325
693-62-9: 25327
693-65-2: 18264
693-67-4: 25271
693-72-1: 17495
693-89-0: 10714
693-93-6: 17939
693-95-8: 24526
693-98-1: 14778
694-05-3: 22746
694-28-0: 10662
694-31-5: 22402
694-33-7: 10770
694-35-9: 10710
694-48-4: 22401
694-51-9: 10322
694-53-1: 24083
694-59-7: 22709
694-80-4: 02590
694-81-5: 23067
694-85-9: 22990
695-02-3: 25132
695-06-7: 13096
695-12-5: 09854
695-19-2: 22995
695-28-3: 10310
695-34-1: 22549
695-53-4: 17950
695-77-2: 10490
695-84-1: 19349
695-96-5: 19097
695-98-7: 22754
695-99-8: 09702
696-04-8: 23133
696-07-1: 23139
696-23-1: 14781
696-28-6: 01446
696-29-7: 09899
696-30-0: 22685
696-33-3: 03555
696-41-3: 01738
696-44-6: 02189
696-54-8: 22794
696-59-3: 12927
696-62-8: 03540
696-63-9: 05553
696-71-9: 10458
696-74-2: 10820
696-86-6: 10443
697-11-0: 09587
697-22-3: 11300
697-82-5: 19663
697-86-9: 02027
697-91-6: 09711
698-00-0: 02032
698-01-1: 02075
698-19-1: 04886
698-24-8: 14975
698-26-0: 14968
698-27-1: 01719
698-29-3: 23090
698-31-7: 04723
698-42-0: 05763
698-63-5: 12950
698-67-9: 01824
698-69-1: 02077
698-71-5: 19369
698-87-3: 04854
698-88-4: 03069
699-02-5: 04858
699-06-9: 01701
699-08-1: 03381
699-09-2: 05550
699-17-2: 08650
699-30-9: 13036
699-83-2: 12362
699-98-9: 13149
700-12-9: 03872
700-13-0: 04773
700-17-4: 02496
700-38-9: 19503
700-47-0: 22135
700-57-2: 25128
700-58-3: 25129
700-63-0: 04105
700-75-4: 02316
700-85-6: 15346
700-87-8: 03557
700-88-9: 02984
701-34-8: 05342
701-35-9: 24111
701-46-2: 06535
701-47-3: 05064
701-56-4: 02317
701-57-5: 03786
701-97-3: 10095
701-99-5: 00940
702-23-8: 04852
702-28-3: 10822
702-54-5: 17923
702-96-5: 23376
703-32-2: 23621
703-49-1: 23623
703-59-3: 06788
703-61-7: 23611
703-66-2: 23622
703-80-0: 12444

703-95-7: 23099
704-01-8: 04448
704-13-2: 01730
704-14-3: 19421
704-77-8: 09447
704-78-9: 09448
704-79-0: 08917
704-80-3: 08916
704-96-1: 21807
705-15-7: 12417
705-16-8: 10807
705-54-4: 21804
705-60-2: 03818
705-76-0: 05038
705-86-2: 22287
706-03-6: 20996
706-14-9: 13101
706-78-5: 10718
706-79-6: 10702
707-29-9: 11304
707-55-1: 09872
708-06-5: 16596
708-08-7: 22793
708-76-9: 01703
708-79-2: 22174
708-82-7: 21822
709-09-1: 03183
709-60-4: 05524
710-14-5: 22244
710-24-7: 05289
710-43-0: 10821
710-77-0: 23199
711-41-1: 03211
711-44-4: 03869
711-55-7: 16503
711-79-5: 12427
712-32-3: 03212
712-33-4: 00168
712-34-5: 00270
712-48-1: 01417
712-50-5: 15907
713-36-0: 03737
713-46-2: 12149
713-68-8: 19555
713-95-1: 22280
715-51-5: 18835
716-39-2: 17056
716-43-8: 04250
716-53-0: 01129
716-76-7: 07591
716-79-0: 05630
717-21-5: 22013
717-74-8: 04080
718-25-2: 03484
719-22-2: 09697
719-54-0: 00981
719-59-5: 15822
719-79-9: 02761
720-44-5: 05101
722-27-0: 02247
722-96-3: 04107
723-46-6: 05326
723-62-6: 01169
725-00-8: 20392
725-12-2: 17900
725-92-8: 22261
726-25-0: 24891
726-42-1: 02311
726-44-3: 03890
727-51-5: 01246
728-84-7: 20955
728-86-9: 16042
728-87-0: 05097
728-90-5: 01948
729-43-1: 12503
729-62-4: 22258
730-46-1: 13095
732-26-3: 19674
736-30-1: 03309
736-31-2: 03329
737-89-3: 16183
739-27-5: 01054
742-20-1: 06979
742-29-0: 23554
744-38-7: 01953
744-45-6: 04600
744-80-9: 23177
746-47-4: 12735
747-90-0: 09349
747-98-8: 23552
748-10-7: 23553
751-38-2: 16543
753-58-2: 15603
753-89-9: 20521
753-90-2: 11735
754-03-0: 12046
754-05-2: 24058
754-10-9: 20418
754-14-3: 24237
754-17-6: 11801
755-25-9: 14404
756-48-9: 08597
756-79-6: 19835
757-45-9: 20573
757-58-4: 24407
758-08-7: 00285
758-10-1: 12053
758-12-3: 00852
758-18-9: 11701
758-24-7: 12546
758-28-1: 21011
758-39-4: 20549
758-46-3: 24235
758-95-2: 22061
758-96-3: 20419
759-05-7: 08477
759-36-4: 20856
759-54-6: 14454
759-73-9: 25443
760-03-2: 21825
760-20-3: 18642
760-21-4: 18259
760-23-6: 08730
760-30-5: 00666
760-40-7: 21429
760-41-8: 00068
760-54-3: 20839
760-55-4: 20866
760-67-8: 14368
760-78-1: 17351
760-79-2: 07851
760-90-7: 08117
760-92-9: 20679
760-93-0: 21887
761-35-3: 13895
761-65-9: 12817
762-04-9: 19816
762-21-0: 09005
762-42-5: 09006
762-49-2: 11765
762-50-5: 11783
762-62-9: 18635
762-63-0: 18672
762-72-1: 24147
762-84-5: 00222
763-20-2: 13605
763-29-1: 18641
763-30-4: 18102
763-35-9: 20424
763-50-8: 00821
763-69-9: 21174
763-89-3: 18726
763-93-9: 14507
764-01-2: 09014
764-13-6: 13949
764-29-4: 00863

764-33-0: 14553
764-35-2: 14541
764-42-1: 08792
764-47-6: 20594
764-48-7: 12162
764-49-8: 24650
764-56-7: 13457
764-67-0: 13893
764-73-8: 17553
764-78-3: 12577
764-85-2: 17289
764-93-2: 11018
764-96-5: 25330
764-97-6: 25329
764-99-8: 12588
765-03-7: 11565
765-05-9: 10919
765-09-3: 25143
765-13-9: 18084
765-14-0: 11495
765-27-5: 11604
765-30-0: 10765
765-34-4: 18031
765-42-4: 10832
765-43-5: 12336
765-47-9: 10703
765-58-2: 24681
765-69-5: 10632
765-70-8: 10631
765-87-7: 10066
766-00-7: 10634
766-07-4: 14723
766-08-5: 24079
766-09-6: 20165
766-20-1: 11268
766-36-9: 23440
766-38-1: 13106
766-39-2: 13033
766-46-1: 02648
766-51-8: 02857
766-53-0: 07416
766-77-8: 24050
766-79-0: 20186
766-80-3: 02671
766-81-4: 02649
766-84-7: 06678
766-85-8: 03539
766-90-5: 03898
766-92-7: 03784
766-93-8: 12816
766-94-9: 03365
766-95-0: 23285
766-96-1: 02650
767-00-0: 06712
767-03-3: 10725
767-05-5: 10638
767-10-2: 23347
767-12-4: 10127
767-15-7: 23025
767-54-4: 10215
767-58-8: 15016
767-59-9: 15046
767-60-2: 15048
767-88-4: 22778
767-99-7: 03753
768-00-3: 03752
768-22-9: 15106
768-29-6: 17897
768-32-1: 24145
768-33-2: 23990
768-39-8: 07729
768-49-0: 03755
768-52-5: 02368
768-56-9: 02747
768-66-1: 20222
768-90-1: 25112
768-94-5: 25107
768-95-6: 25127
769-06-2: 02500
769-11-9: 02106
769-25-5: 03367
769-39-1: 19615
769-40-4: 05561
769-68-6: 04319
769-92-6: 02199
770-09-2: 24146
770-12-7: 19965
770-35-4: 21418
770-71-8: 25124
771-13-1: 16213
771-51-7: 15186
771-56-2: 03868
771-60-8: 02447
771-61-9: 19551
771-62-0: 05560
771-63-1: 04769
771-97-1: 16703
771-98-2: 02968
771-99-3: 20200
772-00-9: 11273
772-01-0: 11272
772-03-2: 23643
772-13-4: 05181
772-17-8: 05240
772-33-8: 19116
772-55-4: 25447
772-68-9: 00864
772-79-2: 19962
773-47-7: 19853
773-64-8: 05522
773-76-2: 23812
773-82-0: 06729
774-40-3: 04192
774-74-3: 00934
775-12-2: 24052
775-16-6: 23431
775-51-9: 09688
775-56-4: 24037
776-11-4: 16945
776-19-2: 14749
776-34-1: 16190
776-35-2: 18837
776-74-9: 02676
776-75-0: 20136
776-76-1: 24075
776-99-8: 21545
777-22-0: 03678
777-37-7: 02917
777-52-6: 19964
777-59-3: 25448
777-95-7: 11263
778-18-7: 10104
778-22-3: 03670
778-24-5: 24048
778-25-6: 24164
778-28-9: 05459
778-48-3: 18949
778-66-5: 03900
778-97-2: 23105
779-02-2: 01156
779-03-3: 01121
779-51-1: 03663
779-76-0: 02070
779-84-0: 19570
779-90-8: 12273
780-20-1: 02054
780-24-5: 15593
780-25-6: 04950
780-69-8: 24127
781-17-9: 22510
781-35-1: 21547
781-43-1: 01141
781-73-7: 12391
782-05-8: 03220
782-06-9: 03219
782-45-6: 01815

782-75-2: 19382
782-76-3: 19515
782-78-5: 19383
783-95-9: 22775
784-04-3: 12272
784-50-9: 12767
784-62-3: 07143
785-80-8: 02437
785-86-4: 24957
787-84-8: 05978
788-25-0: 19541
789-02-6: 02952
789-24-2: 12750
789-25-3: 24153
790-69-2: 24867
790-90-9: 09188
791-28-6: 19764
791-29-7: 24081
791-31-1: 24168
792-57-4: 15941
793-23-7: 02579
793-39-5: 05601
795-18-6: 00452
796-30-5: 16384
797-20-6: 04475
797-67-1: 05602
799-53-1: 11385
806-71-3: 02738
806-98-4: 08107
807-75-0: 13268
811-93-8: 20751
811-96-1: 00033
811-97-2: 11885
811-98-3: 15814
812-01-1: 09332
812-03-3: 20694
812-25-9: 21609
813-44-5: 18588
813-56-9: 20888
813-74-1: 20998
813-76-3: 19772
813-78-5: 19920
814-49-3: 19949
814-67-5: 20960
814-68-6: 22034
814-75-5: 08629
814-78-8: 08964
815-16-7: 21738
815-17-8: 08396
815-24-7: 18590
815-52-1: 18575
815-58-7: 22039
815-60-1: 18579
815-66-7: 14360
815-68-9: 18353
816-11-5: 08409
816-39-7: 21539
816-50-2: 21203
816-51-3: 00813
816-63-7: 13625
816-77-3: 22041
816-79-5: 18673
817-11-8: 00997
817-76-5: 08352
817-80-1: 09220
817-91-4: 13647
817-95-8: 00616
818-23-5: 18077
818-38-2: 18297
818-41-7: 20420
818-42-8: 09078
818-81-5: 17712
818-92-8: 21713
819-07-8: 11903
819-43-2: 19948
819-44-3: 14529
819-93-2: 08375
819-97-6: 08478
820-29-1: 10990
820-71-3: 21987
821-00-1: 13930
821-06-7: 08766
821-07-8: 14377
821-08-9: 13969
821-09-0: 18717
821-10-3: 08999
821-11-4: 08840
821-25-0: 13502
821-38-5: 24360
821-41-0: 14479
821-48-7: 11692
821-55-6: 17279
821-76-1: 13504
821-95-4: 25322
821-96-5: 25326
821-98-7: 25328
821-99-8: 17209
822-06-0: 14033
822-19-5: 17462
822-23-1: 00709
822-26-4: 11459
822-28-6: 13866
822-35-5: 09585
822-36-6: 14779
822-38-8: 11440
822-50-4: 10519
822-51-5: 23293
822-66-2: 10398
822-67-3: 10397
822-68-4: 19735
822-73-1: 07366
822-84-4: 24760
822-85-5: 10219
822-87-7: 10222
823-09-6: 23075
823-17-6: 10311
823-18-7: 10062
823-22-3: 22286
823-39-2: 22542
823-40-5: 04465
823-69-8: 07284
823-76-7: 12334
823-78-9: 02584
823-86-9: 02086
824-11-3: 25194
824-22-6: 15018
824-40-8: 22835
824-46-4: 04758
824-47-5: 02878
824-55-5: 02870
824-63-5: 15017
824-69-1: 04749
824-72-6: 19863
824-75-9: 01897
824-94-2: 02882
825-25-2: 10663
825-41-2: 02107
825-44-5: 07099
825-51-4: 16974
825-55-8: 24763
825-92-3: 23985
825-94-5: 24099
826-03-9: 12890
826-18-6: 03876
826-36-8: 20301
826-39-1: 07261
826-62-0: 15802
826-73-3: 05691
826-74-4: 16393
826-77-7: 23638
826-81-3: 23816
827-08-7: 03030
827-15-6: 03866
827-16-7: 24999
827-23-6: 02136
827-52-1: 02972

827-54-3: 16394
827-94-1: 02137
828-01-3: 05224
828-39-7: 18569
828-51-3: 25120
829-14-1: 05693
829-26-5: 16559
829-83-4: 01395
829-84-5: 19736
830-03-5: 00707
830-09-1: 21882
830-13-7: 09610
830-79-5: 01793
830-81-9: 16903
830-89-7: 14899
830-96-6: 15232
831-61-8: 06619
831-67-4: 10859
831-81-2: 02931
831-82-3: 19556
831-91-4: 03885
832-01-9: 21884
832-64-4: 18879
832-66-6: 22206
832-69-9: 18876
832-70-2: 18833
832-71-3: 18878
832-92-8: 04815
832-97-3: 15234
833-48-7: 11107
834-25-3: 03736
834-28-6: 14949
834-99-1: 18871
835-06-3: 18873
835-11-0: 15863
835-64-3: 19066
836-30-6: 02432
836-41-9: 02345
836-43-1: 05144
837-08-1: 19388
837-45-6: 18910
837-56-9: 23295
838-40-4: 23297
838-41-5: 17935
838-45-9: 03433
838-95-9: 02223
840-10-8: 01807
840-65-3: 16727
840-97-1: 25247
841-32-7: 04293
841-73-6: 23183
843-33-4: 04673
844-20-2: 18891
844-51-9: 09727
846-48-0: 01046
848-53-3: 11084
848-83-9: 16383
849-99-0: 14126
852-38-0: 17898
853-39-4: 15871
865-49-6: 15721
866-55-7: 24224
867-49-2: 12057
868-59-7: 10873
868-77-9: 21902
868-78-0: 14586
868-85-9: 19987
869-19-2: 15533
869-24-9: 11693
869-29-4: 21752
869-56-7: 19984
870-23-5: 21768
870-24-6: 11689
870-63-3: 08753
870-74-6: 18560
870-93-9: 08399
871-22-7: 07994
871-28-3: 18752
871-76-1: 11734
871-83-0: 17211
871-84-1: 17552
872-05-9: 10998
872-10-6: 18273
872-31-1: 24677
872-36-6: 11342
872-50-4: 23425
872-53-7: 10591
872-55-9: 24723
872-85-5: 22787
872-93-5: 24776
873-32-5: 06677
873-49-4: 02989
873-62-1: 06711
873-66-5: 03897
873-69-8: 22790
873-73-4: 02833
873-74-5: 06663
873-75-6: 04991
873-76-7: 05004
873-83-6: 23117
873-94-9: 10293
873-99-4: 11296
874-14-6: 23135
874-23-7: 10217
874-35-1: 15019
874-41-9: 03399
874-42-0: 01627
874-52-2: 02193
874-60-2: 07226
874-63-5: 03576
874-79-3: 03911
874-86-2: 06688
874-90-8: 06717
875-30-9: 15134
875-40-1: 03949
875-51-4: 02051
875-59-2: 12425
875-62-7: 16695
875-79-6: 15133
876-02-8: 12426
876-05-1: 10614
877-06-5: 07258
877-08-7: 03942
877-09-8: 03943
877-10-1: 03944
877-11-2: 03860
877-22-5: 06358
877-43-0: 23631
877-44-1: 04036
877-48-5: 19266
877-53-2: 19263
877-65-6: 05053
877-82-7: 19274
877-95-2: 00363
878-93-3: 17118
879-12-9: 16546
879-18-5: 16587
879-37-8: 15178
879-39-0: 03951
879-95-8: 04888
879-97-0: 19297
880-09-1: 20184
880-36-4: 24761
880-93-3: 16450
881-03-8: 16454
881-07-2: 23681
881-51-6: 00169
881-67-4: 03182
881-68-5: 01569
881-87-8: 00167
882-06-4: 21947
882-23-5: 18159
882-25-7: 09485
882-26-8: 03800
882-33-7: 11420
882-35-9: 22965

882-36-0: 07864
883-20-5: 18880
883-40-9: 02995
883-93-2: 07064
883-99-8: 16682
884-35-5: 06347
884-74-2: 15809
884-89-9: 11294
885-77-8: 05122
885-81-4: 00238
886-06-6: 21524
886-08-8: 19581
886-38-4: 10837
886-58-8: 03674
886-66-8: 02741
886-74-8: 20900
886-77-1: 18128
888-71-1: 17895
890-51-7: 16197
891-32-7: 16234
892-78-4: 00004
894-18-8: 05235
896-32-2: 04843
897-06-3: 01043
897-78-9: 10218
898-03-3: 16230
898-22-6: 25031
900-02-7: 04742
900-91-4: 05219
900-95-8: 24215
903-46-8: 01326
905-56-6: 11167
909-39-7: 20121
910-19-0: 09366
910-31-6: 09373
910-86-1: 24851
912-60-7: 15340
913-59-7: 11161
913-76-8: 23548
915-35-5: 24206
916-51-8: 11158
916-52-9: 11157
916-55-2: 06962
917-92-0: 08993
917-93-1: 20382
917-95-3: 20664
918-00-3: 21586
918-07-0: 07942
918-22-9: 14020
918-71-8: 18475
918-78-5: 21150
918-79-6: 18141
918-82-1: 18773
918-84-3: 18204
919-12-0: 18776
919-19-7: 19799
919-30-2: 20470
919-31-3: 21001
919-94-8: 07993
920-37-6: 21756
920-46-7: 22036
920-54-7: 00220
920-66-1: 21400
921-03-9: 21587
921-08-4: 08358
921-13-1: 15645
921-47-1: 14093
921-74-4: 15421
922-08-7: 14128
922-09-8: 17640
922-28-1: 13515
922-55-4: 10870
922-56-5: 10866
922-59-8: 18778
922-62-3: 18680
922-65-6: 18116
923-06-8: 08116
923-26-2: 21903
923-28-4: 17762
923-32-0: 10878
923-42-2: 08280
923-99-9: 19995
924-16-3: 07870
924-41-4: 13962
924-42-5: 21650
924-43-6: 17151
924-49-2: 08309
924-50-5: 08910
924-52-7: 17137
924-88-9: 08110
925-06-4: 08112
925-15-5: 08158
925-21-3: 08819
925-44-0: 17441
925-52-0: 13730
925-78-0: 17280
926-02-3: 20597
926-04-5: 07850
926-05-6: 17133
926-06-7: 15756
926-26-1: 08109
926-39-6: 12093
926-41-0: 21323
926-55-6: 18753
926-56-7: 18100
926-63-6: 20448
926-64-7: 00884
926-65-8: 20595
926-82-9: 13516
927-18-4: 20946
927-49-1: 25314
927-54-8: 14392
927-61-7: 21711
927-62-8: 07874
927-68-4: 12107
927-70-8: 12056
927-73-1: 08718
927-74-2: 09015
927-80-0: 12642
928-04-1: 09007
928-28-9: 17283
928-40-5: 14155
928-45-0: 17131
928-49-4: 14542
928-50-7: 18628
928-51-8: 08536
928-56-3: 12551
928-65-4: 24106
928-68-7: 13706
928-73-4: 00459
928-80-3: 10988
928-90-5: 14558
928-91-6: 14495
928-92-7: 14494
928-94-9: 14485
928-95-0: 14484
928-96-1: 14489
928-97-2: 14488
929-17-9: 13613
929-20-4: 17772
929-37-3: 12163
929-55-5: 17560
929-61-3: 17614
929-62-4: 17613
929-77-1: 11457
930-02-9: 17378
930-18-7: 10772
930-21-2: 01526
930-22-3: 18009
930-27-8: 12909
930-28-9: 10504
930-30-3: 10748
930-35-8: 11442
930-36-9: 22415
930-37-0: 18017
930-46-1: 10626

930-55-2: 23362
930-60-9: 10738
930-62-1: 14772
930-66-5: 10301
930-68-7: 10409
930-72-3: 24841
930-87-0: 23320
930-88-1: 23343
930-89-2: 10534
930-90-5: 10535
931-10-2: 09783
931-17-9: 10052
931-20-4: 20291
931-56-6: 09892
931-59-9: 05283
931-71-5: 10064
931-77-1: 09804
931-87-3: 10472
931-89-5: 10471
931-97-5: 10004
932-01-4: 10130
932-16-1: 12479
932-17-2: 23418
932-32-1: 02094
932-33-2: 02913
932-39-8: 10514
932-40-1: 10515
932-43-4: 10555
932-44-5: 10556
932-52-5: 23116
932-53-6: 25005
932-56-9: 09649
932-66-1: 12333
932-77-4: 02596
932-78-5: 02914
932-87-6: 02647
932-88-7: 03537
932-96-7: 02100
932-98-9: 02915
933-05-1: 10645
933-17-5: 09650
933-21-1: 10445
933-36-8: 10286
933-48-2: 10214
933-52-8: 09576
933-67-5: 15159
933-75-5: 19640
933-78-8: 19639
933-88-0: 07224
933-90-4: 22800
933-98-2: 03398
934-00-9: 04663
934-32-7: 05604
934-34-9: 07081
934-53-2: 02874
934-56-5: 24233
934-60-1: 22834
934-74-7: 03400
934-80-5: 03403
934-85-0: 03339
935-05-7: 17155
935-07-9: 00049
935-08-0: 11064
935-12-6: 13112
935-28-4: 22621
935-30-8: 01548
935-31-9: 09603
935-44-4: 10790
935-79-5: 15308
935-95-5: 19610
936-02-7: 06305
936-44-7: 23365
936-53-8: 22400
936-58-3: 05148
936-59-4: 21459
936-72-1: 03757
937-05-3: 10133
937-14-4: 04383
937-20-2: 12313
937-26-8: 23333
937-30-4: 12390
937-32-6: 05286
937-33-7: 02200
937-38-2: 21538
937-39-3: 04187
937-54-2: 03668
937-58-6: 08937
938-16-9: 21473
938-18-1: 07239
938-22-7: 03946
938-24-9: 15071
938-25-0: 16696
938-33-0: 23666
938-39-6: 23828
938-45-4: 21490
938-46-5: 21470
938-55-6: 22143
938-67-0: 21455
938-73-8: 01880
938-86-3: 03945
938-95-4: 04158
939-16-2: 23827
939-19-5: 06844
939-23-1: 22717
939-26-4: 16257
939-27-5: 16398
939-48-0: 06487
939-52-6: 08606
939-55-9: 12118
939-58-2: 21806
939-61-7: 02063
939-80-0: 06691
939-83-3: 06722
939-87-7: 10792
940-07-8: 22368
940-31-8: 21271
940-41-0: 24117
940-49-8: 03764
940-54-5: 07822
940-62-5: 21812
940-64-7: 00688
941-43-5: 11330
941-60-6: 15025
941-69-5: 23344
941-81-1: 01552
941-98-0: 12484
942-01-8: 09160
942-06-3: 15315
942-92-7: 14338
943-14-6: 06020
943-15-7: 03697
943-27-1: 12376
943-59-9: 04155
943-88-4: 08963
943-89-5: 21883
945-51-7: 03916
946-31-6: 19162
946-80-5: 03881
946-89-4: 09612
947-04-6: 01501
947-42-2: 24160
947-58-0: 15816
947-72-8: 18836
947-73-9: 18826
947-84-2: 07590
947-91-1: 04102
948-32-3: 04079
948-65-2: 15167
948-95-8: 19953
948-97-0: 03658
948-98-1: 02806
948-99-2: 02807
949-36-0: 04661
949-53-1: 00982
949-67-7: 25253
949-87-1: 11072

949-90-6: 09041
949-98-4: 21928
950-47-0: 19708
951-82-6: 04298
951-86-0: 03059
952-47-6: 19510
952-80-7: 03305
952-81-8: 19514
952-97-6: 03817
953-21-9: 02476
954-46-1: 18888
955-03-3: 19080
955-83-9: 12892
956-05-8: 21996
956-09-2: 14208
956-34-3: 01097
956-90-1: 20202
957-75-5: 25503
958-09-8: 00987
958-60-1: 08648
958-96-3: 07499
959-02-4: 16140
959-27-3: 08843
959-28-4: 08842
959-33-1: 22026
959-36-4: 01694
959-38-6: 24668
959-66-0: 05184
959-68-2: 19540
960-56-5: 25080
960-57-6: 25075
960-71-4: 07732
961-68-2: 02234
961-71-7: 11947
962-05-0: 05174
963-96-2: 01274
964-53-4: 08646
964-68-1: 06036
965-01-5: 05179
967-13-5: 16195
967-65-7: 23796
968-46-7: 04237
975-17-7: 25552
975-53-1: 20363
976-71-6: 20311
977-86-6: 11153
981-18-0: 11078
981-48-6: 17339
982-43-4: 20155
983-30-2: 11660
984-71-4: 06645
985-25-1: 23687
987-42-8: 11156
987-59-7: 23523
990-15-8: 06647
991-17-3: 11162
991-41-3: 06961
992-94-9: 24074
993-13-5: 19829
994-05-8: 08016
994-25-2: 18197
994-30-9: 24009
994-49-0: 11379
996-50-9: 23967
996-98-5: 11984
997-49-9: 24626
998-29-8: 24154
998-30-1: 24121
998-40-3: 19749
998-41-4: 24095
998-93-6: 13482
998-95-8: 13489
999-21-3: 08817
999-52-0: 13488
999-55-3: 21972
999-78-0: 18780
999-97-3: 23971
1000-49-3: 23983
1000-50-6: 23980
1000-82-4: 25451
1000-86-8: 18096
1001-53-2: 00071
1001-89-4: 13487
1002-09-1: 12055
1002-28-4: 14557
1002-35-3: 17773
1002-36-4: 13838
1002-57-9: 17669
1002-69-3: 10911
1002-84-2: 18067
1003-03-8: 10496
1003-09-4: 24676
1003-10-7: 24827
1003-29-8: 23322
1003-30-1: 12899
1003-31-2: 24786
1003-41-4: 24834
1003-56-1: 22991
1003-64-1: 09859
1003-67-4: 22681
1003-73-2: 22679
1003-78-7: 24771
1003-90-3: 23318
1004-04-2: 24964
1004-29-1: 12869
1004-36-0: 22290
1004-38-2: 23172
1004-39-3: 23165
1004-66-6: 03577
1004-77-9: 10256
1005-38-5: 23031
1005-51-2: 25130
1005-56-7: 09241
1005-64-7: 02744
1005-67-0: 16087
1006-06-0: 05069
1006-20-8: 22218
1006-31-1: 03950
1006-32-2: 03948
1006-47-9: 23662
1006-50-4: 23660
1006-64-0: 23364
1006-94-6: 15151
1006-99-1: 07053
1007-01-8: 07288
1007-26-7: 03267
1007-32-5: 08669
1007-36-9: 25471
1008-19-1: 16522
1008-73-7: 13090
1008-88-4: 22716
1008-89-5: 22715
1009-11-6: 08614
1009-14-9: 18542
1009-61-6: 12505
1009-67-2: 05238
1009-81-0: 03508
1009-93-4: 10852
1010-48-6: 05217
1010-60-2: 16778
1010-90-8: 23129
1011-12-7: 10227
1011-50-3: 23763
1011-54-7: 21877
1011-92-3: 21817
1012-84-6: 06540
1013-01-0: 23470
1013-04-3: 16626
1013-08-7: 18893
1013-83-8: 16663
1013-88-3: 04964
1013-96-3: 21939
1014-60-4: 02561
1014-66-0: 24143
1015-38-9: 19785
1015-80-1: 11106

1015-89-0: 18954
1016-05-3: 11214
1016-77-9: 15877
1016-78-0: 15901
1016-94-0: 18970
1017-56-7: 15810
1018-72-0: 05703
1018-73-1: 05704
1018-97-9: 15867
1019-52-9: 06350
1019-57-4: 00037
1020-00-4: 12133
1020-32-2: 04776
1020-33-3: 04779
1020-39-9: 01844
1020-84-4: 10840
1022-13-5: 15884
1022-45-3: 23502
1023-85-4: 24954
1024-34-6: 19847
1024-64-2: 16023
1025-15-6: 25001
1025-86-1: 06873
1029-94-3: 22291
1029-95-4: 22301
1029-97-6: 24844
1032-21-9: 05706
1033-18-7: 05699
1036-16-4: 05705
1038-66-0: 07612
1038-95-5: 19757
1043-91-0: 01736
1045-69-8: 01082
1046-51-1: 11163
1048-05-1: 13176
1048-08-4: 24092
1054-20-2: 23555
1055-23-8: 07255
1055-26-1: 01553
1056-77-5: 12934
1057-07-4: 01065
1057-09-6: 11154
1057-10-9: 11152
1057-11-0: 11166
1059-67-2: 11151
1066-40-6: 24166
1066-45-1: 24217
1067-08-9: 18244
1067-14-7: 20304
1067-20-5: 18225
1067-53-4: 24403
1067-74-9: 00596
1068-19-5: 13517
1068-57-1: 00636
1068-87-7: 18242
1068-90-2: 20791
1069-48-3: 18711
1069-53-0: 14094
1069-72-3: 11686
1069-93-8: 12245
1070-14-0: 13961
1070-19-5: 09186
1070-32-2: 13646
1070-70-8: 21794
1070-71-9: 22114
1070-74-2: 12646
1070-78-6: 20695
1070-83-3: 08387
1070-87-7: 18271
1071-22-3: 21000
1071-26-7: 13509
1071-73-4: 18558
1071-81-4: 14082
1071-93-8: 14133
1071-94-9: 13812
1071-98-3: 09003
1072-05-5: 13513
1072-16-8: 17606
1072-21-5: 14097
1072-36-2: 17364
1072-39-5: 11463
1072-43-1: 24625
1072-44-2: 01531
1072-47-5: 11321
1072-52-2: 01537
1072-53-3: 11306
1072-67-9: 15491
1072-71-5: 24469
1072-72-6: 24840
1072-83-9: 12517
1072-85-1: 02651
1072-90-8: 23368
1072-91-9: 22425
1072-97-5: 22535
1072-98-6: 22539
1073-06-9: 02652
1073-07-0: 10448
1073-23-0: 22629
1073-47-8: 19877
1073-67-2: 02811
1074-11-9: 03073
1074-12-0: 04099
1074-17-5: 03769
1074-24-4: 03006
1074-36-8: 06435
1074-43-7: 03770
1074-55-1: 03771
1074-58-4: 00876
1074-89-1: 22156
1074-92-6: 03233
1074-95-9: 10268
1074-98-2: 22693
1075-06-5: 12365
1075-22-5: 15015
1075-35-0: 15125
1075-38-3: 03234
1075-77-0: 21620
1076-26-2: 16987
1076-38-6: 06845
1076-43-3: 04404
1076-47-7: 06631
1076-55-7: 19435
1076-59-1: 15502
1076-61-5: 16519
1076-88-6: 06637
1077-16-3: 03520
1077-18-5: 07885
1077-58-3: 06226
1078-04-2: 15026
1078-17-7: 19716
1078-19-9: 17012
1078-21-3: 05205
1078-30-4: 23743
1078-61-1: 05213
1078-71-3: 03496
1079-21-6: 07633
1079-56-7: 24141
1079-66-9: 19792
1079-71-6: 01159
1079-96-5: 02572
1080-02-0: 21573
1080-12-2: 08959
1080-16-6: 11059
1080-25-7: 07384
1080-32-6: 19852
1081-15-8: 12803
1081-34-1: 24324
1081-75-0: 03889
1081-77-2: 03824
1083-30-3: 21474
1083-44-9: 11297
1083-56-3: 02742
1084-17-9: 02908
1084-32-8: 23815
1086-80-2: 06776
1087-02-1: 03126

1087-21-4: 04601
1087-49-6: 03514
1088-11-5: 05708
1090-13-7: 16151
1093-58-9: 01084
1094-85-5: 12278
1095-90-5: 13711
1096-43-1: 05392
1098-45-9: 20335
1098-60-8: 19702
1099-01-0: 05197
1104-93-4: 11387
1105-31-3: 11168
1108-82-3: 23557
1109-51-9: 06648
1111-74-6: 24046
1111-92-8: 19784
1111-97-3: 08992
1112-37-4: 19783
1112-39-6: 24043
1112-48-7: 23979
1112-55-6: 24089
1112-56-7: 24228
1112-94-3: 21154
1113-12-8: 24049
1113-30-0: 07728
1113-68-4: 00057
1113-74-2: 18582
1114-00-7: 08127
1114-09-6: 08130
1114-16-5: 15573
1114-17-6: 14744
1114-51-8: 20416
1114-76-7: 07845
1115-08-8: 18103
1115-11-3: 08694
1115-12-4: 20783
1115-30-6: 08106
1115-47-5: 16053
1115-69-1: 01014
1115-86-2: 13702
1116-24-1: 00235
1116-39-8: 07735
1116-40-1: 20461
1116-43-4: 00891
1116-61-6: 07733
1116-76-3: 17567
1117-19-7: 20825
1117-55-1: 17679
1117-59-5: 18427
1117-60-8: 17707
1117-61-9: 17831
1117-65-3: 17817
1117-71-1: 08865
1117-86-8: 17644
1118-09-8: 24140
1118-12-3: 25425
1118-15-6: 11386
1118-58-7: 18097
1118-68-9: 13306
1118-71-4: 13600
1118-84-9: 08495
1119-06-8: 13630
1119-19-3: 08986
1119-29-5: 18158
1119-33-1: 13254
1119-40-0: 18298
1119-46-6: 18411
1119-49-9: 00114
1119-51-3: 18624
1119-60-4: 13771
1119-62-6: 21171
1119-64-8: 14543
1119-65-9: 13826
1119-67-1: 14544
1119-72-8: 13952
1120-07-6: 17194
1120-16-7: 11473
1120-21-4: 25270
1120-28-1: 11593
1120-29-2: 11471
1120-32-7: 17518
1120-34-9: 11466
1120-36-1: 24390
1120-48-5: 17571
1120-55-4: 09589
1120-56-5: 09536
1120-62-3: 10715
1120-64-5: 17930
1120-71-4: 17909
1120-72-5: 10670
1120-73-6: 10753
1120-87-2: 22588
1120-89-4: 04403
1120-90-7: 22670
1120-93-0: 22300
1120-94-1: 24842
1120-97-4: 11270
1121-07-9: 23397
1121-18-2: 10416
1121-24-0: 19399
1121-31-9: 22947
1121-37-5: 15924
1121-55-7: 22643
1121-58-0: 22576
1121-60-4: 22785
1121-76-2: 22598
1121-78-4: 22973
1121-89-7: 20253
1121-92-2: 01538
1121-95-5: 10657
1122-20-9: 10411
1122-25-4: 10244
1122-39-0: 22756
1122-42-5: 03533
1122-54-9: 12516
1122-58-3: 22573
1122-60-7: 09931
1122-61-8: 22707
1122-62-9: 12513
1122-69-6: 22656
1122-72-1: 22789
1122-81-2: 22738
1122-82-3: 09891
1122-83-4: 02489
1122-91-4: 01586
1123-00-8: 10588
1123-09-7: 10413
1123-27-9: 12401
1123-34-8: 10196
1123-63-3: 19153
1123-64-4: 09705
1123-84-8: 03067
1123-85-9: 04855
1123-86-0: 21460
1123-94-0: 19370
1123-96-2: 22651
1124-04-5: 19148
1124-05-6: 03055
1124-06-7: 19152
1124-07-8: 19234
1124-11-4: 22356
1124-13-6: 09562
1124-19-2: 24232
1124-20-5: 03691
1124-25-0: 09907
1124-26-1: 10344
1124-27-2: 09915
1124-33-0: 22708
1124-35-2: 22652
1125-28-6: 12536
1125-52-6: 02727
1125-60-6: 15433
1125-70-8: 04112
1125-74-2: 04675
1125-78-6: 17004

1125-80-0: 15454
1125-90-2: 15611
1126-18-7: 10220
1126-23-4: 22217
1126-39-2: 10171
1126-41-6: 10173
1126-46-1: 06064
1126-65-4: 10576
1126-75-6: 03762
1126-78-9: 02062
1126-79-0: 02750
1126-80-3: 02770
1126-81-4: 00290
1127-45-3: 23824
1127-59-9: 15215
1127-76-0: 16397
1127-97-5: 21457
1128-08-1: 10759
1128-28-5: 19031
1128-54-7: 22422
1128-59-2: 23654
1128-61-6: 23655
1128-74-1: 23656
1128-76-3: 06056
1129-29-9: 03665
1129-30-2: 12512
1129-46-0: 21272
1129-47-1: 21227
1129-50-6: 07865
1129-64-2: 03631
1129-69-7: 22666
1129-89-1: 09619
1130-80-9: 16431
1131-16-4: 22405
1131-18-6: 22377
1131-31-3: 23171
1131-60-8: 19198
1131-62-0: 12370
1131-63-1: 16693
1131-94-8: 04208
1132-21-4: 06197
1132-39-4: 03913
1132-66-7: 03523
1133-80-8: 12723
1134-10-7: 07188
1134-35-6: 07697
1134-36-7: 07676
1134-62-9: 16269
1134-94-7: 02480
1135-12-2: 02471
1135-14-4: 02481
1135-23-5: 05232
1135-66-6: 15817
1135-99-5: 24225
1136-22-7: 11234
1136-58-9: 15908
1136-86-3: 12531
1137-41-3: 15833
1137-42-4: 15994
1137-68-4: 05632
1137-73-1: 13295
1138-52-9: 19072
1138-60-9: 06621
1138-80-3: 13340
1139-52-2: 19095
1139-82-8: 11114
1140-14-3: 15955
1140-16-5: 16012
1140-28-9: 07579
1141-06-6: 12414
1141-37-3: 05980
1141-38-4: 16718
1141-61-3: 01853
1141-88-4: 02246
1142-19-4: 11391
1142-20-7: 01025
1142-39-8: 06292
1142-70-7: 23181
1142-97-8: 05487
1143-38-0: 01342
1143-72-2: 16043
1144-20-3: 23795
1144-74-7: 16028
1145-01-3: 22410
1145-56-8: 25249
1145-75-1: 04490
1146-39-0: 03815
1146-42-5: 05488
1146-44-7: 05486
1146-98-1: 15058
1146-99-2: 15059
1147-43-9: 05887
1148-11-4: 23380
1148-79-4: 24322
1149-16-2: 19346
1149-23-1: 22904
1149-24-2: 22906
1149-26-4: 25539
1149-42-4: 21119
1150-59-0: 16813
1150-62-5: 09175
1151-15-1: 06449
1151-97-9: 22636
1152-48-3: 22030
1152-51-8: 01913
1152-54-1: 24466
1152-61-0: 01470
1153-05-5: 01410
1154-59-2: 01859
1155-00-6: 11403
1155-64-2: 15551
1158-19-6: 06237
1158-23-2: 06564
1159-86-0: 16030
1162-54-5: 20312
1162-82-9: 06915
1164-16-5: 25261
1165-39-5: 13135
1165-48-6: 06920
1166-18-3: 23448
1166-32-1: 24874
1166-52-5: 06618
1168-25-8: 06027
1171-95-5: 19732
1172-76-5: 11359
1172-95-8: 19759
1174-72-7: 24175
1174-92-1: 09375
1179-06-2: 19748
1182-02-1: 19746
1183-84-2: 15560
1184-59-4: 21750
1184-60-7: 21712
1184-78-7: 15608
1185-33-7: 08540
1185-55-3: 24137
1185-81-5: 24220
1186-31-8: 13781
1186-53-4: 18270
1186-73-8: 15770
1187-03-7: 25495
1187-13-9: 08895
1187-42-4: 08794
1187-58-2: 20427
1187-82-2: 18439
1188-01-8: 13276
1188-04-1: 13612
1188-33-6: 15602
1188-37-0: 13250
1188-58-5: 17418
1189-08-8: 21908
1189-23-7: 25370
1189-24-8: 19806
1189-71-5: 24302
1189-99-7: 13569
1190-22-3: 07965

1190-34-7: 13821
1190-39-2: 20815
1190-63-2: 17411
1190-73-4: 00286
1190-83-6: 13561
1190-93-8: 00450
1191-04-4: 14450
1191-08-8: 08245
1191-24-8: 17580
1191-41-9: 17471
1191-43-1: 14186
1191-75-9: 13919
1191-95-3: 09581
1191-96-4: 10775
1191-99-7: 12881
1192-18-3: 10518
1192-30-9: 12867
1192-37-6: 09895
1192-58-1: 23324
1192-62-7: 12394
1192-88-7: 10362
1192-89-8: 15594
1193-02-8: 05541
1193-11-9: 11322
1193-12-0: 20151
1193-16-4: 09784
1193-17-5: 09785
1193-18-6: 10417
1193-21-1: 23056
1193-24-4: 23263
1193-35-7: 12880
1193-46-0: 10122
1193-47-1: 10229
1193-55-1: 10075
1193-59-5: 23323
1193-70-0: 10681
1193-72-2: 02609
1193-74-4: 23024
1193-79-9: 12464
1193-80-2: 03337
1193-81-3: 10088
1193-82-4: 03781
1193-92-6: 22792
1194-02-1: 06707
1194-21-4: 23239
1194-65-6: 06694
1194-98-5: 01646
1195-14-8: 07107
1195-31-9: 10342
1195-32-0: 03692
1195-33-1: 05290
1195-34-2: 23057
1195-42-2: 09793
1195-59-1: 22907
1195-91-1: 10072
1196-01-6: 07393
1196-11-8: 09759
1196-31-2: 10267
1196-39-0: 15455
1196-57-2: 23902
1196-58-3: 03437
1196-67-4: 08697
1196-79-8: 15138
1197-03-1: 22612
1197-09-7: 12363
1197-18-8: 10009
1197-19-9: 06702
1197-37-1: 04439
1197-40-6: 12910
1197-55-3: 04125
1198-14-7: 23808
1198-27-2: 16959
1198-30-7: 15471
1198-34-1: 10679
1198-37-4: 23630
1198-40-9: 23526
1198-47-6: 22141
1198-55-6: 04679
1198-69-2: 13191
1198-96-5: 06781
1199-46-8: 19011
1199-67-3: 05297
1199-77-5: 21929
1199-98-0: 04356
1200-03-9: 02589
1200-14-2: 01599
1200-22-2: 11439
1200-73-3: 11271
1200-89-1: 15820
1201-26-9: 21560
1201-27-0: 07131
1201-38-3: 12369
1201-99-6: 21827
1202-08-0: 12469
1202-34-2: 22558
1202-39-7: 21829
1202-66-0: 00277
1202-71-7: 01498
1203-38-9: 16502
1203-39-0: 16776
1204-06-4: 21872
1204-21-3: 12296
1204-28-0: 15305
1204-72-4: 16455
1204-78-0: 07578
1204-79-1: 07677
1205-02-3: 01015
1205-39-6: 02403
1205-71-6: 02117
1205-91-0: 04746
1206-49-1: 23339
1206-79-7: 25180
1207-69-8: 00962
1207-71-2: 19684
1207-72-3: 19681
1207-75-6: 23473
1207-92-7: 02339
1208-03-3: 01580
1208-64-6: 05218
1208-86-2: 02340
1209-83-2: 18838
1210-12-4: 01164
1210-34-0: 11109
1210-35-1: 11112
1210-39-5: 21624
1211-40-1: 07581
1212-08-4: 05492
1214-24-0: 25559
1214-32-0: 12717
1214-39-7: 22144
1214-47-7: 22020
1216-15-5: 21636
1217-45-4: 01179
1218-34-4: 25229
1218-80-0: 06874
1219-46-1: 15546
1222-98-6: 22033
1223-20-7: 02362
1223-31-0: 03983
1225-65-6: 22999
1226-42-2: 12013
1226-43-3: 05977
1226-46-6: 15767
1228-73-5: 11671
1229-38-5: 20443
1229-71-6: 12728
1230-14-4: 22129
1230-64-4: 01967
1233-36-9: 16391
1233-73-4: 06037
1234-21-5: 25372
1234-30-6: 24849
1234-35-1: 01389
1235-82-1: 20272
1236-49-3: 01055
1240-37-5: 24873

1241-28-7: 04492
1241-65-2: 16123
1241-94-7: 19922
1247-92-3: 19733
1249-81-6: 09377
1250-95-9: 09365
1254-38-2: 15293
1259-22-9: 01073
1260-17-9: 01177
1271-42-7: 12665
1272-08-8: 12658
1272-44-2: 12648
1273-75-2: 12667
1273-91-2: 12675
1273-95-6: 12660
1274-02-8: 12653
1274-08-4: 12670
1284-05-5: 23923
1284-06-6: 17875
1284-10-2: 23922
1287-13-4: 23925
1291-47-0: 12674
1291-48-1: 12659
1295-36-9: 12677
1295-38-1: 12683
1298-66-4: 12666
1300-94-3: 19506
1322-13-0: 21849
1322-67-4: 18861
1323-03-1: 21195
1323-75-7: 23933
1330-61-6: 21875
1330-92-3: 10385
1331-81-3: 04986
1399-22-0: 11574
1399-92-4: 15359
1400-02-8: 15503
1400-05-1: 15504
1400-20-0: 15522
1400-29-9: 15587
1405-86-3: 13227
1415-73-2: 01348
1416-04-2: 24179
1420-07-1: 19298
1420-27-5: 05087
1421-63-2: 08625
1421-86-9: 24260
1421-89-2: 00604
1423-60-5: 09018
1424-14-2: 24855
1424-66-4: 01605
1427-07-2: 03477
1430-97-3: 12740
1432-42-4: 12264
1433-78-9: 19348
1435-50-3: 02612
1435-53-6: 03014
1435-55-8: 09431
1435-88-7: 19508
1436-43-7: 23724
1437-15-6: 22641
1438-16-0: 24560
1438-25-1: 24572
1438-94-4: 23305
1439-07-2: 18008
1439-12-9: 00111
1441-87-8: 07217
1443-80-7: 06660
1444-65-1: 10278
1444-72-0: 06755
1445-23-4: 19293
1445-32-5: 09580
1445-45-0: 11905
1445-73-4: 20298
1445-75-6: 19831
1445-91-6: 05107
1446-34-0: 07439
1447-26-3: 13811
1447-76-3: 23792
1448-98-2: 00067
1449-05-4: 17867
1450-01-7: 15972
1450-14-2: 11361
1450-18-6: 11365
1450-23-3: 11362
1450-31-3: 15768
1450-48-2: 12396
1450-49-3: 12395
1450-58-4: 12953
1450-63-1: 02737
1450-72-2: 12424
1450-74-4: 12318
1450-81-3: 18417
1451-95-2: 24810
1452-77-3: 22795
1453-17-4: 09677
1453-24-3: 10318
1453-25-4: 09669
1453-50-5: 10793
1453-52-7: 15925
1453-58-3: 22416
1453-82-3: 22797
1454-80-4: 07598
1454-84-8: 17173
1454-85-9: 13420
1455-13-6: 15813
1455-18-1: 07108
1455-20-5: 24685
1455-21-6: 17238
1455-77-2: 25055
1457-46-1: 24588
1457-85-8: 00714
1458-99-7: 18662
1459-00-3: 02680
1459-09-2: 03504
1459-10-5: 03954
1459-11-6: 04077
1459-14-9: 02641
1459-15-0: 02825
1459-93-4: 04598
1460-02-2: 04078
1460-05-5: 04096
1460-16-8: 09634
1460-88-4: 10732
1460-98-6: 03627
1461-25-2: 24227
1461-27-4: 10337
1461-96-7: 10609
1462-03-9: 10648
1462-11-9: 14164
1462-12-0: 20832
1462-27-7: 08608
1462-33-5: 11774
1462-34-6: 11773
1462-84-6: 22755
1463-17-8: 23633
1464-53-5: 07441
1466-67-7: 25373
1466-76-8: 06193
1466-82-6: 05856
1466-83-7: 15532
1466-88-2: 21632
1467-05-6: 02876
1467-79-4: 09504
1468-39-9: 08452
1468-55-9: 01518
1468-83-3: 12520
1469-20-1: 11194
1469-70-1: 09206
1469-94-9: 20958
1470-57-1: 15981
1470-79-7: 15933
1470-94-6: 15080
1471-03-0: 21734
1471-96-1: 16806
1472-90-8: 17386

1472-93-1: 17388
1475-11-2: 05029
1475-13-4: 05027
1476-03-5: 11380
1476-07-9: 03390
1476-11-5: 08773
1476-23-9: 21717
1477-42-5: 07042
1477-50-5: 15194
1477-55-0: 04631
1479-46-5: 24240
1481-93-2: 06808
1482-01-5: 18148
1482-13-9: 17759
1482-15-1: 18795
1482-91-3: 17770
1482-97-9: 01020
1483-42-7: 15724
1483-60-9: 03251
1484-08-8: 09163
1484-10-2: 09177
1484-12-4: 09170
1484-13-5: 09165
1484-19-1: 15148
1484-26-0: 02468
1484-84-0: 20260
1484-85-1: 05750
1485-07-0: 16841
1485-45-6: 22091
1485-70-7: 01955
1486-28-8: 19744
1486-75-5: 09618
1486-91-5: 24197
1487-18-9: 12894
1489-47-0: 17265
1489-50-5: 10673
1489-57-2: 09680
1489-60-7: 09588
1490-04-6: 10175
1490-05-7: 15588
1490-33-1: 21021
1492-24-6: 08300
1493-01-2: 15677
1493-02-3: 12858
1493-03-4: 15671
1493-13-6: 15757
1493-15-8: 15766
1493-27-2: 03481
1494-97-9: 22997
1497-17-2: 23400
1498-51-7: 19963
1498-65-3: 19906
1498-69-7: 01125
1498-96-0: 06026
1499-10-1: 01144
1499-55-4: 13257
1501-04-8: 05163
1501-05-9: 05162
1501-06-0: 18295
1501-26-4: 18414
1501-27-5: 18315
1501-60-6: 18098
1501-82-2: 09617
1502-00-7: 25119
1502-03-0: 09605
1502-05-2: 09599
1502-06-3: 09600
1502-22-3: 10226
1502-24-5: 10123
1502-38-1: 10454
1503-53-3: 05862
1504-54-7: 08936
1504-58-1: 22126
1504-70-7: 21793
1504-75-2: 21631
1504-76-3: 21633
1504-96-7: 02973
1508-07-2: 20777
1508-76-5: 23413
1509-34-8: 01031
1509-92-8: 16052
1510-24-3: 07518
1511-62-2: 15631
1512-12-5: 24239
1512-14-7: 14950
1515-16-8: 08634
1515-76-0: 07805
1515-80-6: 13958
1515-95-3: 03420
1516-24-1: 18115
1516-58-1: 02542
1516-59-2: 02543
1516-60-5: 02544
1516-80-9: 24064
1517-05-1: 12101
1517-22-2: 18920
1517-49-3: 21749
1517-51-7: 21751
1517-53-9: 12609
1517-63-1: 05128
1517-66-4: 08591
1517-67-5: 08583
1517-69-7: 05106
1518-75-8: 24766
1518-86-1: 19048
1518-89-4: 19050
1519-36-4: 01278
1519-40-0: 03443
1519-59-1: 22719
1520-21-4: 02256
1520-42-9: 03318
1520-45-2: 03434
1520-46-3: 03748
1521-38-6: 06187
1521-41-1: 01872
1521-51-3: 10299
1522-20-9: 18747
1522-22-1: 18357
1522-46-9: 08296
1522-88-9: 20580
1523-06-4: 00706
1523-11-1: 16957
1523-20-2: 06243
1526-73-4: 16781
1527-89-5: 06716
1527-95-3: 10905
1527-96-4: 08693
1527-97-5: 13976
1527-98-6: 07815
1528-30-9: 10547
1528-74-1: 07501
1529-17-5: 24144
1529-41-5: 04312
1529-68-6: 08047
1530-03-6: 03908
1530-04-7: 03522
1530-05-8: 03498
1530-12-7: 07252
1530-20-7: 03495
1532-84-9: 15431
1532-97-4: 15438
1533-20-6: 08610
1534-26-5: 25170
1534-27-6: 11541
1536-78-3: 00823
1538-74-5: 09101
1538-87-0: 11788
1539-42-0: 22920
1540-28-9: 18396
1540-29-0: 14201
1540-34-7: 18356
1541-23-7: 13437
1544-68-9: 03465
1546-79-8: 14793
1547-87-1: 00398
1549-45-7: 00797

1550-09-0: 14007
1551-06-0: 23299
1551-09-3: 23316
1551-16-2: 23300
1551-21-9: 20669
1551-27-5: 24765
1551-32-2: 24739
1551-44-6: 08366
1551-88-8: 10230
1551-89-9: 10231
1552-12-1: 10450
1552-67-6: 14456
1552-91-6: 00558
1552-92-7: 00559
1552-94-9: 18113
1552-95-0: 18114
1555-80-2: 04133
1556-00-9: 00534
1556-18-9: 10545
1556-99-6: 12742
1558-17-4: 23068
1558-24-3: 24101
1558-25-4: 24098
1558-31-2: 24015
1558-33-4: 24014
1559-02-0: 10819
1559-81-5: 16523
1559-87-1: 03031
1560-06-1: 02746
1560-09-4: 02745
1560-11-8: 10729
1560-68-5: 14384
1560-75-4: 17158
1560-89-0: 13410
1560-93-6: 18064
1560-96-9: 25151
1561-10-0: 14253
1561-11-1: 14250
1561-18-8: 13656
1561-20-2: 21826
1562-93-2: 06552
1562-94-3: 11039
1563-01-5: 05436
1563-83-3: 00056
1563-87-7: 00062
1564-53-0: 15835
1565-71-5: 05031
1565-80-6: 08551
1565-86-2: 05131
1566-30-9: 24967
1566-42-3: 25455
1567-89-1: 21546
1568-70-3: 19419
1569-01-3: 21420
1569-02-4: 21396
1569-43-3: 14499
1569-44-4: 14505
1569-59-1: 18736
1569-60-4: 13790
1569-69-3: 10099
1570-45-2: 22869
1570-48-5: 21526
1570-64-5: 19176
1570-65-6: 19235
1570-67-8: 19232
1570-76-9: 19151
1571-33-1: 19845
1571-72-8: 05919
1572-52-7: 18293
1573-17-7: 09010
1573-28-0: 13671
1573-58-6: 07803
1573-92-8: 12765
1574-10-3: 14701
1574-18-1: 16921
1574-33-0: 18756
1574-41-0: 18094
1575-37-7: 04423
1575-57-1: 08626
1575-61-7: 18595
1575-96-8: 16640
1576-43-8: 05371
1576-48-3: 16864
1576-67-6: 18855
1576-69-8: 18853
1576-85-8: 18733
1576-86-9: 18614
1576-87-0: 18613
1576-95-0: 18720
1576-96-1: 18719
1577-22-6: 14452
1577-52-2: 17367
1578-21-8: 21821
1578-58-1: 13627
1578-63-8: 04183
1579-40-4: 03845
1584-02-7: 14441
1585-06-4: 03376
1585-07-5: 02644
1585-40-6: 05158
1586-00-1: 05145
1587-04-8: 03754
1587-15-1: 08170
1587-17-3: 15558
1587-18-4: 08168
1587-20-8: 21029
1587-21-9: 21030
1587-26-4: 08775
1587-29-7: 08774
1587-41-3: 15665
1588-44-9: 03519
1588-49-4: 03326
1588-60-9: 13069
1588-83-6: 05948
1589-47-5: 21332
1589-60-2: 10192
1589-61-3: 14683
1590-02-9: 21032
1590-08-5: 17014
1590-22-3: 16580
1590-87-0: 11355
1591-17-9: 08850
1591-30-6: 07617
1591-31-7: 07516
1591-44-2: 15508
1592-31-0: 02559
1592-38-7: 16846
1592-83-2: 07825
1594-69-0: 01247
1595-04-6: 02763
1595-05-7: 02764
1595-11-5: 02762
1595-14-8: 01614
1595-15-9: 01728
1595-16-0: 03707
1596-84-5: 08190
1599-41-3: 20725
1599-67-3: 11461
1600-37-9: 21732
1601-00-9: 12460
1601-82-7: 10798
1603-40-3: 22548
1603-41-4: 22550
1603-61-8: 02975
1603-79-8: 04282
1603-91-4: 24494
1604-01-9: 22747
1604-02-0: 10654
1604-11-1: 08183
1604-42-8: 01285
1604-43-9: 01286
1605-18-1: 02567
1605-65-8: 19955
1606-08-2: 09818
1606-67-3: 22494
1607-57-4: 02634

1608-51-1: 22012
1608-52-2: 22011
1609-47-8: 11244
1609-86-5: 20632
1609-93-4: 21800
1610-23-7: 09883
1610-24-8: 09902
1610-29-3: 18722
1610-39-5: 25197
1611-50-3: 09072
1611-56-9: 25302
1611-83-2: 21653
1613-32-7: 23702
1613-34-9: 23645
1613-45-2: 20607
1613-46-3: 08043
1613-49-6: 24689
1613-51-0: 24828
1615-00-5: 21927
1615-14-1: 14824
1615-70-9: 18105
1615-75-4: 11782
1615-80-1: 14610
1615-82-3: 14634
1616-40-6: 06958
1616-86-0: 16975
1617-17-0: 20970
1617-18-1: 08922
1617-90-9: 11573
1618-26-4: 15623
1619-28-9: 10756
1619-34-7: 01489
1619-56-3: 08365
1619-57-4: 08381
1619-62-1: 20822
1620-14-0: 21543
1620-75-3: 22772
1620-98-0: 01582
1621-55-2: 06888
1621-56-3: 06932
1621-91-6: 22435
1622-10-2: 13498
1623-14-9: 19923
1623-19-4: 19932
1624-26-6: 05651
1624-62-0: 11680
1626-00-2: 13177
1626-09-1: 17652
1627-73-2: 14678
1628-89-3: 22671
1629-58-9: 18737
1630-09-7: 03125
1630-77-9: 12575
1630-78-0: 12574
1630-79-1: 11243
1630-94-0: 10771
1631-82-9: 23999
1631-84-1: 24028
1632-16-2: 13544
1632-68-4: 07306
1632-76-4: 22524
1632-83-3: 05619
1632-99-1: 11907
1633-82-5: 21007
1633-88-1: 18376
1633-89-2: 18375
1633-90-5: 14191
1633-97-2: 18378
1634-02-2: 24663
1634-04-4: 20645
1634-09-9: 16268
1634-69-1: 13713
1635-02-5: 14515
1635-61-6: 02111
1635-84-3: 02207
1636-33-5: 16428
1636-34-6: 08228
1636-39-1: 07421
1636-82-4: 18551
1637-31-6: 07802
1638-22-8: 19131
1638-26-2: 10517
1638-63-7: 04350
1638-86-4: 19874
1638-94-4: 10642
1638-95-5: 10641
1639-01-6: 08277
1639-03-8: 18381
1639-04-9: 18380
1639-05-0: 18379
1639-09-4: 13604
1640-39-7: 15177
1640-89-7: 10526
1641-40-3: 05712
1641-41-4: 15079
1643-15-8: 00687
1643-30-7: 05210
1643-49-8: 12726
1645-55-2: 13557
1646-26-0: 12274
1646-27-1: 05812
1647-08-1: 14401
1647-16-1: 10893
1647-23-0: 07914
1647-26-3: 09801
1649-08-7: 11814
1649-25-8: 01088
1649-46-3: 01066
1653-19-6: 07791
1653-31-2: 25163
1653-32-3: 24384
1653-34-5: 18071
1653-40-3: 13648
1653-57-2: 17142
1654-86-0: 10961
1655-41-0: 13720
1655-42-1: 18565
1655-70-5: 03840
1655-71-6: 03591
1655-72-7: 03592
1655-74-9: 03842
1656-44-6: 05508
1656-48-0: 20995
1657-45-0: 03735
1657-46-1: 03485
1657-47-2: 03115
1657-49-4: 02930
1657-50-7: 02929
1657-51-8: 02925
1657-52-9: 02924
1657-53-0: 03616
1657-56-3: 03323
1657-69-8: 03116
1660-04-4: 12523
1662-01-7: 18962
1663-35-0: 12583
1663-45-2: 19739
1663-67-8: 20959
1665-59-4: 11936
1666-13-3: 11354
1666-28-0: 06013
1666-96-2: 23696
1667-00-1: 02990
1667-01-2: 12534
1667-04-5: 02959
1667-11-4: 07462
1668-10-6: 00064
1670-14-0: 04397
1670-46-8: 10661
1670-81-1: 15195
1671-11-0: 23267
1671-15-4: 23266
1671-75-6: 13698
1672-46-4: 09267
1673-59-2: 12982
1674-10-8: 10305

1674-18-6: 03888
1674-30-2: 05011
1674-33-5: 18213
1674-37-9: 17755
1674-38-0: 11545
1674-56-2: 21353
1675-69-0: 17218
1676-63-7: 12388
1677-46-9: 23843
1678-25-7: 05389
1678-45-1: 11526
1678-53-1: 22692
1678-80-4: 09977
1678-81-5: 09974
1678-82-6: 09912
1678-91-7: 09856
1678-92-8: 09952
1678-93-9: 09810
1678-97-3: 09973
1678-98-4: 09929
1679-06-7: 14190
1679-07-8: 10640
1679-08-9: 21016
1679-09-0: 08278
1679-36-3: 14564
1679-41-0: 08692
1679-47-6: 13083
1679-64-7: 04627
1680-51-9: 16526
1680-53-1: 16445
1680-58-6: 16463
1681-10-3: 22140
1681-15-8: 22152
1681-22-7: 16534
1681-37-4: 22577
1685-82-1: 15013
1686-14-2: 17918
1686-28-8: 07329
1686-30-2: 17530
1686-41-5: 09653
1687-29-2: 10035
1689-64-1: 12776
1689-71-0: 18007
1689-73-2: 19572
1689-78-7: 24715
1689-79-8: 24716
1689-82-3: 19561
1689-83-4: 06713
1689-84-5: 06693
1690-75-1: 20248
1694-19-5: 03615
1694-20-8: 03813
1694-31-1: 08488
1694-83-3: 14242
1694-92-4: 05516
1695-04-1: 03611
1696-20-4: 16084
1697-34-3: 16110
1698-80-2: 19704
1699-46-3: 15476
1699-51-0: 15449
1700-30-7: 05143
1700-37-4: 01772
1701-81-1: 10823
1701-95-7: 06167
1703-46-4: 05047
1703-52-2: 12898
1703-58-8: 08260
1703-88-4: 02748
1703-90-8: 08609
1704-15-0: 22014
1705-84-6: 25199
1705-85-7: 09409
1705-89-1: 12794
1706-11-2: 03578
1706-12-3: 03731
1706-73-6: 04765
1707-03-5: 19777
1708-29-8: 12882
1708-39-0: 11329
1709-03-1: 00158
1709-14-4: 01519
1709-39-3: 05314
1709-44-0: 01573
1710-13-0: 13015
1710-98-1: 07210
1711-02-0: 07220
1711-05-3: 07222
1711-06-4: 07225
1711-07-5: 07213
1712-64-7: 17136
1712-70-5: 02872
1712-84-1: 18810
1717-00-6: 11817
1718-50-9: 03875
1719-53-5: 24017
1719-57-9: 23984
1719-58-0: 23992
1720-32-7: 03516
1721-89-7: 23629
1721-93-3: 15453
1722-10-7: 22522
1722-12-9: 23048
1722-62-9: 20237
1723-00-8: 20243
1723-94-0: 16099
1724-08-9: 07296
1725-74-2: 14442
1726-02-9: 11394
1726-12-1: 03880
1726-14-3: 02749
1726-77-8: 17558
1726-80-3: 08178
1729-67-5: 21131
1729-99-3: 16624
1730-37-6: 12739
1730-86-5: 02945
1730-89-8: 14252
1730-91-2: 08447
1730-92-3: 18445
1731-79-9: 11504
1731-84-6: 17253
1731-86-8: 25296
1731-88-0: 25159
1731-92-6: 13417
1731-94-8: 17170
1732-08-7: 13579
1732-09-8: 17638
1732-10-1: 17227
1733-12-6: 19064
1733-63-7: 12537
1733-76-2: 16246
1735-17-7: 10029
1735-84-8: 22611
1736-71-6: 24896
1738-25-6: 20978
1738-36-9: 00889
1739-84-0: 14769
1740-54-1: 10930
1740-57-4: 04520
1740-97-2: 19475
1741-01-1: 14659
1741-41-9: 20574
1741-83-9: 18260
1741-84-0: 20667
1745-16-0: 03877
1745-81-9: 19588
1746-11-8: 05784
1746-13-0: 03901
1746-23-2: 03225
1746-77-6: 09090
1747-60-0: 07041
1747-92-8: 24158
1749-19-5: 02467
1751-97-9: 07628
1752-24-5: 19392

1754-49-0: 19848
1754-62-7: 21961
1754-74-1: 02907
1755-01-7: 15789
1755-05-1: 18132
1758-10-7: 03758
1758-25-4: 04168
1758-33-4: 18001
1758-64-1: 01233
1758-68-5: 01229
1758-73-2: 15746
1758-85-6: 03150
1758-88-9: 03402
1759-53-1: 10797
1759-58-6: 10521
1759-64-4: 10309
1759-71-3: 10057
1759-81-5: 10716
1759-88-2: 24076
1760-24-3: 11966
1760-46-9: 04285
1761-56-4: 19381
1761-61-1: 01594
1761-71-3: 09789
1762-15-8: 12487
1765-40-8: 02701
1767-46-0: 14285
1768-31-6: 21576
1768-59-8: 14667
1768-64-5: 22230
1771-18-2: 19680
1772-10-7: 03705
1772-43-6: 17932
1772-76-5: 21943
1773-44-0: 01753
1773-99-5: 25123
1774-34-1: 19598
1774-35-2: 03919
1774-37-4: 03917
1775-95-7: 15825
1776-17-6: 09327
1776-18-7: 22018
1776-20-1: 09322
1776-22-3: 09323
1776-26-7: 22015
1776-30-3: 22006
1776-37-0: 14974
1777-82-8: 05022
1777-84-0: 00239
1777-87-3: 02265
1778-02-5: 20361
1778-08-1: 01904
1778-09-2: 12481
1779-48-2: 19781
1779-60-8: 08496
1779-81-3: 24490
1780-31-0: 23059
1780-40-1: 23078
1781-78-8: 10475
1783-25-1: 15773
1783-96-6: 01459
1785-02-0: 25000
1785-65-5: 16770
1785-74-6: 11113
1785-98-4: 15064
1786-05-6: 06853
1786-32-9: 13483
1790-22-3: 08069
1792-17-2: 25411
1792-40-1: 05624
1792-41-2: 05615
1792-81-0: 10055
1794-45-2: 21619
1794-48-5: 02726
1794-84-9: 15650
1794-89-4: 17451
1795-01-3: 24724
1795-04-6: 24781
1795-05-7: 24782
1795-09-1: 24774
1795-15-9: 09937
1795-16-0: 09823
1795-17-1: 09849
1795-18-2: 09958
1795-19-3: 24309
1795-20-6: 10566
1795-21-7: 10509
1795-22-8: 10573
1795-26-2: 09982
1795-27-3: 09981
1795-48-8: 20630
1796-05-0: 03280
1798-04-5: 00607
1798-09-0: 04246
1801-42-9: 15102
1801-72-5: 25437
1804-15-5: 20395
1805-61-4: 19450
1806-26-4: 19545
1806-29-7: 07631
1807-02-9: 01086
1807-36-9: 06862
1808-12-4: 11687
1809-02-5: 07787
1809-04-7: 17623
1809-05-8: 18253
1809-10-5: 18183
1809-19-4: 19815
1809-20-7: 19805
1809-26-3: 18657
1810-66-8: 23830
1810-72-6: 23612
1812-50-6: 14194
1812-51-7: 07506
1813-77-0: 10888
1816-00-8: 22760
1816-78-0: 20349
1817-47-6: 03673
1817-57-8: 09019
1817-73-8: 02036
1817-74-9: 03656
1817-75-0: 03803
1817-76-1: 03651
1818-07-1: 03830
1818-27-5: 12528
1818-96-8: 04623
1820-09-3: 07390
1820-59-3: 07552
1820-80-0: 22374
1821-02-9: 18460
1821-12-1: 04362
1821-27-8: 02429
1821-36-9: 02127
1821-39-2: 02485
1822-62-4: 24847
1822-74-8: 12585
1823-31-0: 03671
1823-52-5: 17981
1823-90-1: 08656
1823-91-2: 04334
1824-81-3: 22551
1825-23-6: 07231
1825-30-5: 16334
1825-31-6: 16333
1825-62-3: 24063
1825-82-7: 24104
1826-14-8: 24517
1826-56-8: 00244
1829-32-9: 06084
1829-81-8: 05388
1830-54-2: 18319
1832-53-7: 19837
1833-51-8: 23996
1835-96-7: 09320
1835-98-9: 09318
1835-99-0: 22016

1836-00-6: 09319
1836-06-2: 12298
1836-87-9: 12751
1838-59-1: 12856
1838-77-3: 14506
1839-84-5: 07448
1839-88-9: 09975
1843-05-6: 15987
1843-21-6: 07045
1843-24-9: 19526
1844-13-9: 17655
1845-30-3: 07389
1845-59-6: 10172
1846-13-5: 20745
1846-70-4: 17320
1846-76-0: 06795
1849-18-9: 19088
1849-29-2: 15620
1849-36-1: 05558
1850-14-2: 15698
1850-15-3: 04389
1852-58-0: 01089
1852-61-5: 01071
1859-03-6: 05292
1859-76-3: 16008
1860-27-1: 08020
1860-60-2: 05103
1861-00-3: 03637
1861-21-8: 23217
1864-92-2: 02262
1866-16-6: 11742
1866-31-5: 21968
1866-39-3: 21932
1867-37-4: 20785
1868-53-7: 15660
1869-63-2: 00859
1871-52-9: 10469
1871-57-4: 21681
1871-58-5: 20723
1871-76-7: 04355
1871-96-1: 10932
1873-25-2: 08572
1873-29-6: 20631
1873-54-7: 23646
1873-88-7: 25203
1873-92-3: 24027
1874-23-3: 12984
1874-24-4: 21938
1875-48-5: 15366
1875-63-4: 16379
1875-87-2: 04835
1875-88-3: 04832
1875-89-4: 04857
1877-72-1: 06126
1877-73-2: 04276
1877-77-6: 04967
1878-49-5: 00686
1878-65-5: 04152
1878-66-6: 04153
1878-67-7: 04137
1878-68-8: 04138
1878-83-7: 00646
1878-84-8: 00647
1878-85-9: 00676
1878-87-1: 00704
1878-91-7: 00498
1879-06-7: 12458
1879-07-8: 09906
1879-09-0: 19294
1879-56-7: 00497
1883-32-5: 04868
1883-78-9: 12911
1884-68-0: 24777
1885-14-9: 09236
1885-29-6: 06661
1885-35-4: 06741
1885-38-7: 21765
1886-57-3: 03239
1886-75-5: 20585
1888-57-9: 14352
1888-71-7: 21714
1888-85-3: 07987
1890-28-4: 18118
1892-29-1: 12154
1892-43-9: 12128
1892-54-2: 18824
1892-72-4: 05030
1893-99-8: 00811
1895-97-2: 21930
1896-62-4: 08970
1897-45-6: 04511
1898-14-2: 03613
1899-24-7: 12945
1899-94-1: 05377
1901-26-4: 08965
1904-72-9: 23455
1904-73-0: 23453
1904-98-9: 22189
1906-82-7: 13274
1907-13-7: 24214
1907-65-9: 05346
1912-21-6: 21270
1912-43-2: 15184
1914-00-7: 15123
1914-65-4: 16652
1915-77-1: 23503
1916-07-0: 06629
1916-08-1: 06344
1916-95-6: 17353
1917-15-3: 12977
1917-30-2: 06990
1918-00-9: 06159
1918-02-1: 22830
1918-77-0: 24783
1918-79-2: 24808
1918-82-7: 24722
1920-81-6: 20023
1921-70-6: 18065
1927-25-9: 14576
1927-94-2: 01609
1928-30-9: 25102
1928-38-7: 00590
1928-40-1: 00789
1928-44-5: 00593
1928-57-0: 00586
1929-29-9: 05237
1929-73-3: 00587
1930-72-9: 06680
1931-60-8: 08244
1931-78-8: 20938
1931-83-5: 01509
1932-04-3: 20162
1932-36-1: 24865
1937-54-8: 17186
1937-62-8: 17507
1938-17-6: 04411
1938-22-3: 05157
1939-99-7: 04958
1940-18-7: 10137
1940-19-8: 10136
1940-27-8: 02029
1940-57-4: 12724
1942-39-8: 16565
1942-45-6: 17846
1942-46-7: 11022
1942-52-5: 12069
1943-16-4: 15652
1943-54-0: 19233
1943-83-5: 11786
1943-95-9: 19197
1945-53-5: 18911
1945-92-2: 12151
1946-82-3: 15543
1947-00-8: 14264
1948-31-8: 01018
1948-33-0: 04755

1948-40-9: 01702
1949-07-1: 20084
1949-89-9: 14534
1951-11-7: 08680
1952-38-1: 19380
1953-33-9: 23182
1953-54-4: 15266
1955-45-9: 17980
1955-46-0: 04612
1956-39-4: 12328
1964-45-0: 07096
1965-25-9: 02535
1967-25-5: 25378
1967-31-3: 04620
1969-43-3: 13520
1969-73-9: 04094
1971-46-6: 15172
1971-49-9: 15365
1971-55-7: 24307
1973-09-7: 25017
1973-22-4: 02642
1974-04-5: 13480
1975-43-5: 16648
1975-44-6: 16649
1975-45-7: 16650
1975-50-4: 06493
1975-52-6: 06494
1975-78-6: 10950
1977-72-6: 05618
1981-38-0: 01160
1981-90-4: 09356
1984-58-3: 03085
1984-59-4: 03082
1984-65-2: 03083
1985-12-2: 02662
1985-59-7: 16505
1985-97-3: 03424
1986-06-7: 19422
1986-43-2: 24569
1986-81-8: 22815
1986-90-9: 18267
1989-33-9: 12763
1990-29-0: 01035
1990-90-5: 22575
1992-48-9: 24173
1995-76-2: 00399
1998-64-7: 22151
2001-93-6: 23159
2002-16-6: 13378
2002-24-6: 12092
2002-59-7: 22219
2004-03-7: 22157
2004-69-5: 18754
2004-70-8: 18093
2005-43-8: 23561
2006-14-6: 21936
2007-97-8: 05715
2009-00-9: 07397
2009-24-7: 13130
2009-74-7: 13806
2009-83-8: 14272
2010-15-3: 14901
2012-29-5: 19256
2013-12-9: 17350
2014-83-7: 03047
2016-42-4: 24351
2016-57-1: 10907
2017-68-7: 16840
2018-61-3: 19722
2018-75-9: 13065
2018-90-8: 16844
2019-16-1: 11695
2021-28-5: 05220
2022-78-8: 23236
2024-83-1: 06700
2025-40-3: 21818
2026-16-6: 01146
2027-17-0: 16444
2027-19-2: 16498
2027-46-5: 13884
2028-52-6: 18794
2028-63-9: 09016
2029-94-9: 01640
2031-62-1: 24036
2031-67-6: 24125
2032-13-5: 14874
2032-15-7: 14886
2032-17-9: 14801
2032-35-1: 11761
2033-24-1: 11282
2034-22-2: 14792
2034-25-5: 18430
2034-28-8: 25049
2034-72-2: 24253
2035-89-4: 12802
2035-94-1: 04844
2036-41-1: 23074
2037-19-6: 15349
2037-31-2: 05545
2038-03-1: 16130
2038-72-4: 01424
2039-66-9: 19018
2039-76-1: 12497
2039-77-2: 12498
2039-82-9: 02630
2039-83-0: 03063
2039-85-2: 02810
2039-86-3: 02629
2039-87-4: 02809
2039-88-5: 02628
2039-89-6: 03345
2039-90-9: 03344
2039-91-0: 03366
2039-93-2: 03660
2039-95-4: 18938
2040-07-5: 12533
2040-14-4: 21513
2040-95-1: 10503
2040-96-2: 10572
2042-14-0: 19502
2042-37-7: 06671
2043-24-5: 23428
2043-38-1: 06983
2043-61-0: 10006
2044-08-8: 10298
2044-27-1: 22951
2044-72-6: 07844
2045-00-3: 01423
2045-70-7: 24667
2046-18-6: 04359
2047-21-4: 12535
2049-66-3: 20834
2049-67-4: 18686
2049-73-2: 03133
2049-74-3: 12166
2049-80-1: 20883
2049-84-5: 23469
2049-86-7: 08126
2049-92-5: 02226
2049-94-7: 03635
2049-95-8: 03262
2049-96-9: 06544
2050-01-3: 21240
2050-03-5: 03269
2050-07-9: 18541
2050-08-0: 06318
2050-14-8: 19054
2050-15-9: 19055
2050-16-0: 19056
2050-18-2: 08925
2050-20-6: 13578
2050-23-9: 17637
2050-24-0: 03148
2050-25-1: 12236
2050-32-0: 09679
2050-33-1: 09678

2050-37-5: 15937
2050-43-3: 00226
2050-46-6: 03132
2050-47-7: 03835
2050-48-8: 03922
2050-49-9: 16918
2050-54-6: 09500
2050-60-4: 11980
2050-61-5: 11979
2050-67-1: 07471
2050-68-2: 07474
2050-69-3: 16331
2050-72-8: 16335
2050-73-9: 16336
2050-74-0: 16337
2050-75-1: 16338
2050-76-2: 16920
2050-77-3: 10922
2050-78-4: 17132
2050-85-3: 00361
2050-87-5: 25213
2050-89-7: 07600
2050-92-2: 18172
2050-94-4: 09201
2050-95-5: 08557
2051-00-5: 11974
2051-04-9: 11397
2051-05-0: 24300
2051-07-2: 18123
2051-10-7: 09413
2051-25-4: 10892
2051-30-1: 17605
2051-31-2: 10982
2051-32-3: 17735
2051-33-4: 14287
2051-49-2: 14207
2051-53-8: 02373
2051-59-4: 06825
2051-60-7: 07457
2051-61-8: 07458
2051-62-9: 07459
2051-78-7: 08517
2051-79-8: 04480
2051-84-5: 02357
2051-85-6: 04728
2051-90-3: 03094
2051-95-8: 04373
2051-97-0: 23314
2051-98-1: 00012
2052-01-9: 21085
2052-06-4: 02031
2052-07-5: 07449
2052-13-3: 02753
2052-14-4: 06298
2052-15-5: 18464
2054-35-5: 07137
2055-40-5: 03362
2055-46-1: 23168
2057-43-4: 22712
2057-49-0: 22729
2057-84-3: 18139
2058-62-0: 02386
2058-74-4: 15213
2062-77-3: 08611
2065-23-8: 00727
2065-37-4: 16772
2065-70-5: 16339
2067-33-6: 18404
2067-47-2: 15511
2067-48-3: 15510
2067-52-9: 15971
2067-84-7: 23014
2073-32-7: 01521
2074-05-7: 13460
2076-36-0: 05814
2077-13-6: 02875
2077-46-5: 04022
2078-54-8: 19086
2079-95-0: 24366
2081-44-9: 22269
2082-59-9: 18400
2082-61-3: 03751
2082-76-0: 10957
2083-91-2: 23968
2084-13-1: 04144
2084-18-6: 08279
2084-19-7: 18372
2084-45-9: 03056
2085-88-3: 18021
2086-96-6: 05683
2088-24-6: 00677
2088-71-3: 01083
2088-87-1: 16711
2089-06-7: 20336
2089-36-3: 05746
2089-87-4: 16714
2089-89-6: 16719
2089-91-0: 16715
2090-14-4: 09411
2090-15-5: 09875
2091-29-4: 13920
2093-20-1: 12089
2094-72-6: 25118
2094-98-6: 10003
2095-17-2: 19698
2095-57-0: 23155
2099-26-5: 01074
2099-63-0: 05868
2100-17-6: 18611
2101-86-2: 02540
2101-88-4: 02533
2103-57-3: 01791
2104-04-3: 24493
2104-09-8: 24498
2104-64-5: 19869
2106-54-9: 14268
2106-93-6: 12630
2107-69-9: 15084
2107-70-2: 05214
2107-76-8: 06833
2108-92-1: 09833
2109-66-2: 16132
2110-78-3: 21205
2111-75-3: 10364
2113-47-5: 00094
2113-51-1: 07514
2113-54-4: 00095
2113-56-6: 07520
2113-57-7: 07450
2113-58-8: 07535
2113-68-0: 07682
2114-00-3: 21445
2114-11-6: 09119
2114-42-3: 09951
2116-36-1: 02717
2116-62-3: 22722
2116-64-5: 22723
2116-65-6: 22726
2117-11-5: 18793
2117-24-0: 24086
2121-66-6: 21010
2121-67-7: 18300
2122-61-4: 05911
2122-70-5: 16572
2122-77-2: 12247
2123-27-5: 03070
2123-28-6: 03062
2124-31-4: 12372
2124-47-2: 02206
2127-09-5: 22952
2128-93-0: 15841
2129-29-5: 06251
2129-59-1: 15437
2130-23-6: 17815
2131-18-2: 03863
2131-38-6: 16554

2131-39-7: 16552
2131-41-1: 16556
2131-42-2: 16557
2131-43-3: 16540
2131-55-7: 02854
2131-62-6: 06432
2131-63-7: 06431
2131-64-8: 02308
2132-80-1: 07481
2134-29-4: 20394
2136-70-1: 12241
2136-71-2: 12180
2136-81-4: 02954
2136-89-2: 02953
2136-95-0: 03859
2138-20-7: 04690
2138-22-9: 04648
2138-43-4: 04668
2138-48-9: 04667
2140-49-0: 14223
2141-42-6: 01163
2141-62-0: 20980
2142-01-0: 15401
2142-06-5: 23405
2142-29-2: 03040
2142-63-4: 12303
2142-66-7: 12279
2142-73-6: 12379
2144-08-3: 01789
2146-37-4: 10531
2146-38-5: 10709
2147-34-4: 17084
2150-18-7: 22655
2150-37-0: 06199
2150-38-1: 06196
2150-41-6: 06190
2150-42-7: 06188
2150-44-9: 06179
2150-47-2: 06174
2153-08-4: 00152
2155-30-8: 21188
2155-42-2: 03066
2155-94-4: 21661
2156-29-8: 22963
2156-97-0: 21844
2157-31-5: 18332
2157-39-3: 04594
2157-50-8: 21522
2157-52-0: 12797
2157-56-4: 18355
2157-57-5: 14180
2158-55-6: 09632
2160-93-2: 12147
2160-94-3: 10387
2161-40-2: 09667
2161-85-5: 12366
2162-63-2: 15913
2162-92-7: 14021
2162-98-3: 10915
2162-99-4: 17597
2163-00-0: 14023
2163-44-2: 20810
2163-48-6: 20885
2163-77-1: 01422
2164-08-1: 10691
2164-09-2: 21644
2164-17-2: 25429
2164-19-4: 09782
2167-39-7: 17977
2168-13-0: 22972
2168-93-6: 08044
2169-64-4: 25007
2169-87-1: 16717
2170-03-8: 13030
2171-74-6: 05759
2171-82-6: 19866
2173-56-0: 18471
2173-69-5: 04840
2173-82-2: 08393
2174-64-3: 04707
2175-90-8: 02983
2175-91-9: 10487
2176-62-7: 22711
2176-98-9: 24231
2177-18-6: 21845
2177-32-4: 14512
2177-47-1: 15047
2177-77-7: 18441
2177-78-8: 18447
2179-57-9: 11421
2179-60-4: 11426
2180-43-0: 14327
2180-68-9: 18106
2180-69-0: 18107
2182-66-3: 24159
2188-15-0: 22107
2189-60-8: 03828
2192-35-0: 15894
2194-88-9: 04800
2196-13-6: 22781
2196-23-8: 13728
2196-25-0: 14379
2197-57-1: 16794
2198-20-1: 09602
2198-51-8: 19082
2198-61-0: 14256
2198-64-3: 01021
2198-65-4: 07838
2198-66-5: 19105
2198-72-3: 13693
2198-74-5: 14716
2198-75-6: 16332
2198-77-8: 16340
2199-44-2: 23328
2199-51-1: 23330
2199-52-2: 23331
2199-53-3: 23332
2199-60-2: 23335
2203-14-7: 04637
2203-36-3: 08535
2203-80-7: 14546
2205-27-8: 10377
2206-38-4: 02981
2206-48-6: 02977
2206-89-5: 21802
2207-01-4: 09839
2207-27-4: 10491
2207-50-3: 17926
2208-25-5: 07150
2210-07-3: 01522
2210-25-5: 21652
2210-28-8: 21917
2210-79-9: 18020
2211-67-8: 07968
2211-68-9: 08758
2211-69-0: 08759
2211-70-3: 08716
2211-81-6: 15345
2211-94-1: 18018
2212-05-7: 17998
2212-06-8: 17992
2212-17-1: 24864
2213-08-3: 01004
2213-23-2: 13511
2213-32-3: 18632
2213-43-6: 20127
2213-63-0: 23870
2213-71-0: 19966
2214-14-4: 03636
2215-20-5: 13339
2215-21-6: 06337
2215-63-6: 14984
2215-77-2: 06547
2216-12-8: 03809
2216-13-9: 16651
2216-15-1: 02163

2216-16-2: 02162
2216-30-0: 13512
2216-32-2: 13522
2216-35-5: 17199
2216-36-6: 17203
2216-45-7: 05113
2216-51-5: 10180
2216-52-6: 10184
2216-67-3: 16218
2216-68-4: 16180
2216-69-5: 16429
2216-70-8: 17532
2216-75-3: 12954
2216-77-5: 21378
2216-81-1: 21177
2216-84-4: 10079
2216-87-7: 25311
2216-92-4: 13337
2216-93-5: 14859
2216-94-6: 22121
2216-99-1: 06370
2217-02-9: 07313
2217-06-3: 03984
2217-07-4: 02244
2217-08-5: 23196
2217-14-3: 08137
2217-15-4: 08135
2217-25-6: 23472
2217-31-4: 23839
2217-41-6: 16199
2217-45-0: 02497
2217-46-1: 02499
2217-55-2: 11393
2217-58-5: 16542
2217-59-6: 18178
2217-60-9: 04761
2217-65-4: 03849
2217-81-4: 23944
2218-94-2: 25078
2219-30-9: 25532
2219-66-1: 08998
2219-72-9: 19307
2219-82-1: 19304
2223-67-8: 07263
2223-82-7: 21842
2224-98-8: 00984
2227-79-4: 04386
2230-69-5: 10346
2230-70-8: 10290
2231-05-2: 05756
2231-57-4: 09246
2233-00-3: 21743
2233-29-6: 22753
2234-13-1: 16470
2234-14-2: 12539
2234-16-4: 12345
2234-20-0: 03342
2234-26-6: 07292
2235-15-6: 00027
2236-11-5: 10017
2236-60-4: 22136
2237-30-1: 06662
2237-36-7: 06359
2239-78-3: 13901
2240-21-3: 14881
2240-25-7: 23228
2240-41-7: 19850
2240-53-1: 19780
2242-78-6: 05895
2243-24-5: 17192
2243-27-8: 17237
2243-30-3: 02448
2243-32-5: 06542
2243-33-6: 09684
2243-35-8: 12257
2243-42-7: 06545
2243-43-8: 08257
2243-44-9: 02821
2243-47-2: 07558
2243-53-0: 08926
2243-54-1: 16423
2243-55-2: 14643
2243-56-3: 14644
2243-57-4: 14645
2243-58-5: 14646
2243-61-0: 16698
2243-62-1: 16699
2243-63-2: 16700
2243-64-3: 16701
2243-67-6: 16704
2243-69-8: 00058
2243-71-2: 01268
2243-73-4: 01946
2243-74-5: 04721
2243-76-7: 06402
2243-79-0: 16026
2243-80-3: 16027
2243-81-4: 16601
2243-82-5: 16602
2243-83-6: 16588
2243-89-2: 23715
2243-90-5: 23720
2243-93-8: 16561
2243-94-9: 16562
2243-95-0: 16564
2243-98-3: 25343
2244-05-5: 25268
2244-07-7: 25282
2244-16-8: 10421
2245-25-2: 18014
2245-38-7: 16558
2245-52-5: 04912
2245-90-1: 15479
2246-44-8: 16176
2247-71-4: 00098
2247-78-1: 00396
2251-50-5: 07232
2251-65-2: 07237
2252-63-3: 20069
2253-67-0: 03977
2253-73-8: 20637
2254-94-6: 07074
2256-01-1: 07820
2257-35-4: 21975
2257-52-5: 14633
2258-58-4: 18708
2259-14-5: 10809
2259-96-3: 06976
2268-77-1: 07075
2270-20-4: 05160
2270-60-2: 17819
2272-04-0: 06252
2272-45-9: 02409
2273-08-7: 25070
2273-09-8: 25081
2273-93-0: 22164
2277-23-8: 10962
2279-76-7: 24219
2279-99-4: 14204
2280-28-6: 25534
2282-84-0: 06636
2283-08-1: 16634
2283-82-1: 01094
2288-18-8: 03659
2291-80-7: 09107
2292-83-3: 09582
2294-43-1: 02720
2294-71-5: 04179
2294-72-6: 25347
2294-76-0: 22713
2294-94-2: 03319
2295-31-0: 24543
2295-58-1: 21528
2297-30-5: 01075
2298-07-9: 16154
2300-47-2: 10763

2300-51-8: 10764
2300-53-0: 10762
2302-88-7: 00472
2303-25-5: 03717
2304-94-1: 01006
2305-13-7: 05270
2305-31-9: 10039
2305-32-0: 10037
2305-36-4: 05930
2306-88-9: 17692
2307-06-4: 24192
2307-10-0: 12062
2307-69-9: 05465
2307-81-5: 05723
2308-38-5: 08392
2310-98-7: 20505
2311-05-9: 19305
2311-46-8: 14257
2311-59-3: 10974
2312-15-4: 06258
2313-65-7: 14299
2313-87-3: 04462
2314-17-2: 07050
2314-36-5: 01638
2314-61-6: 11443
2314-78-5: 23392
2314-97-8: 15710
2315-36-8: 00143
2315-37-9: 00151
2315-68-6: 06572
2316-50-9: 19107
2316-64-5: 04994
2317-91-1: 12554
2318-08-3: 12702
2319-61-1: 09860
2320-30-1: 10238
2321-07-5: 24196
2321-08-6: 12701
2322-00-1: 00453
2323-81-1: 18412
2324-41-6: 14673
2325-10-2: 12001
2326-77-4: 22811
2327-69-7: 08379
2328-24-7: 04254
2338-12-7: 07154
2338-18-3: 14987
2338-20-7: 06625
2338-27-4: 05633
2344-70-9: 05273
2344-71-0: 04377
2344-80-1: 24000
2344-83-4: 24007
2345-26-8: 21229
2345-27-9: 24387
2345-28-0: 18075
2345-32-6: 09035
2345-34-8: 05861
2345-61-1: 21799
2346-00-1: 24510
2346-81-8: 14004
2348-77-8: 16818
2348-81-4: 16771
2348-82-5: 16814
2348-91-6: 10739
2349-07-7: 21236
2349-13-5: 21235
2349-67-9: 24459
2349-71-5: 04760
2350-01-8: 04494
2350-19-8: 13493
2350-89-2: 07503
2351-13-5: 11367
2351-97-5: 18703
2352-37-6: 25016
2358-84-1: 21914
2359-07-1: 19885
2361-27-5: 24806
2362-10-9: 11369
2362-12-1: 19113
2362-18-7: 03837
2362-57-4: 19560
2363-71-5: 13392
2364-46-7: 16563
2364-47-8: 16956
2364-50-3: 07860
2364-54-7: 19484
2364-59-2: 08508
2365-44-8: 15212
2365-48-2: 00670
2366-52-1: 07998
2366-96-3: 07244
2367-22-8: 07511
2367-82-0: 03960
2370-12-9: 18487
2370-13-0: 14273
2370-14-1: 17704
2370-88-9: 10850
2371-19-9: 13704
2373-51-5: 15649
2379-55-7: 23874
2379-90-0: 01213
2380-94-1: 15265
2381-15-9: 01987
2381-16-0: 01995
2381-19-3: 06749
2381-21-7: 22507
2381-31-9: 01994
2381-39-7: 06947
2382-40-3: 11561
2382-80-1: 25230
2382-96-9: 07181
2384-70-5: 11019
2384-75-0: 13736
2384-85-2: 11020
2384-86-3: 11021
2384-90-9: 13435
2385-27-5: 01945
2385-70-8: 08327
2385-77-5: 17791
2386-25-6: 12383
2386-26-7: 23327
2386-60-9: 08259
2387-23-7: 25415
2388-14-9: 03666
2391-71-1: 19391
2393-18-2: 21777
2393-23-9: 04926
2393-53-5: 11675
2396-43-2: 25189
2396-53-4: 03853
2396-60-3: 11069
2396-63-6: 13458
2396-65-8: 17189
2396-76-1: 10056
2396-77-2: 14459
2396-78-3: 14465
2396-80-7: 14476
2396-83-0: 14464
2396-84-1: 13957
2396-85-2: 17816
2398-09-6: 09532
2398-10-9: 09533
2398-12-1: 09566
2398-16-5: 09565
2398-37-0: 02664
2398-64-3: 03525
2398-65-4: 03829
2398-66-5: 03723
2398-68-7: 03299
2398-81-4: 22854
2399-48-6: 21973
2400-00-2: 03397
2400-01-3: 03521
2400-02-4: 03429
2400-03-5: 03907

2400-04-6: 02759
2401-22-1: 02847
2401-24-3: 02088
2401-73-2: 11373
2402-06-4: 10773
2402-13-3: 20329
2402-77-9: 22616
2402-78-0: 22618
2402-79-1: 22745
2402-97-3: 22587
2403-27-2: 22004
2403-35-2: 20452
2403-55-6: 16109
2403-88-5: 20287
2404-18-4: 20229
2404-35-5: 09628
2404-73-1: 19832
2404-80-0: 19795
2404-81-1: 19794
2404-87-7: 24764
2404-89-9: 07717
2407-68-3: 21759
2408-20-0: 21279
2408-37-9: 10287
2409-52-1: 08185
2409-55-4: 19302
2410-48-2: 04788
2411-36-1: 13469
2411-58-7: 25278
2411-74-7: 14689
2411-95-2: 10468
2412-73-9: 06128
2412-80-8: 18450
2414-29-1: 20313
2415-72-7: 10784
2415-85-2: 07862
2416-30-0: 21392
2416-94-6: 19664
2416-98-0: 07672
2417-04-1: 07650
2417-90-5: 20967
2418-14-6: 08148
2418-31-7: 08835
2419-73-0: 08224
2420-16-8: 01610
2420-17-9: 14861
2420-18-0: 14848
2420-26-0: 01611
2421-28-5: 15309
2422-79-9: 01989
2423-01-0: 10700
2423-10-1: 17476
2423-71-4: 19319
2424-01-3: 21666
2424-92-2: 11585
2424-98-8: 21232
2425-28-7: 04995
2425-54-9: 24354
2425-66-3: 20537
2425-79-8: 17993
2426-02-0: 15326
2426-08-6: 17994
2426-54-2: 21830
2426-87-1: 01751
2428-04-8: 25028
2430-00-4: 00389
2431-24-5: 13757
2431-30-3: 14549
2431-50-7: 08743
2431-52-9: 08040
2431-54-1: 08787
2432-11-3: 24320
2432-12-4: 19237
2432-63-5: 08818
2432-87-3: 10941
2432-89-5: 10937
2432-90-8: 04556
2432-99-7: 25286
2433-56-9: 18942
2433-97-8: 25103
2435-53-2: 09750
2435-54-3: 09749
2435-82-7: 00009
2436-79-5: 23334
2436-84-2: 13440
2436-85-3: 16207
2436-90-0: 17527
2436-92-2: 23634
2436-93-3: 23642
2436-96-6: 07496
2437-03-8: 16208
2437-05-0: 21941
2437-17-4: 16636
2437-18-5: 16683
2437-20-9: 15202
2437-25-4: 11506
2437-30-1: 00339
2437-33-4: 05426
2437-39-0: 02375
2437-48-1: 19634
2437-49-2: 04735
2437-56-1: 25174
2437-66-3: 06632
2437-72-1: 23712
2437-73-2: 23713
2437-76-5: 02687
2437-88-9: 04035
2437-98-1: 00365
2438-03-1: 06575
2438-04-2: 06488
2438-05-3: 06573
2438-12-2: 10394
2439-35-2: 21841
2439-54-5: 17569
2439-77-2: 01921
2442-31-1: 06849
2442-49-1: 24339
2443-03-0: 10558
2443-04-1: 10557
2443-58-5: 12775
2443-91-6: 20550
2444-19-1: 04762
2444-29-3: 03716
2444-36-2: 04150
2444-37-3: 00697
2444-68-0: 01147
2444-89-5: 03836
2445-69-4: 21239
2445-76-3: 21178
2445-77-4: 08459
2445-78-5: 08449
2445-81-0: 06856
2445-82-1: 06857
2445-83-2: 06861
2446-69-7: 19379
2446-84-6: 11083
2449-10-7: 10939
2450-26-2: 01717
2450-71-7: 22095
2451-00-5: 16403
2451-01-6: 10086
2451-91-4: 09499
2452-70-2: 19830
2453-35-2: 20474
2455-24-5: 21918
2455-43-8: 02575
2456-28-2: 10929
2456-43-1: 24308
2457-47-8: 22619
2457-76-3: 05898
2458-12-0: 05933
2458-23-3: 23729
2459-05-4: 08807
2459-09-8: 22873
2459-24-7: 16639
2459-25-8: 16685

2460-77-7: 09696
2462-85-3: 17366
2462-94-4: 11601
2463-53-8: 17292
2463-63-0: 13731
2465-29-4: 25562
2466-76-4: 14758
2467-02-9: 19454
2467-03-0: 19385
2467-09-6: 07688
2467-10-9: 18268
2467-15-4: 07739
2469-99-0: 08256
2470-10-2: 04927
2470-11-3: 10082
2470-68-0: 13470
2471-92-3: 07124
2473-01-0: 17202
2475-44-7: 01218
2475-93-6: 08647
2476-37-1: 12346
2478-38-8: 12404
2479-47-2: 02369
2479-49-4: 04546
2479-64-3: 05696
2483-57-0: 00703
2484-88-0: 05479
2486-07-9: 03282
2486-70-6: 05935
2486-75-1: 05934
2487-40-3: 22225
2489-03-4: 03993
2489-77-2: 24897
2489-86-3: 16494
2490-54-2: 13619
2491-52-3: 11077
2491-66-9: 05431
2492-08-2: 09663
2493-02-9: 02661
2495-25-2: 21920
2495-35-4: 21971
2497-18-9: 14486
2497-21-4: 14509
2497-33-8: 19511
2497-91-8: 06074
2498-50-2: 04392
2498-66-0: 02006
2498-75-1: 01991
2498-76-2: 01990
2499-59-4: 21951
2499-95-8: 21860
2501-03-3: 02338
2503-56-2: 25098
2505-06-8: 09670
2506-41-4: 16289
2507-55-3: 24375
2508-29-4: 18480
2510-55-6: 18905
2511-07-1: 19796
2511-22-0: 05983
2511-27-5: 19798
2512-56-3: 06316
2512-81-4: 20145
2513-25-9: 06787
2513-49-7: 07087
2514-91-2: 24910
2516-34-9: 09519
2516-95-2: 06115
2516-96-3: 06105
2517-04-6: 01525
2517-43-3: 08547
2519-24-6: 10740
2523-37-7: 12743
2523-39-9: 12741
2523-42-4: 12736
2523-46-8: 12737
2523-48-0: 12752
2523-55-9: 09787
2523-56-0: 09786
2524-04-1: 19952
2524-09-6: 19972
2524-37-0: 06181
2524-52-9: 22833
2524-64-3: 19951
2524-67-6: 02420
2525-02-2: 04677
2525-16-8: 01515
2525-21-5: 23775
2525-62-4: 14061
2527-96-0: 12978
2528-16-7: 04575
2528-38-3: 19930
2528-61-2: 13726
2529-36-4: 06633
2530-26-9: 22706
2530-87-2: 24006
2531-80-8: 21260
2531-84-2: 18877
2532-58-3: 10520
2534-77-2: 07267
2536-35-8: 13879
2536-38-1: 17397
2536-91-6: 07044
2539-53-9: 01676
2541-69-7: 01993
2541-75-5: 13423
2544-06-1: 21220
2545-26-8: 22155
2545-59-7: 00784
2547-61-7: 15761
2548-87-0: 17788
2549-67-9: 01530
2549-99-7: 24128
2550-04-1: 24129
2550-06-3: 24100
2550-21-2: 14345
2550-22-3: 14520
2550-26-7: 08670
2550-28-9: 14565
2550-36-9: 09803
2553-04-0: 16002
2553-19-7: 24034
2554-06-5: 10847
2554-75-8: 22545
2555-13-7: 21655
2555-49-9: 00726
2556-46-9: 04393
2561-21-9: 00830
2562-38-1: 10561
2562-52-9: 22162
2563-97-5: 00393
2565-39-1: 13798
2565-47-1: 23214
2566-19-0: 13344
2566-22-5: 19724
2567-14-8: 21741
2567-29-5: 07453
2567-51-3: 00131
2567-58-0: 00117
2567-60-4: 00149
2568-30-1: 11314
2568-33-4: 08213
2568-90-3: 08018
2568-91-4: 20650
2568-92-5: 08019
2570-12-9: 10466
2570-26-5: 18058
2571-00-8: 01757
2571-05-3: 13603
2571-39-3: 15957
2571-52-0: 06742
2571-54-2: 06740
2572-40-9: 06250
2573-79-7: 24837
2573-80-0: 24843
2575-20-4: 08649

2576-47-8: 11683
2578-45-2: 22601
2579-04-6: 13430
2579-22-8: 22093
2581-34-2: 19500
2582-30-1: 09196
2583-25-7: 20882
2584-71-6: 20368
2585-18-4: 23836
2586-62-1: 16256
2586-89-2: 13827
2587-02-2: 22571
2587-42-0: 11036
2589-71-1: 18537
2591-66-4: 19904
2591-67-5: 19905
2591-68-6: 19894
2591-70-0: 19902
2591-71-1: 19901
2591-74-4: 19899
2591-86-8: 20233
2592-95-2: 07152
2594-20-9: 09091
2594-21-0: 09050
2595-97-3: 01032
2595-98-4: 24305
2601-13-0: 14255
2601-89-0: 20975
2603-10-3: 09105
2605-18-7: 03617
2606-85-1: 06748
2607-03-6: 09559
2612-02-4: 06360
2612-33-1: 20896
2612-35-3: 11316
2612-46-6: 14378
2613-44-7: 00807
2613-61-8: 13568
2613-65-2: 10537
2613-66-3: 10536
2613-69-6: 10582
2613-72-1: 10585
2613-76-5: 15027
2614-88-2: 13967
2615-05-6: 02320
2615-09-0: 12521
2619-22-9: 05353
2620-44-2: 05736
2620-45-3: 05726
2620-50-0: 05752
2621-46-7: 02877
2621-78-5: 09116
2622-08-4: 19998
2622-21-1: 10314
2622-60-8: 05629
2622-89-1: 07749
2623-23-6: 10186
2623-33-8: 00060
2623-37-2: 16637
2623-40-7: 17079
2623-45-2: 12315
2623-50-9: 23641
2623-51-0: 24958
2623-54-3: 07276
2623-82-7: 17671
2623-83-8: 17241
2623-84-9: 25289
2623-87-2: 08332
2623-95-2: 10960
2624-01-3: 13615
2625-30-1: 09932
2625-31-2: 10562
2626-34-8: 06969
2626-60-0: 09774
2627-86-3: 04930
2627-95-4: 11376
2628-17-3: 19351
2628-87-7: 18321
2631-71-2: 12269
2632-13-5: 12300
2632-87-3: 19907
2635-13-4: 05728
2636-79-5: 16766
2637-34-5: 22948
2637-37-8: 23773
2639-63-6: 08423
2641-01-2: 06253
2641-89-6: 04767
2642-63-9: 12347
2642-81-1: 03355
2642-82-2: 04833
2642-98-0: 09395
2644-93-1: 22433
2645-08-1: 01385
2645-24-1: 24646
2645-36-5: 20235
2648-57-9: 08641
2648-61-5: 12349
2648-69-3: 08637
2650-40-0: 10203
2650-64-8: 13263
2652-13-3: 17290
2653-63-6: 11074
2653-70-5: 11075
2653-77-2: 11076
2654-57-1: 22459
2654-58-2: 22458
2655-27-8: 02450
2655-46-1: 22457
2655-48-3: 22456
2655-83-6: 02818
2655-84-7: 02636
2655-95-0: 03128
2657-20-7: 16039
2657-25-2: 22021
2657-65-0: 08048
2657-67-2: 08049
2661-23-6: 09491
2664-63-3: 19623
2672-58-4: 05580
2673-15-6: 14301
2676-33-7: 21184
2677-33-0: 18646
2677-69-2: 21438
2678-54-8: 12572
2679-14-3: 01003
2679-87-0: 07991
2680-00-4: 12603
2680-01-5: 12605
2680-03-7: 21645
2681-83-6: 14210
2681-92-7: 18402
2681-94-9: 08329
2682-45-3: 17100
2683-43-4: 02148
2683-57-0: 18486
2687-12-9: 02936
2687-25-4: 04442
2687-43-6: 14733
2688-77-9: 15451
2688-84-8: 02452
2689-07-8: 02920
2689-59-0: 15970
2689-88-5: 20826
2690-08-6: 17631
2693-46-1: 12703
2694-54-4: 05575
2696-84-6: 02486
2696-85-7: 02059
2698-38-6: 00537
2698-39-7: 00538
2698-44-4: 20782
2699-13-0: 10323
2700-22-3: 20786
2702-58-1: 06239
2705-87-5: 10096

2706-56-1: 22912
2708-97-6: 03962
2713-09-9: 12643
2714-60-5: 11348
2716-62-3: 02356
2717-42-2: 16547
2717-44-4: 16346
2718-25-4: 23913
2719-07-5: 00612
2719-21-3: 00061
2719-23-5: 00378
2719-27-9: 10005
2719-52-0: 03633
2719-61-1: 03794
2719-62-2: 03879
2719-63-3: 02768
2719-64-4: 03910
2720-41-4: 10101
2721-38-2: 10513
2721-43-9: 22428
2721-59-7: 23845
2722-36-3: 05274
2724-56-3: 25297
2724-57-4: 25160
2724-58-5: 13418
2724-59-6: 17427
2724-69-8: 24885
2725-53-3: 01667
2725-82-8: 02643
2726-21-8: 03217
2726-82-1: 06630
2730-43-0: 21687
2732-18-5: 06829
2732-58-3: 09402
2733-41-7: 07190
2734-70-5: 02177
2735-04-8: 02175
2736-37-0: 21594
2736-40-5: 08685
2736-80-3: 20933
2738-19-4: 14425
2738-38-7: 20315
2739-12-0: 02354
2739-16-4: 23734
2741-16-4: 03667
2743-00-2: 02342
2745-25-7: 12940
2745-26-8: 12935
2746-68-1: 24838
2747-48-0: 18723
2747-53-7: 18731
2748-11-0: 10068
2748-21-2: 22468
2748-22-3: 22462
2750-80-3: 17109
2753-11-9: 16392
2755-07-9: 25277
2756-56-1: 07327
2757-90-6: 13256
2758-18-1: 10754
2759-28-6: 20090
2759-54-8: 20412
2760-31-8: 20425
2761-24-2: 24126
2761-84-4: 11560
2765-04-0: 25216
2765-11-9: 18056
2765-18-6: 16497
2766-42-9: 23960
2767-84-2: 07298
2768-41-4: 09001
2769-64-4: 08008
2774-90-5: 17474
2775-95-3: 19900
2777-65-3: 25348
2777-66-4: 25351
2778-68-9: 07286
2778-96-3: 17438
2779-21-7: 05417
2779-35-3: 19888
2782-07-2: 13158
2782-57-2: 24997
2782-91-4: 24894
2783-17-7: 11502
2784-93-2: 04489
2785-74-2: 04655
2785-75-3: 04654
2785-89-9: 19366
2785-98-0: 06191
2786-05-2: 11144
2786-08-5: 11224
2786-80-3: 19326
2789-92-6: 05902
2790-09-2: 04599
2791-29-9: 13864
2792-42-9: 07354
2798-73-4: 08983
2799-17-9: 21354
2801-26-5: 23721
2801-29-8: 23683
2802-06-4: 13294
2802-07-5: 10074
2804-50-4: 21685
2805-57-4: 00249
2806-85-1: 20393
2807-30-9: 12239
2807-54-7: 08805
2808-71-1: 10319
2808-76-6: 10306
2808-77-7: 10308
2808-79-9: 10307
2808-80-2: 09904
2808-86-8: 22743
2809-64-5: 16525
2809-67-8: 17844
2809-69-0: 13974
2810-04-0: 24805
2810-69-7: 03935
2810-74-4: 07609
2813-62-9: 12610
2815-34-1: 20058
2815-67-0: 03187
2816-57-1: 10235
2817-71-2: 22411
2817-96-1: 08956
2818-66-8: 05644
2819-41-2: 03515
2819-48-9: 09799
2819-86-5: 19552
2820-37-3: 22403
2825-00-5: 22294
2825-01-6: 16432
2825-66-3: 17497
2825-79-8: 17496
2826-26-8: 20780
2834-05-1: 25288
2834-23-3: 00525
2834-78-8: 02381
2834-79-9: 24485
2834-92-6: 16958
2835-04-3: 05413
2835-06-5: 04121
2835-21-4: 21718
2835-30-5: 25490
2835-33-8: 22735
2835-39-4: 08467
2835-53-2: 01208
2835-58-7: 02457
2835-59-8: 02458
2835-61-2: 16185
2835-66-7: 01574
2835-68-9: 01804
2835-69-0: 07686
2835-77-0: 15831
2835-78-1: 15832
2835-81-6: 08299

2835-82-7: 08302
2835-96-3: 19027
2835-97-4: 19021
2835-98-5: 19023
2835-99-6: 19029
2836-00-2: 19025
2836-03-5: 04436
2836-04-6: 04461
2840-26-8: 05928
2844-15-7: 02218
2844-72-6: 16386
2845-85-4: 04593
2845-89-8: 02858
2846-89-1: 23170
2847-30-5: 22342
2848-25-1: 04670
2848-62-6: 13178
2849-98-1: 21915
2850-77-3: 22326
2850-78-4: 22344
2850-79-5: 22313
2855-08-5: 07931
2855-19-8: 17999
2855-27-8: 09969
2856-63-5: 04311
2858-66-4: 21582
2859-67-8: 22935
2859-68-9: 22934
2859-78-1: 02614
2860-50-6: 15380
2861-28-1: 05740
2861-48-5: 08238
2866-82-2: 01845
2866-92-4: 13352
2867-05-2: 07404
2867-47-2: 21891
2867-59-6: 08528
2867-63-2: 10680
2868-37-3: 10813
2869-25-2: 20986
2870-04-4: 03401
2870-37-3: 06565
2871-85-4: 12746
2871-91-2: 25198
2872-48-2: 01240
2873-18-9: 24678
2873-74-7: 18363
2873-89-4: 02072
2873-97-4: 21647
2874-74-0: 11529
2875-32-3: 00476
2875-95-8: 20734
2876-08-6: 05613
2876-22-4: 18966
2876-23-5: 18967
2876-35-9: 16366
2876-44-0: 16474
2876-45-1: 16414
2876-51-9: 16473
2876-52-0: 16413
2876-53-1: 16417
2876-63-3: 21519
2876-70-2: 16574
2876-78-0: 16573
2877-13-6: 00383
2877-26-1: 17383
2879-00-7: 22351
2879-20-1: 12354
2880-05-9: 01755
2880-58-2: 09734
2882-19-1: 04135
2882-98-6: 10564
2883-02-5: 09934
2883-05-8: 17589
2883-07-0: 09820
2883-08-1: 09923
2883-12-7: 02986
2883-45-6: 13450
2883-70-7: 01145
2885-00-9: 17392
2886-52-4: 21489
2887-55-0: 21492
2887-56-1: 22083
2887-61-8: 08613
2887-68-5: 22085
2889-31-8: 18308
2890-28-0: 05698
2890-62-2: 12455
2892-15-1: 22071
2892-18-4: 14513
2892-20-8: 22084
2892-22-0: 22073
2892-25-3: 22072
2892-27-5: 21456
2892-30-0: 21442
2892-33-3: 22086
2892-42-4: 15910
2893-05-2: 08666
2895-21-8: 00159
2896-60-8: 04701
2896-63-1: 04676
2898-54-6: 15942
2899-90-3: 18497
2901-34-0: 04374
2901-75-9: 19719
2902-69-4: 12522
2902-96-7: 21342
2902-98-9: 04352
2903-12-0: 07406
2903-64-2: 04297
2904-86-1: 22082
2905-25-1: 05498
2905-27-3: 05507
2905-30-8: 05506
2905-31-9: 05505
2905-56-8: 20204
2905-65-9: 06057
2908-76-1: 02236
2909-38-8: 02852
2910-85-2: 06483
2912-62-1: 04351
2913-40-8: 09730
2913-43-1: 09733
2915-52-8: 08812
2917-26-2: 13874
2917-47-7: 21351
2917-73-9: 17222
2917-80-8: 20821
2918-13-0: 13793
2918-83-4: 05454
2919-05-3: 18759
2919-23-5: 09579
2920-38-9: 07587
2921-14-4: 00466
2922-51-2: 13425
2922-83-0: 04365
2923-68-4: 17699
2928-43-0: 07659
2929-52-4: 24135
2929-86-4: 19996
2929-91-1: 15741
2929-96-6: 19628
2930-02-1: 18013
2930-80-5: 09374
2933-74-6: 12207
2934-05-6: 19085
2935-23-1: 10802
2935-35-5: 04123
2935-44-6: 14159
2935-90-2: 21214
2938-48-9: 22247
2938-98-9: 08218
2941-19-7: 04953
2941-20-0: 04914
2941-63-1: 07061
2941-71-1: 07056

2941-78-8: 05931
2942-15-6: 07060
2942-40-7: 14977
2942-42-9: 14980
2942-59-8: 22842
2944-49-2: 03574
2944-58-3: 06300
2944-59-4: 06301
2946-61-4: 19875
2947-60-6: 04336
2947-61-7: 04337
2949-26-0: 16410
2955-56-8: 11597
2955-65-9: 17653
2955-88-6: 23409
2958-36-3: 15821
2958-87-4: 23888
2959-96-8: 22250
2961-04-8: 01321
2961-47-9: 22594
2961-49-1: 22593
2961-50-4: 22714
2969-81-5: 08333
2971-22-4: 04011
2971-79-1: 20247
2972-25-0: 05161
2973-00-4: 15659
2973-19-5: 19180
2973-27-5: 23726
2973-50-4: 04303
2973-76-4: 01595
2973-78-6: 01591
2973-80-0: 01590
2974-92-7: 07472
2976-74-1: 00584
2976-75-2: 00699
2977-31-3: 04871
2978-58-7: 08987
2979-19-3: 10236
2980-69-0: 25280
2980-70-3: 10502
2980-71-4: 17296
2983-26-8: 11806
2983-37-1: 14228
2983-38-2: 08408
2984-40-9: 08740
2984-55-6: 11523
2985-59-3: 15966
2986-17-6: 24852
2986-19-8: 09128
2987-68-0: 01868
2987-77-1: 24132
2989-98-2: 25377
2996-62-5: 23423
2996-92-1: 24138
2997-53-7: 04971
2998-04-1: 14139
2998-08-5: 21934
2998-18-7: 21909
2999-40-8: 05593
3000-79-1: 20177
3000-88-2: 17215
3001-72-7: 23443
3002-23-1: 13596
3002-24-2: 14172
3002-33-3: 03804
3002-81-1: 18960
3004-73-7: 00382
3004-93-1: 17683
3005-87-6: 13345
3007-31-6: 11484
3007-91-8: 16675
3007-97-4: 16632
3008-39-7: 09640
3008-40-0: 10629
3008-41-1: 09609
3009-03-8: 23927
3009-97-0: 00897
3010-02-4: 00882
3010-04-6: 00873
3010-53-5: 02161
3010-82-0: 04521
3010-96-6: 09572
3011-34-5: 01732
3012-37-1: 24649
3013-27-2: 03475
3016-19-1: 17783
3016-97-5: 16031
3017-67-2: 11000
3017-68-3: 08749
3017-70-7: 08754
3017-71-8: 08748
3017-95-6: 20504
3017-96-7: 20502
3018-09-5: 12545
3018-12-0: 00880
3019-04-3: 21561
3019-17-8: 23269
3019-20-3: 03677
3019-25-8: 12332
3022-16-0: 02556
3022-41-1: 09604
3023-09-4: 22087
3023-98-1: 17085
3024-72-4: 07206
3024-83-7: 01017
3025-96-5: 08295
3027-00-7: 00055
3027-01-8: 01820
3027-13-2: 21565
3027-21-2: 24045
3027-38-1: 17044
3029-19-4: 22512
3029-32-1: 18045
3029-40-1: 03827
3030-47-5: 11943
3030-90-8: 19873
3031-05-8: 16549
3031-08-1: 16553
3031-15-0: 16537
3031-66-1: 14551
3031-73-0: 14726
3031-74-1: 14725
3032-32-4: 05903
3032-81-3: 03080
3034-03-5: 06894
3034-08-0: 06890
3034-19-3: 14647
3034-38-6: 14787
3034-48-8: 24506
3034-53-5: 24504
3034-97-7: 05533
3035-67-4: 00592
3035-75-4: 25149
3036-66-6: 07797
3039-55-2: 21985
3040-44-6: 20259
3041-16-5: 11289
3041-40-5: 20784
3042-22-6: 23313
3042-50-0: 03266
3042-69-1: 16306
3043-33-2: 18104
3046-82-0: 01758
3047-32-3: 17978
3048-76-8: 11303
3049-24-9: 19851
3049-56-7: 22331
3054-63-5: 11494
3054-92-0: 18533
3054-95-3: 21706
3055-14-9: 03628
3056-60-8: 19358
3058-39-7: 06714
3058-47-7: 03785
3059-97-0: 15488

3061-05-0: 18940
3061-36-7: 03290
3062-35-9: 23771
3062-81-5: 07734
3066-71-5: 21819
3067-19-4: 25533
3068-00-6: 08283
3068-88-0: 17985
3073-05-0: 23910
3073-56-1: 00305
3073-66-3: 09971
3073-77-6: 23030
3073-87-8: 17940
3073-92-5: 08042
3074-61-1: 10721
3074-64-4: 13747
3074-71-3: 13510
3074-75-7: 14055
3074-76-8: 14053
3074-77-9: 14054
3074-89-3: 07539
3075-84-1: 07548
3077-16-5: 16108
3077-85-8: 09172
3077-97-2: 08664
3079-28-5: 10926
3079-30-9: 11500
3079-31-0: 24359
3079-32-1: 24285
3080-42-0: 01856
3081-24-1: 19725
3082-75-5: 01024
3083-23-6: 18028
3085-42-5: 03918
3085-53-8: 12827
3085-54-9: 12828
3087-14-7: 10692
3087-62-5: 13189
3088-15-1: 15932
3088-41-3: 20271
3088-42-4: 13299
3088-78-6: 07765
3095-35-0: 19760
3096-45-5: 12768
3096-47-7: 12790
3096-52-4: 12795
3096-56-8: 12789
3096-57-9: 12785
3096-71-7: 19010
3098-18-8: 15980
3101-04-0: 01604
3101-60-8: 18005
3102-32-7: 18744
3102-33-8: 18743
3102-61-2: 10422
3102-87-2: 04504
3107-04-8: 01030
3108-24-5: 17694
3112-01-4: 12280
3112-80-9: 03928
3112-85-4: 03782
3112-88-7: 03884
3113-71-1: 06497
3113-72-2: 06500
3113-98-2: 10089
3114-62-3: 19320
3118-97-6: 16981
3121-61-7: 21876
3121-82-2: 18337
3122-21-2: 08104
3122-95-0: 19439
3123-97-5: 13074
3127-67-1: 04182
3128-06-1: 14261
3128-07-2: 13635
3128-15-2: 10735
3129-39-3: 16777
3129-91-7: 09769
3131-27-9: 22928
3132-64-7: 17989
3132-99-8: 01585
3133-01-5: 25104
3134-10-9: 04459
3135-54-4: 06954
3137-55-1: 23225
3137-60-8: 23240
3138-86-1: 23860
3139-05-7: 19317
3140-92-9: 24696
3140-93-0: 24695
3141-12-6: 01391
3141-24-0: 24778
3141-26-2: 24698
3141-27-3: 24697
3141-62-6: 19761
3142-58-3: 18354
3142-66-3: 18556
3142-84-5: 13824
3144-04-5: 08258
3144-16-9: 07303
3144-54-5: 14337
3146-66-5: 23185
3147-39-5: 06616
3147-55-5: 06141
3152-68-9: 18741
3153-36-4: 08354
3153-37-5: 08355
3154-36-7: 18765
3155-43-9: 17389
3155-65-5: 14681
3156-41-0: 03801
3156-70-5: 21727
3156-73-8: 21416
3156-74-9: 08594
3158-26-7: 17624
3160-35-8: 08962
3160-40-5: 08955
3160-46-1: 23379
3162-05-8: 06871
3163-07-3: 04720
3163-27-7: 16255
3164-46-3: 01534
3167-49-5: 22838
3171-45-7: 04434
3171-46-8: 04432
3172-34-7: 23975
3172-52-9: 24702
3172-56-3: 07718
3173-53-3: 09890
3173-79-3: 17983
3174-61-6: 15464
3174-67-2: 18324
3174-83-2: 05275
3175-23-3: 20719
3176-62-3: 14973
3176-66-7: 14976
3178-22-1: 09843
3178-23-2: 09896
3178-24-3: 09950
3178-29-8: 13555
3179-09-7: 03607
3179-10-0: 03608
3179-11-1: 06531
3179-16-6: 02574
3179-31-5: 25068
3179-63-3: 21325
3180-09-4: 19129
3182-95-4: 05263
3183-08-2: 10868
3184-13-2: 17872
3184-35-8: 14149
3185-99-7: 03709
3187-19-7: 06343
3187-94-8: 12870
3188-13-4: 11787
3188-92-9: 15423

3189-12-6: 15141
3189-22-8: 15152
3192-94-7: 23373
3194-15-8: 21480
3195-24-2: 20824
3196-15-4: 08331
3197-42-0: 20178
3199-61-9: 05811
3199-62-0: 05818
3200-96-2: 09592
3205-11-6: 11706
3208-16-0: 12897
3209-13-0: 19411
3209-22-1: 03103
3211-48-1: 09567
3212-75-7: 10473
3217-86-5: 20436
3218-02-8: 10081
3221-20-3: 20123
3221-61-2: 17626
3226-37-7: 24629
3228-03-3: 19479
3228-04-4: 19478
3228-99-7: 20562
3229-00-3: 20546
3229-86-5: 16189
3229-88-7: 16192
3229-89-8: 16194
3230-23-7: 20171
3230-33-9: 16225
3230-35-1: 16227
3230-36-2: 16228
3230-45-3: 19579
3232-84-6: 25071
3234-49-9: 18206
3234-54-6: 08891
3234-64-8: 18767
3235-04-9: 01744
3238-38-8: 19617
3238-40-2: 13012
3238-60-6: 20214
3238-75-3: 12153
3240-09-3: 14522
3240-34-4: 15296
3240-60-6: 23272
3241-23-4: 17163
3243-42-3: 16855
3246-65-9: 19384
3246-66-0: 00846
3246-76-2: 06406
3246-78-4: 02346
3246-80-8: 25545
3249-66-9: 24282
3249-68-1: 14260
3251-56-7: 19413
3252-43-5: 00878
3253-39-2: 21907
3253-41-6: 21888
3254-93-1: 14894
3255-90-1: 13230
3256-88-0: 22696
3260-87-5: 19173
3261-33-4: 11960
3261-62-9: 04808
3263-79-4: 23319
3264-21-9: 12511
3265-35-8: 15220
3268-49-3: 20397
3268-79-9: 24628
3271-80-5: 07665
3272-27-3: 25454
3272-91-1: 16191
3272-92-2: 16193
3274-20-2: 01297
3274-29-1: 13624
3274-56-4: 23296
3276-13-9: 12355
3277-26-7: 11382
3278-46-4: 21288
3278-88-4: 03994
3278-89-5: 04000
3279-07-0: 19311
3279-26-3: 19970
3279-27-4: 19327
3279-76-3: 22993
3280-08-8: 04870
3280-51-1: 13964
3282-16-4: 03895
3282-27-7: 25374
3282-30-2: 21606
3283-12-3: 19774
3285-04-9: 10272
3285-97-0: 01277
3285-98-1: 01276
3286-01-9: 01281
3287-99-8: 04921
3288-04-8: 03568
3288-23-1: 25432
3289-28-9: 10015
3290-06-0: 03048
3290-53-7: 03756
3290-92-4: 21899
3296-05-7: 02534
3297-72-1: 15460
3298-16-6: 20303
3298-40-6: 20868
3299-05-6: 03375
3299-38-5: 08643
3299-82-9: 16671
3300-23-0: 01196
3302-10-1: 14269
3306-02-3: 02963
3306-62-5: 05301
3306-93-2: 06955
3310-68-7: 25057
3310-97-2: 18916
3316-09-4: 04672
3316-23-2: 02573
3318-61-4: 18359
3319-15-1: 05100
3320-83-0: 02851
3320-86-3: 03563
3320-87-4: 03564
3320-90-9: 12879
3321-50-4: 09853
3321-64-0: 19809
3326-71-4: 12971
3326-89-4: 17917
3327-24-0: 22019
3328-77-6: 12358
3328-86-7: 24719
3329-48-4: 14482
3329-56-4: 07997
3330-50-5: 01053
3333-23-1: 16631
3333-44-6: 13056
3333-52-6: 08103
3335-34-0: 03787
3335-35-1: 03788
3336-41-2: 06158
3338-15-6: 21488
3338-55-4: 17777
3342-77-6: 06285
3343-19-9: 04653
3343-24-6: 05598
3343-38-2: 16812
3343-45-1: 12307
3343-59-7: 01648
3343-92-8: 17517
3344-19-2: 00608
3344-70-5: 11489
3346-21-2: 23054
3347-22-6: 17048
3347-57-7: 05147
3347-99-7: 04605
3350-30-9: 10441

3350-78-5: 08980
3351-28-8: 09404
3351-30-2: 09407
3351-31-3: 09406
3351-32-4: 09405
3351-86-8: 09274
3352-87-2: 11475
3352-93-0: 10399
3353-12-6: 22509
3354-39-0: 11444
3354-55-0: 19312
3354-58-3: 19517
3355-17-7: 08997
3355-43-9: 18867
3357-55-9: 17945
3358-28-9: 12931
3360-41-6: 04375
3360-55-2: 02873
3363-56-2: 05636
3365-82-0: 07573
3366-65-2: 18823
3366-92-5: 14854
3367-49-5: 20036
3367-95-1: 20238
3368-16-9: 04291
3368-21-6: 21925
3369-37-7: 02344
3372-81-4: 24624
3373-59-9: 24908
3375-23-3: 16947
3375-38-0: 14176
3376-48-5: 10966
3376-95-2: 22823
3377-71-7: 15168
3377-86-4: 13998
3377-87-5: 13999
3378-09-4: 06764
3378-72-1: 04909
3379-37-1: 03288
3379-38-2: 03289
3379-41-7: 03851
3383-21-9: 09744
3385-34-0: 22256
3385-94-2: 11366
3386-31-0: 09335
3386-33-2: 17376
3386-97-8: 08736
3387-26-6: 13013
3391-10-4: 05013
3391-86-4: 17820
3391-87-5: 10269
3393-59-7: 05634
3393-64-4: 08660
3393-78-0: 03975
3393-90-6: 09314
3393-96-2: 01950
3394-04-5: 24887
3395-79-7: 03181
3395-91-3: 21081
3397-62-4: 25015
3398-07-0: 01572
3398-09-2: 02383
3398-16-1: 22390
3398-22-9: 20369
3399-21-1: 10047
3400-45-1: 10592
3402-70-8: 03733
3402-72-0: 03699
3402-74-2: 03718
3402-76-4: 16482
3402-85-5: 03602
3404-57-7: 13764
3404-58-8: 14405
3404-61-3: 14409
3404-63-5: 18644
3404-67-9: 14069
3404-71-5: 18636
3404-72-6: 18631
3404-73-7: 18633
3404-77-1: 14398
3404-78-2: 14419
3404-80-6: 14068
3405-42-3: 20464
3405-88-7: 00546
3406-76-6: 00499
3406-84-6: 07657
3408-30-8: 16670
3410-66-0: 25243
3411-95-8: 19057
3412-44-0: 03893
3412-99-5: 20806
3416-24-8: 13233
3416-59-9: 05656
3417-91-2: 25259
3419-18-9: 01576
3419-71-4: 10354
3420-02-8: 15158
3422-70-6: 20741
3422-71-7: 00434
3422-75-1: 05182
3422-79-5: 05185
3424-93-9: 01923
3426-01-5: 22452
3428-24-8: 04652
3430-10-2: 22567
3430-27-1: 22568
3430-33-9: 22565
3430-95-3: 11479
3433-37-2: 20265
3433-80-5: 02583
3435-28-7: 23039
3437-89-6: 18574
3437-95-4: 24745
3438-16-2: 06094
3438-46-8: 23073
3438-48-0: 23077
3439-97-2: 24080
3440-02-6: 24047
3442-78-2: 22508
3443-83-2: 24376
3445-11-2: 23424
3446-98-8: 23348
3451-02-3: 05532
3452-07-1: 11600
3452-09-3: 17313
3452-97-9: 14288
3453-64-3: 15350
3454-07-7: 03350
3455-21-8: 07161
3456-99-3: 09180
3457-13-4: 15986
3457-17-8: 15973
3457-46-3: 12012
3457-48-5: 12014
3457-58-7: 03713
3457-66-7: 01517
3458-28-4: 15581
3460-02-4: 17125
3460-11-5: 01947
3460-18-2: 03027
3460-20-6: 03997
3460-24-0: 03003
3460-67-1: 22871
3463-26-1: 22420
3463-27-2: 22421
3463-40-9: 03999
3463-92-1: 09290
3464-18-4: 23403
3464-50-4: 09381
3466-80-6: 20199
3468-99-3: 04286
3469-00-9: 04287
3469-26-9: 16354
3471-31-6: 15183
3473-63-0: 15776
3475-74-9: 23338

3476-89-9: 23886
3478-88-4: 15860
3478-90-8: 15845
3479-10-5: 05400
3479-30-9: 19855
3481-20-7: 02494
3482-34-6: 14730
3482-37-9: 05837
3483-82-7: 25248
3484-18-2: 15147
3487-99-8: 21964
3494-76-6: 11269
3497-00-5: 19871
3505-67-7: 11301
3506-09-0: 21411
3508-00-7: 13404
3508-78-9: 18361
3508-94-9: 04268
3510-70-1: 24148
3512-80-9: 22572
3513-69-7: 18915
3516-95-8: 21500
3517-90-6: 03603
3518-66-9: 12045
3521-91-3: 13776
3522-94-9: 14091
3524-62-7: 12448
3524-68-3: 21864
3524-73-0: 14411
3524-75-2: 10571
3524-87-6: 23268
3527-63-7: 09991
3528-17-4: 07140
3529-87-1: 03570
3530-15-2: 25037
3531-24-6: 21761
3532-31-8: 07189
3533-10-6: 21640
3535-67-9: 03584
3536-29-6: 05276
3540-05-4: 17723
3540-84-9: 10300
3540-95-2: 20158
3541-14-8: 05490
3543-02-0: 22910
3544-24-9: 01803
3544-25-0: 04304
3547-04-4: 03408
3548-85-4: 03200
3552-01-0: 10157
3553-11-5: 01867
3553-78-4: 23981
3554-74-3: 20280
3555-84-8: 15943
3555-85-9: 15855
3555-86-0: 16044
3556-56-7: 25263
3558-19-8: 05938
3558-60-9: 03590
3558-69-8: 22637
3565-26-2: 23823
3566-52-7: 18712
3567-38-2: 05072
3568-81-8: 19707
3570-46-5: 05722
3570-55-6: 12075
3570-58-9: 12120
3570-93-2: 10103
3572-06-3: 08627
3574-40-1: 07790
3574-43-4: 12103
3575-31-3: 06537
3577-87-5: 19989
3581-87-1: 24525
3581-91-7: 24516
3582-71-6: 25202
3586-12-7: 02453
3586-14-9: 03730
3587-57-3: 20527
3587-60-8: 02861
3587-64-2: 21407
3587-75-5: 21405
3588-17-8: 13951
3588-30-5: 07798
3588-63-4: 25535
3592-12-9: 11290
3593-00-8: 13770
3593-85-9: 01076
3594-37-4: 08233
3597-91-9: 07660
3598-14-9: 00896
3600-24-6: 25211
3607-78-1: 20619
3609-48-1: 05234
3609-53-8: 05842
3610-02-4: 07134
3610-36-4: 15225
3611-76-5: 13058
3613-42-1: 01124
3615-21-2: 05611
3615-56-3: 24182
3616-56-6: 11698
3616-57-7: 11697
3623-23-2: 02709
3623-51-6: 10178
3623-52-7: 10179
3623-53-8: 10177
3625-52-3: 11570
3626-94-6: 09451
3634-92-2: 17450
3637-01-2: 12380
3637-61-4: 10635
3638-35-5: 10783
3638-64-0: 12586
3638-73-1: 02130
3642-89-5: 01072
3643-64-9: 10353
3644-61-9: 21510
3646-50-2: 05202
3651-02-3: 16948
3651-23-8: 24020
3651-67-0: 16106
3652-91-3: 09168
3653-23-4: 14906
3654-49-7: 18121
3657-07-6: 00762
3658-48-8: 19803
3658-79-5: 18224
3658-80-8: 25212
3658-93-3: 14029
3658-94-4: 07980
3658-95-5: 07978
3661-77-6: 09621
3663-20-5: 00119
3663-34-1: 00245
3663-82-9: 05724
3669-52-1: 12430
3669-80-5: 25293
3673-79-8: 21133
3674-09-7: 21146
3674-10-0: 21381
3674-13-3: 21130
3674-65-5: 18850
3674-66-6: 18852
3674-69-9: 18857
3674-72-4: 18862
3674-73-5: 18900
3674-75-7: 18868
3675-68-1: 08061
3675-69-2: 08062
3678-62-4: 22604
3679-78-5: 14630
3680-02-2: 12584
3681-71-8: 14490
3681-78-5: 11534
3682-35-7: 24992

3682-56-2: 06550
3683-19-0: 14428
3683-22-5: 14427
3685-84-5: 00545
3688-92-4: 16833
3689-24-5: 24654
3690-04-8: 20417
3690-05-9: 19389
3693-22-9: 11138
3694-52-8: 04445
3695-48-5: 24546
3695-77-0: 04962
3695-89-4: 21642
3695-93-0: 04314
3696-23-9: 24862
3696-28-4: 22639
3696-36-4: 20781
3697-24-3: 09408
3697-27-6: 09400
3697-30-1: 01984
3697-42-5: 24325
3698-06-4: 23914
3698-12-2: 24566
3698-54-2: 02502
3699-01-2: 03746
3699-67-0: 21153
3706-26-1: 04802
3709-08-8: 13073
3710-23-4: 16442
3710-30-3: 17522
3710-31-4: 13586
3710-41-6: 17521
3710-43-8: 12885
3710-84-7: 11717
3711-34-0: 14587
3711-37-3: 14591
3717-73-5: 00231
3718-53-4: 14835
3719-43-5: 11094
3721-95-7: 09550
3722-72-3: 06785
3722-73-4: 06784
3722-74-5: 06783
3724-10-5: 06508
3725-05-1: 13970
3725-07-3: 17551
3725-30-2: 10470
3726-23-6: 22813
3728-34-5: 05755
3728-43-6: 24139
3728-54-9: 09863
3730-56-1: 18918
3730-60-7: 14295
3731-51-9: 22916
3731-52-0: 22917
3731-53-1: 22918
3732-81-8: 19960
3735-90-8: 09138
3736-81-0: 12967
3737-33-5: 19703
3737-38-0: 24809
3739-30-8: 08435
3739-38-6: 06546
3739-94-4: 22885
3740-52-1: 04275
3741-00-2: 10570
3741-38-6: 11307
3741-91-1: 09761
3742-34-5: 10525
3742-38-9: 10711
3742-42-5: 10320
3743-22-4: 19276
3743-28-0: 01917
3746-39-2: 00661
3748-84-3: 22749
3749-44-8: 20491
3756-41-0: 02826
3757-32-2: 18468
3760-11-0: 17303
3760-98-3: 24386
3761-94-2: 09644
3763-33-5: 01042
3763-39-1: 10842
3763-55-1: 09286
3764-01-0: 23079
3764-20-3: 06977
3765-62-6: 01875
3768-43-2: 20292
3768-55-6: 23969
3768-60-3: 24301
3768-63-6: 24274
3769-23-1: 14410
3769-41-3: 19574
3770-80-7: 16047
3772-26-7: 23427
3773-93-1: 11328
3774-03-6: 11287
3774-99-0: 24552
3775-01-7: 14869
3775-72-2: 08304
3775-73-3: 08303
3775-90-4: 21892
3779-29-1: 09557
3779-61-1: 17776
3780-50-5: 19546
3780-58-3: 14248
3780-72-1: 23398
3785-34-0: 00485
3785-90-8: 20779
3788-32-7: 10559
3788-61-2: 18863
3789-74-0: 04682
3789-77-3: 02118
3789-78-4: 12373
3790-71-4: 11553
3804-70-4: 13124
3806-42-6: 24585
3807-77-0: 00017
3808-87-5: 11408
3810-26-2: 10760
3810-51-3: 06348
3811-20-9: 14672
3811-29-8: 22299
3814-34-4: 18190
3814-93-5: 03340
3816-62-4: 05940
3819-88-3: 03463
3834-18-2: 01048
3835-64-1: 05052
3837-38-5: 01280
3839-50-7: 10483
3840-31-1: 05153
3842-03-3: 07981
3842-20-4: 15409
3846-50-2: 02271
3846-73-9: 23818
3848-12-2: 00506
3848-24-6: 14171
3849-33-0: 20614
3850-30-4: 07890
3852-09-3: 21221
3853-80-3: 21059
3856-25-5: 25131
3856-37-9: 08131
3857-25-8: 13048
3858-47-7: 10174
3858-78-4: 07877
3859-41-4: 10630
3862-11-1: 19272
3862-12-2: 19264
3864-08-2: 11098
3864-18-4: 07485
3867-18-3: 02254
3870-70-0: 13271
3871-23-6: 07681
3871-31-6: 19934

3873-54-9: 01037
3874-54-2: 08605
3875-51-2: 10548
3875-52-3: 10522
3875-68-1: 25553
3876-97-9: 16551
3877-15-4: 20668
3877-19-8: 16524
3878-45-3: 19769
3878-55-5: 08191
3879-08-1: 08433
3881-57-0: 00983
3882-50-6: 06756
3882-98-2: 24492
3883-64-5: 15329
3883-95-2: 06011
3885-72-1: 21439
3885-88-9: 04532
3886-69-9: 04929
3891-07-4: 15383
3891-33-6: 07905
3891-79-0: 08094
3898-08-6: 24876
3898-41-7: 03293
3899-36-3: 14446
3900-46-7: 00294
3901-77-7: 10855
3905-64-4: 16247
3908-48-3: 09706
3908-79-0: 22589
3913-02-8: 17703
3913-23-3: 02683
3913-71-1: 10995
3913-81-3: 10997
3913-85-7: 11006
3916-64-1: 13453
3917-15-5: 21709
3920-53-4: 13608
3922-28-9: 11490
3927-72-8: 01234
3933-94-6: 07540
3934-20-1: 23055
3934-92-7: 06588
3937-40-4: 01411
3937-45-9: 10090
3937-46-0: 10091
3937-56-2: 17230
3938-95-2: 21166
3938-96-3: 00674
3943-74-6: 06366
3943-77-9: 06195
3944-00-1: 12009
3944-36-3: 21410
3944-37-4: 21336
3946-29-0: 21449
3949-36-8: 06820
3951-96-0: 24188
3953-65-9: 19701
3955-26-8: 04005
3955-58-6: 24880
3958-03-0: 24767
3958-38-1: 10444
3958-57-4: 02691
3958-60-9: 02689
3959-05-5: 04883
3962-77-4: 03585
3963-62-0: 04812
3963-78-8: 01207
3963-79-9: 02384
3964-04-3: 23563
3964-52-1: 18995
3964-56-5: 19099
3964-58-7: 06086
3964-61-2: 19142
3965-53-5: 06584
3965-81-9: 16117
3966-30-1: 04212
3966-32-3: 04244
3966-58-3: 10824
3967-54-2: 11332
3968-85-2: 03634
3969-54-8: 01436
3969-57-1: 01434
3970-35-2: 06103
3970-40-9: 02899
3970-41-0: 02103
3970-62-5: 18524
3971-33-3: 17639
3971-58-2: 01038
3972-26-7: 21294
3972-40-5: 08121
3972-56-3: 02798
3972-65-4: 02621
3974-99-0: 00772
3976-34-9: 07489
3976-36-1: 07547
3977-23-9: 23241
3977-87-5: 10796
3978-81-2: 22634
3982-64-7: 03265
3982-66-9: 25563
3982-67-0: 03453
3984-76-7: 11941
3985-61-3: 20022
3988-03-2: 15848
3991-61-5: 03729
3991-73-9: 19925
3993-57-5: 06864
3995-37-7: 02156
3995-38-8: 02157
3999-78-8: 22658
4001-61-0: 05548
4001-73-4: 01822
4004-51-7: 16951
4007-00-5: 00782
4008-48-4: 23822
4009-55-6: 18085
4010-33-7: 12276
4010-51-9: 15978
4013-94-9: 11929
4013-95-0: 11932
4016-11-9: 18010
4018-65-9: 04647
4021-75-4: 08538
4023-34-1: 10791
4023-65-8: 21769
4023-79-4: 02564
4024-81-1: 10430
4025-64-3: 06123
4025-77-8: 00932
4028-23-3: 23991
4028-56-2: 07969
4032-80-8: 11399
4032-86-4: 13514
4032-92-2: 13566
4032-93-3: 13565
4032-94-4: 17603
4033-07-2: 22112
4036-30-0: 08198
4036-43-5: 07549
4038-04-4: 18637
4038-13-5: 15944
4038-14-6: 15945
4038-15-7: 15940
4038-17-9: 06768
4038-18-0: 06767
4038-90-8: 20114
4039-92-3: 02905
4041-09-2: 10665
4043-71-4: 05737
4044-60-4: 15956
4044-65-9: 03165
4045-30-1: 20153
4046-70-2: 25032
4046-86-0: 21340
4048-30-0: 13788

4048-31-1: 13785
4048-32-2: 13791
4048-33-3: 14271
4049-81-4: 13944
4050-45-7: 14413
4051-56-3: 01878
4052-30-6: 06506
4053-34-3: 23847
4054-38-0: 09624
4056-73-9: 10070
4057-42-5: 17803
4058-51-9: 18087
4064-06-6: 13160
4065-80-9: 10156
4068-78-4: 06091
4072-22-4: 15979
4074-22-0: 09686
4074-24-2: 08159
4074-43-5: 19130
4074-88-8: 21954
4075-79-0: 00096
4076-40-8: 06750
4076-43-1: 06746
4079-14-5: 11150
4079-15-6: 11159
4079-52-1: 12454
4079-68-9: 22096
4081-35-0: 19378
4082-55-7: 17161
4085-18-1: 07580
4088-60-2: 08931
4088-81-7: 18931
4089-71-8: 14165
4090-55-5: 11292
4091-26-3: 12611
4091-39-8: 08632
4093-28-1: 05851
4093-29-2: 05853
4095-09-4: 24094
4095-22-1: 24753
4096-20-2: 20198
4096-21-3: 23363
4096-34-8: 10410
4097-48-7: 19524
4097-63-6: 19409
4099-63-2: 19364
4100-09-8: 18334
4100-80-5: 13029
4101-68-2: 10913
4103-26-8: 11175
4104-01-2: 18629
4104-67-0: 10825
4104-77-2: 15428
4105-38-8: 25506
4105-92-4: 05579
4106-66-5: 11139
4107-62-4: 21121
4107-64-6: 22511
4107-65-7: 06697
4110-44-5: 17607
4110-50-3: 20605
4110-77-4: 02812
4113-12-6: 13870
4113-72-8: 02539
4114-28-7: 14715
4114-31-2: 14706
4117-15-1: 25265
4117-31-1: 23961
4117-33-3: 15548
4117-50-4: 09348
4119-41-9: 03553
4120-64-3: 24581
4120-77-8: 00355
4120-78-9: 00356
4124-30-5: 00571
4124-31-6: 00761
4124-88-3: 17304
4125-54-6: 19589
4126-78-7: 09630
4127-45-1: 10786
4127-47-3: 10785
4128-31-8: 17719
4128-37-4: 25367
4128-71-6: 00360
4130-08-9: 24161
4130-42-1: 19075
4131-03-7: 15858
4131-04-8: 16006
4132-48-3: 03597
4132-49-4: 03236
4132-70-1: 03097
4132-71-2: 03177
4132-72-3: 03249
4132-76-7: 03176
4132-77-8: 03246
4135-93-7: 24432
4136-11-2: 11164
4136-21-4: 12421
4136-97-4: 05924
4140-70-9: 00211
4140-71-0: 05607
4140-83-4: 05309
4142-85-2: 24078
4143-41-3: 11053
4144-54-1: 10977
4145-94-2: 24513
4145-95-3: 24591
4145-98-6: 21014
4145-99-7: 08267
4146-00-3: 21015
4146-01-4: 17663
4146-02-5: 08275
4146-05-8: 18479
4146-06-9: 01910
4146-09-2: 00263
4146-13-8: 09204
4146-14-9: 18374
4146-15-0: 08268
4146-16-1: 21019
4146-18-3: 08276
4146-19-4: 24509
4146-20-7: 24593
4146-21-8: 24511
4146-43-4: 08133
4147-13-1: 11663
4147-89-1: 23970
4148-97-4: 08180
4151-25-1: 19778
4152-09-4: 11959
4152-90-3: 04892
4155-89-9: 22675
4156-16-5: 15501
4157-77-1: 02751
4158-12-7: 17439
4159-03-9: 11436
4160-80-9: 22251
4160-82-1: 22248
4161-24-4: 08534
4161-32-4: 21348
4161-40-4: 07962
4161-60-8: 18557
4162-62-3: 21731
4164-28-7: 15614
4164-39-0: 20108
4164-43-6: 03655
4164-92-5: 13981
4165-98-4: 18322
4166-46-5: 14309
4166-53-4: 22249
4166-86-3: 19140
4167-08-2: 17410
4167-74-2: 19523
4167-77-5: 10608
4169-04-4: 21345
4170-24-5: 08346
4170-46-1: 24108

4170-90-5: 05155
4171-83-9: 03816
4171-93-1: 25183
4173-44-8: 13966
4175-53-5: 15011
4175-54-6: 16508
4175-66-0: 24515
4179-19-5: 03173
4180-23-8: 03620
4181-97-9: 06448
4182-44-9: 11925
4183-86-2: 19569
4185-53-9: 02734
4185-55-1: 16258
4185-62-0: 16265
4185-63-1: 16298
4185-79-9: 02624
4186-30-5: 10794
4186-39-4: 10795
4187-86-4: 18789
4187-87-5: 05071
4188-17-4: 23856
4192-56-7: 13232
4192-77-2: 21959
4198-33-8: 08122
4199-88-6: 18964
4200-95-7: 13402
4202-38-4: 11499
4203-18-3: 23616
4205-23-6: 13385
4206-75-1: 24003
4208-49-5: 12992
4208-64-4: 13047
4209-02-3: 12306
4209-15-8: 02405
4209-16-9: 02406
4209-24-9: 12321
4209-44-3: 22279
4209-90-9: 14308
4209-91-0: 14313
4214-28-2: 03530
4214-74-8: 22541
4214-75-9: 22554
4214-76-0: 22555
4214-79-3: 22981
4215-80-9: 23972
4215-86-5: 03727
4219-24-3: 14451
4219-48-1: 11525
4219-49-2: 13894
4220-52-4: 14179
4221-98-1: 09683
4223-03-4: 21656
4224-70-8: 14216
4225-26-7: 07063
4225-85-8: 23586
4225-86-9: 23599
4228-00-6: 11531
4228-89-1: 05612
4229-91-8: 12923
4230-36-8: 06426
4230-91-5: 07057
4232-27-3: 19333
4233-81-2: 24592
4235-09-0: 00357
4236-05-9: 16260
4236-41-3: 23857
4236-65-1: 20115
4237-37-0: 19159
4237-40-5: 03775
4237-70-1: 03271
4237-83-6: 13092
4238-71-5: 14791
4240-92-0: 21072
4242-18-6: 16653
4242-46-0: 04306
4244-84-2: 01002
4245-37-8: 21896
4245-76-5: 13369
4246-51-9: 20467
4250-38-8: 17159
4250-94-6: 11160
4251-16-5: 24651
4251-21-2: 04774
4253-34-3: 24162
4253-76-3: 20763
4253-89-8: 11398
4253-91-2: 20685
4254-18-6: 12288
4254-29-9: 15076
4254-59-5: 08118
4255-61-2: 17098
4255-62-3: 10239
4258-93-9: 09925
4259-00-1: 10579
4259-43-2: 20724
4261-68-1: 20473
4264-29-3: 03811
4265-25-2: 05798
4265-55-8: 21123
4265-97-8: 17678
4268-36-4: 09047
4269-15-2: 12787
4271-27-6: 11483
4271-30-1: 13251
4273-98-7: 02478
4276-09-9: 08531
4277-08-1: 10124
4279-22-5: 07967
4279-70-3: 03703
4279-76-9: 03457
4280-28-8: 03586
4282-24-0: 13010
4282-28-4: 13011
4282-31-9: 24812
4282-32-0: 13016
4282-40-0: 13533
4282-90-0: 19687
4283-80-1: 18189
4285-42-1: 09127
4286-47-9: 14039
4286-49-1: 07985
4288-84-0: 21370
4289-95-6: 19720
4291-79-6: 09926
4291-80-9: 09927
4291-81-0: 09928
4291-98-9: 10720
4292-19-7: 11498
4292-75-5: 09886
4292-92-6: 09948
4293-58-7: 00256
4293-61-2: 24521
4293-62-3: 24590
4295-06-1: 23589
4300-46-3: 25491
4301-39-7: 12965
4303-17-7: 12963
4303-44-0: 20523
4303-70-2: 17477
4309-66-4: 02463
4311-12-0: 13261
4313-03-5: 13433
4313-32-0: 11326
4316-23-8: 04571
4316-56-7: 02079
4316-93-2: 23063
4316-94-3: 23032
4316-98-7: 23113
4318-37-0: 11087
4318-76-7: 22879
4318-78-9: 22882
4319-49-7: 16082
4320-29-0: 05340
4323-69-7: 17115
4324-37-2: 21217

4324-39-4: 21218
4325-48-8: 18200
4325-49-9: 13492
4325-74-0: 07431
4325-77-3: 18890
4325-82-0: 18732
4325-85-3: 24167
4325-97-7: 12067
4328-94-3: 18385
4337-62-6: 25525
4338-94-7: 24869
4339-05-3: 14560
4340-76-5: 11595
4341-00-8: 06646
4341-76-8: 09013
4342-61-4: 11357
4344-55-2: 02058
4346-18-3: 08510
4347-61-9: 12275
4348-19-0: 02270
4350-09-8: 25232
4350-41-8: 22610
4351-54-6: 12843
4352-89-0: 17086
4354-89-6: 20870
4355-11-7: 10030
4356-43-8: 15783
4358-59-2: 08900
4358-75-2: 07889
4358-87-6: 04194
4359-46-0: 11319
4360-12-7: 00992
4360-96-7: 20846
4361-06-2: 20860
4362-01-0: 09455
4362-40-7: 11315
4363-04-6: 16962
4363-93-3: 23733
4363-97-7: 09441
4364-06-1: 03194
4370-59-6: 09707
4370-81-4: 04231
4371-31-7: 07684
4371-48-6: 19474
4371-64-6: 20840
4372-10-5: 11302
4373-13-1: 16347
4375-11-5: 14944
4375-14-8: 15164
4376-18-5: 04573
4376-23-2: 14514
4376-29-8: 12285
4376-30-1: 08325
4377-37-1: 22751
4377-73-5: 09692
4382-53-0: 15185
4382-54-1: 15199
4383-05-5: 05084
4383-06-6: 05085
4383-07-7: 05088
4383-18-0: 05130
4383-22-6: 04952
4383-93-1: 14221
4385-35-7: 06867
4385-98-2: 05221
4388-01-6: 06973
4388-87-8: 14183
4388-88-9: 14184
4389-09-7: 01977
4389-45-1: 05929
4389-50-8: 05932
4389-53-1: 06386
4390-75-4: 08252
4390-78-7: 08255
4392-24-9: 02721
4392-30-7: 02964
4393-05-9: 05154
4393-06-0: 05057
4393-09-3: 04903
4394-11-0: 07695
4394-85-8: 16128
4395-73-7: 04935
4395-79-3: 02887
4395-80-6: 02888
4395-81-7: 03548
4396-02-5: 08561
4396-19-4: 11853
4396-98-9: 05002
4397-05-1: 04988
4397-13-1: 05083
4397-53-9: 01773
4402-32-8: 21384
4403-69-4: 04880
4404-98-2: 18503
4405-42-9: 24004
4406-41-1: 20429
4406-66-0: 11070
4406-87-5: 00015
4407-36-7: 21989
4408-60-0: 04299
4408-78-0: 00737
4409-83-0: 04409
4409-84-1: 04407
4409-85-2: 04408
4410-99-5: 04817
4411-26-1: 25114
4412-52-6: 09208
4413-16-5: 02974
4413-21-2: 10532
4413-31-4: 04014
4414-88-4: 05637
4416-71-1: 15492
4418-61-5: 24422
4418-84-2: 11049
4420-74-0: 21018
4421-09-4: 05742
4422-95-1: 05570
4423-74-9: 12831
4424-17-3: 01814
4425-82-5: 12744
4426-47-5: 07757
4426-75-9: 07136
4426-83-9: 13537
4427-56-9: 19480
4428-13-1: 04842
4429-77-0: 09620
4432-34-2: 16133
4433-10-7: 07487
4433-11-8: 07492
4433-63-0: 07758
4433-79-8: 07839
4435-14-7: 09998
4435-18-1: 00877
4435-50-1: 08282
4436-59-3: 09654
4436-81-1: 13075
4436-96-8: 05056
4437-18-7: 12866
4437-20-1: 12893
4437-22-3: 12917
4437-23-4: 12919
4437-50-7: 14182
4437-51-8: 14174
4439-20-7: 12160
4439-24-1: 12211
4441-57-0: 10001
4441-63-8: 09999
4441-90-1: 20857
4442-79-9: 10078
4443-55-4: 09852
4443-56-5: 09919
4443-57-6: 11578
4443-58-7: 09953
4443-59-8: 09812
4443-60-1: 09888
4443-61-2: 09938

4444-12-6: 13916
4445-06-1: 09936
4445-07-2: 03826
4445-08-3: 17385
4445-09-4: 07766
4445-59-4: 07623
4447-21-6: 13971
4447-60-3: 20654
4447-64-7: 18474
4447-67-0: 25521
4447-69-2: 25520
4447-74-9: 25524
4447-75-0: 18473
4447-76-1: 25522
4447-80-7: 25523
4448-77-5: 09641
4450-68-4: 05136
4453-79-6: 04067
4455-26-9: 17570
4457-00-5: 10543
4457-32-3: 09234
4457-67-4: 07918
4458-31-5: 20445
4458-39-3: 07601
4460-86-0: 01792
4461-39-6: 12097
4462-09-3: 20183
4462-96-8: 17881
4463-03-0: 05595
4463-33-6: 03170
4463-42-7: 07768
4466-59-5: 15314
4467-06-5: 22695
4467-88-3: 15037
4468-42-2: 03396
4468-48-8: 04301
4469-89-0: 10946
4471-05-0: 05141
4475-42-7: 20308
4475-72-3: 24412
4476-28-2: 04269
4476-77-1: 24385
4477-76-3: 24948
4478-10-8: 02688
4480-83-5: 11281
4481-08-7: 13166
4481-28-1: 05893
4481-30-5: 03263
4482-03-5: 07513
4482-75-1: 13198
4482-83-1: 09765
4483-47-0: 15034
4484-72-4: 24102
4484-74-6: 03546
4485-08-9: 03808
4485-09-0: 17281
4485-13-6: 13766
4485-16-9: 13762
4486-75-3: 00547
4488-22-6: 07436
4488-40-8: 16643
4488-44-2: 16686
4489-22-9: 04993
4489-23-0: 02719
4489-33-2: 00182
4489-90-1: 15516
4489-91-2: 15513
4489-96-7: 15515
4489-97-8: 15514
4491-19-4: 00949
4492-01-7: 22406
4493-42-9: 10899
4494-45-5: 05175
4494-46-6: 05183
4494-47-7: 00426
4495-66-3: 21523
4497-29-4: 15685
4497-40-9: 21622
4498-72-0: 12443
4506-36-9: 16350
4506-71-2: 07129
4506-96-1: 24825
4507-84-0: 16647
4507-97-5: 19630
4508-39-8: 12622
4510-34-3: 21990
4511-42-6: 11284
4515-21-3: 01466
4516-69-2: 10580
4518-10-9: 05875
4518-16-5: 15517
4518-17-6: 15512
4518-98-3: 11363
4519-35-1: 19861
4520-29-0: 24085
4521-28-2: 04370
4521-30-6: 07090
4521-61-3: 07238
4523-36-8: 16235
4523-45-9: 16178
4523-46-0: 16215
4524-77-0: 19102
4525-36-4: 13113
4525-44-4: 24022
4530-31-8: 23959
4531-79-7: 02431
4533-97-5: 10861
4534-09-2: 10862
4534-53-6: 03645
4534-74-1: 10139
4535-61-9: 13813
4535-66-4: 12105
4536-23-6: 14246
4536-30-5: 12155
4536-88-3: 03638
4537-00-2: 16629
4537-05-7: 03979
4537-13-7: 03724
4539-51-9: 09155
4540-44-7: 21365
4541-14-4: 08564
4542-61-4: 07731
4543-96-8: 20767
4546-48-9: 23759
4547-57-3: 04141
4548-45-2: 22607
4549-31-9: 13499
4549-32-0: 17595
4549-33-1: 17208
4549-74-0: 18099
4551-03-5: 23176
4553-07-5: 04161
4556-23-4: 22946
4560-87-6: 24996
4562-36-1: 09262
4563-33-1: 04956
4565-32-6: 07093
4573-05-1: 09675
4573-09-5: 10684
4573-50-6: 10426
4584-46-7: 11694
4584-57-0: 02217
4584-63-8: 09698
4584-68-3: 09661
4584-82-1: 18870
4586-22-5: 15801
4588-18-5: 17798
4589-12-2: 01958
4589-93-9: 01064
4589-97-3: 04087
4590-52-7: 23840
4591-39-3: 05714
4593-90-2: 05239
4594-78-9: 10094
4595-60-2: 23047
4597-87-9: 22552

4599-94-4: 11127
4604-28-8: 02813
4604-72-2: 22359
4606-07-9: 10810
4606-65-9: 20266
4607-38-9: 04210
4609-87-4: 10928
4609-89-6: 18262
4611-18-1: 22822
4613-11-0: 03216
4614-01-1: 15054
4619-49-2: 20006
4619-74-3: 19637
4620-70-6: 12146
4620-79-5: 24157
4621-04-9: 10159
4621-66-3: 22780
4627-10-5: 11262
4628-21-1: 08757
4630-20-0: 09169
4630-82-4: 10019
4630-83-5: 10598
4631-12-3: 01468
4631-98-5: 10204
4632-01-3: 10189
4634-87-1: 13445
4635-59-0: 08681
4635-87-4: 18692
4636-16-2: 12447
4637-24-5: 15604
4638-92-0: 10808
4641-34-3: 14936
4643-25-8: 13795
4645-15-2: 09942
4648-14-0: 03610
4648-33-3: 03609
4652-44-2: 09062
4653-13-8: 16579
4654-39-1: 04827
4655-34-9: 21906
4656-80-8: 01413
4657-97-0: 00007
4659-47-6: 01784
4662-96-8: 03306
4663-22-3: 10782
4664-00-0: 23442
4667-38-3: 24016
4667-76-9: 15404
4667-99-6: 24008
4669-01-6: 10568
4669-02-7: 18069
4669-06-1: 16184
4670-56-8: 06376
4671-03-8: 25054
4672-38-2: 19856
4675-18-7: 17937
4676-51-1: 08182
4684-12-2: 16157
4685-47-6: 03579
4692-94-8: 07834
4694-12-6: 10686
4695-31-2: 21216
4695-62-9: 07349
4696-61-1: 11256
4698-30-0: 19028
4701-17-1: 24793
4701-36-4: 03435
4702-13-0: 15407
4702-32-3: 18601
4702-34-5: 06819
4703-15-5: 05365
4703-96-2: 24570
4705-06-0: 10400
4705-34-4: 03324
4705-94-6: 16980
4706-81-4: 24383
4706-89-2: 03250
4706-90-5: 03248
4707-71-5: 18906
4709-20-0: 16923
4709-68-6: 12733
4712-36-1: 21432
4712-55-4: 19820
4712-70-3: 19709
4714-23-2: 02928
4717-96-8: 22240
4720-64-3: 20293
4720-82-5: 12286
4721-34-0: 19841
4721-37-3: 19838
4726-96-9: 23963
4727-17-7: 10478
4727-72-4: 20286
4730-22-7: 13663
4731-34-4: 03843
4731-84-4: 19132
4732-72-3: 05816
4733-39-5: 18961
4733-50-0: 04510
4734-58-1: 23948
4734-90-1: 10474
4735-50-6: 16839
4736-48-5: 09615
4737-41-1: 18202
4740-00-5: 24775
4740-51-6: 23964
4741-80-4: 17975
4743-17-3: 07171
4743-82-2: 18704
4744-08-5: 20571
4744-10-9: 20581
4744-11-0: 20589
4746-61-6: 00273
4747-05-1: 07794
4747-15-3: 03587
4747-24-4: 11708
4748-78-1: 01682
4749-27-3: 21745
4749-28-4: 21728
4749-31-9: 20683
4749-35-3: 21289
4751-49-9: 13582
4752-10-7: 20003
4753-75-7: 13039
4753-80-4: 24618
4754-26-1: 09729
4755-33-3: 07283
4755-36-6: 09671
4755-77-5: 00539
4756-10-9: 12939
4756-45-0: 06368
4766-33-0: 16155
4766-57-8: 24169
4767-03-7: 21201
4767-55-9: 15354
4767-56-0: 15355
4769-73-7: 21373
4771-08-8: 01949
4771-47-5: 06107
4771-50-0: 15193
4771-80-6: 10371
4773-53-9: 21382
4775-09-1: 19773
4775-86-4: 18349
4780-79-4: 16845
4781-76-4: 22263
4784-14-9: 25022
4784-77-4: 08747
4785-62-0: 20886
4786-20-3: 08846
4789-40-6: 12863
4792-30-7: 15319
4795-29-3: 13040
4795-86-2: 07282
4798-44-1: 14477
4798-46-3: 14483

4798-60-1: 14500
4798-62-3: 13786
4798-64-5: 17826
4799-68-2: 21347
4799-69-3: 08596
4801-58-5: 20175
4802-84-0: 14923
4802-88-4: 23461
4802-89-5: 14920
4802-93-1: 14926
4803-35-4: 14925
4806-11-5: 24742
4806-12-6: 24687
4806-13-7: 24741
4806-14-8: 24707
4806-58-0: 10787
4806-59-1: 10788
4806-61-5: 09534
4807-55-0: 24578
4808-64-4: 22635
4810-09-7: 13739
4813-57-4: 21950
4814-74-8: 23341
4814-79-3: 11956
4814-80-6: 00188
4815-57-0: 03294
4819-75-4: 11829
4822-40-6: 15145
4822-44-0: 00288
4823-47-6: 21785
4824-12-8: 13342
4825-53-0: 19248
4829-04-3: 11438
4830-56-2: 05032
4830-99-3: 15040
4831-43-0: 23420
4832-17-1: 17027
4832-52-4: 03685
4835-11-4: 14104
4835-90-9: 21200
4837-88-1: 03598
4838-00-0: 14982
4839-46-7: 18302
4843-42-9: 22118
4845-05-0: 14722
4845-58-3: 07078
4850-28-6: 10586
4850-32-2: 10553
4850-68-4: 09990
4853-66-1: 08500
4856-97-7: 05640
4857-01-6: 05638
4857-51-6: 00261
4860-03-1: 13861
4861-58-9: 24762
4861-59-0: 24755
4861-61-4: 24721
4861-63-6: 24747
4868-08-0: 09573
4869-59-4: 06434
4870-65-9: 04134
4872-69-9: 23915
4873-56-7: 05467
4875-10-9: 05557
4877-77-4: 02714
4877-93-4: 07479
4880-33-5: 01110
4884-24-6: 07423
4885-02-3: 15668
4887-30-3: 14259
4887-83-6: 05621
4890-85-1: 06555
4891-29-6: 24717
4891-44-5: 24728
4894-61-5: 08756
4894-62-6: 13442
4894-75-1: 10279
4895-65-2: 19913
4897-84-1: 08334
4897-95-4: 14582
4900-30-5: 17183
4900-63-4: 16433
4901-51-3: 19608
4901-75-1: 11090
4904-61-4: 09613
4911-44-8: 20498
4911-46-0: 07951
4911-55-1: 18130
4911-65-3: 02950
4911-70-0: 18505
4912-92-9: 15009
4913-13-7: 02501
4914-89-0: 14447
4914-91-4: 18670
4914-92-5: 18669
4915-21-3: 12937
4915-22-4: 12938
4915-23-5: 12936
4916-57-8: 22640
4916-63-6: 14381
4919-33-9: 04176
4919-43-1: 16654
4919-51-1: 10838
4920-77-8: 19499
4920-80-3: 06453
4920-84-7: 03186
4920-95-0: 07550
4923-53-9: 16809
4923-77-7: 09864
4923-84-6: 19818
4926-55-0: 12221
4926-76-5: 09866
4926-78-7: 09867
4926-90-3: 09862
4928-41-0: 02158
4930-98-7: 22985
4934-55-8: 15824
4935-80-2: 18978
4937-86-4: 20952
4938-74-3: 18746
4942-47-6: 25115
4944-94-9: 05650
4945-28-2: 23714
4945-47-5: 20129
4945-48-6: 20137
4947-16-4: 01215
4949-20-6: 18117
4949-44-4: 18462
4949-55-7: 23281
4957-14-6: 03653
4958-56-9: 16007
4960-55-8: 03894
4964-71-0: 23564
4964-77-6: 23620
4965-26-8: 07119
4965-30-4: 16817
4965-33-7: 23593
4965-36-0: 23566
4972-29-6: 05298
4972-49-0: 12004
4974-21-4: 00764
4974-27-0: 17523
4974-44-1: 00555
4975-21-7: 13118
4977-62-2: 00445
4979-79-7: 23604
4981-66-2: 01188
4984-01-4: 17797
4984-85-4: 14356
4985-46-0: 05173
4985-70-0: 01127
4985-85-7: 12098
4986-89-4: 21781
4988-33-4: 24829
4992-28-3: 24583
4992-29-4: 24559

4992-30-7: 24584
4994-16-5: 02970
4996-48-9: 12987
4997-87-9: 14924
4998-57-6: 14567
4998-76-9: 09780
5001-37-6: 08474
5006-22-4: 09547
5006-66-6: 22847
5008-69-5: 20660
5008-72-0: 07996
5008-73-1: 20652
5008-87-7: 22314
5008-88-8: 22315
5008-89-9: 22316
5008-90-2: 22317
5009-11-0: 14526
5012-80-6: 14669
5013-77-4: 04900
5014-47-1: 01846
5014-86-8: 02382
5018-27-9: 12815
5018-87-1: 06512
5020-21-3: 15067
5020-41-7: 04851
5021-43-2: 23885
5022-29-7: 15381
5022-65-1: 15230
5023-02-9: 06803
5023-60-9: 03741
5024-21-5: 24362
5026-76-6: 13742
5029-67-4: 22669
5030-67-1: 09714
5031-78-7: 12499
5034-76-4: 15121
5035-52-9: 24238
5037-60-5: 15086
5039-61-2: 14655
5042-08-0: 25554
5043-19-6: 16658
5049-61-6: 22304
5053-43-0: 23072
5057-96-5: 08146
5057-98-7: 10623
5057-99-8: 10625
5058-19-5: 22590
5061-21-2: 13064
5069-23-8: 24706
5069-26-1: 24756
5070-13-3: 09195
5071-96-5: 04925
5072-45-7: 20150
5075-33-2: 14561
5075-52-5: 24084
5075-92-3: 23419
5076-20-0: 18026
5076-82-4: 20112
5077-57-6: 13703
5077-67-8: 08651
5079-90-3: 06553
5081-36-7: 06454
5085-74-5: 18875
5090-16-4: 13805
5090-41-5: 17475
5096-57-1: 05684
5097-12-1: 03929
5097-92-7: 22721
5097-93-8: 22720
5100-98-1: 24916
5101-15-5: 07562
5103-42-4: 07254
5106-98-9: 06087
5107-18-6: 15540
5108-29-2: 24744
5110-08-7: 23761
5111-73-9: 16630
5112-21-0: 20898
5113-93-9: 10345
5119-88-0: 20179
5122-82-7: 12310
5129-40-8: 08385
5129-41-9: 09030
5131-60-2: 04451
5131-66-8: 21366
5132-75-2: 13634
5133-12-0: 02007
5136-45-8: 11478
5137-45-1: 11839
5144-11-6: 24427
5145-42-6: 09709
5145-64-2: 23315
5145-99-3: 20606
5146-68-9: 23395
5149-15-5: 01376
5150-42-5: 19258
5150-80-1: 13456
5150-93-6: 08189
5153-67-3: 03797
5154-00-7: 22978
5154-01-8: 22989
5159-41-1: 05090
5160-99-6: 03706
5161-04-6: 03708
5161-13-7: 24772
5161-14-8: 24773
5161-16-0: 24830
5161-17-1: 24832
5162-03-8: 15900
5162-44-7: 08710
5162-48-1: 18532
5162-63-0: 10841
5162-82-3: 06098
5166-35-8: 08722
5166-53-0: 14516
5167-18-0: 22223
5171-84-6: 14087
5175-48-4: 12210
5177-27-5: 23041
5178-20-1: 01041
5182-44-5: 04831
5183-77-7: 09829
5183-79-9: 09835
5191-80-0: 10872
5192-03-0: 15112
5193-57-7: 20276
5194-50-3: 13946
5194-51-4: 13945
5195-37-9: 16507
5195-40-4: 16506
5204-87-5: 17463
5206-09-7: 24727
5206-83-7: 10264
5208-49-1: 09270
5209-18-7: 03892
5209-31-4: 00011
5209-33-6: 00013
5210-87-7: 15819
5211-62-1: 21564
5216-25-1: 02955
5216-32-0: 03060
5217-05-0: 10597
5221-17-0: 20384
5221-40-9: 01959
5221-42-1: 00367
5221-44-3: 01960
5223-59-6: 03404
5223-61-0: 03773
5227-53-2: 01516
5228-61-5: 02053
5228-65-9: 04425
5231-87-8: 09591
5232-99-5: 21815
5233-04-5: 04427
5233-42-1: 06986
5240-32-4: 10145

5240-72-2: 07304
5244-34-8: 12159
5247-95-0: 22281
5256-79-1: 18055
5257-20-5: 11260
5258-50-4: 24831
5259-28-9: 11629
5259-65-4: 10436
5259-98-3: 18485
5260-04-8: 25362
5263-87-6: 23665
5265-18-9: 21437
5267-39-0: 04897
5271-26-1: 20088
5271-38-5: 12214
5271-39-6: 03409
5271-67-0: 24790
5272-02-6: 13495
5274-50-0: 19880
5274-70-4: 01726
5281-18-5: 01692
5283-66-9: 24114
5284-07-1: 19776
5284-66-2: 21004
5285-18-7: 18109
5285-74-5: 24643
5285-87-0: 24648
5287-64-9: 14538
5292-21-7: 09985
5292-43-3: 00484
5292-53-5: 20879
5293-97-0: 15849
5306-96-7: 04482
5307-00-6: 04495
5307-14-2: 04501
5307-19-7: 22915
5307-40-4: 08537
5307-46-0: 11257
5308-25-8: 20068
5309-18-2: 16353
5309-46-6: 13886
5311-96-6: 14714
5312-97-0: 06698
5314-37-4: 07653
5314-41-0: 00038
5314-55-6: 24066
5315-79-7: 22517
5319-77-7: 24449
5320-98-9: 16112
5323-65-9: 19161
5324-00-5: 15038
5324-19-6: 15394
5325-88-2: 16838
5325-97-3: 18889
5326-10-3: 19945
5326-23-8: 22844
5326-27-2: 06294
5326-34-1: 02695
5326-42-1: 15982
5326-47-6: 05927
5326-53-4: 11032
5326-81-8: 00156
5326-82-9: 00130
5326-87-4: 00107
5327-44-6: 03583
5328-01-8: 16395
5328-76-7: 16261
5329-12-4: 14658
5329-50-0: 01372
5331-08-8: 10247
5331-28-2: 03242
5331-43-1: 14709
5331-91-9: 07071
5331-93-1: 01637
5332-06-9: 08247
5332-24-1: 23562
5332-25-2: 23565
5332-26-3: 15370
5332-35-4: 23401
5332-52-5: 25283
5332-73-0: 20458
5332-98-9: 18120
5333-00-6: 22027
5333-02-8: 23627
5333-03-9: 09700
5333-40-4: 00682
5333-99-3: 15393
5334-39-4: 22419
5334-43-0: 22431
5334-99-6: 22491
5335-24-0: 07671
5335-28-4: 04100
5335-75-1: 22592
5336-22-1: 19624
5336-57-2: 04316
5336-76-5: 05223
5337-03-1: 22255
5337-30-4: 09132
5337-31-5: 16125
5337-36-0: 07741
5337-41-7: 07744
5337-70-2: 17873
5337-72-4: 10126
5337-93-9: 21515
5339-85-5: 04821
5340-31-8: 25309
5340-36-3: 17738
5340-41-0: 14329
5340-96-5: 07135
5341-07-1: 23570
5341-58-2: 16681
5341-76-4: 18339
5341-95-7: 08221
5342-31-4: 15736
5342-87-0: 04859
5343-16-8: 13041
5343-65-7: 00105
5343-96-4: 08592
5343-99-7: 17055
5344-18-3: 08622
5344-20-7: 18169
5344-33-2: 18338
5344-44-5: 02108
5344-49-0: 06106
5344-55-8: 08519
5344-68-3: 19565
5344-82-1: 24861
5344-90-1: 04966
5345-01-7: 18307
5345-09-5: 19321
5345-42-6: 03599
5345-46-0: 00022
5345-47-1: 22837
5345-89-1: 21828
5346-37-2: 14685
5346-90-7: 13200
5347-15-9: 23568
5347-68-2: 20147
5347-82-0: 15750
5348-42-5: 04429
5348-74-3: 06363
5349-24-6: 00116
5349-55-3: 21194
5349-60-0: 05067
5349-62-2: 18570
5349-78-0: 23837
5350-47-0: 16018
5350-51-6: 23582
5350-57-2: 15962
5350-67-4: 20977
5351-17-7: 05881
5351-69-9: 14677
5351-85-9: 14663
5354-04-1: 04366
5355-54-4: 21309
5355-81-7: 21485

5356-83-2: 24055
5357-38-0: 24077
5359-04-6: 12462
5362-55-0: 18724
5364-87-4: 10209
5367-26-0: 02895
5367-28-2: 02903
5367-32-8: 03601
5368-28-5: 20126
5368-75-2: 02576
5368-81-0: 06442
5369-17-5: 02060
5369-19-7: 02198
5369-22-2: 02465
5369-55-1: 12891
5370-19-4: 24795
5370-25-2: 12309
5371-48-2: 18761
5372-81-6: 04614
5378-49-4: 24421
5378-52-9: 24431
5379-20-4: 03343
5380-40-5: 12991
5381-25-9: 07128
5381-92-0: 04869
5382-16-1: 20278
5382-42-3: 23735
5382-77-4: 22262
5384-03-2: 22080
5384-06-5: 21486
5384-08-7: 22078
5384-09-8: 21487
5384-13-4: 21483
5384-19-0: 22079
5384-54-3: 21484
5385-37-5: 22498
5386-09-4: 00028
5386-11-8: 00030
5388-62-5: 02078
5390-04-5: 18792
5390-71-6: 20530
5390-72-7: 20528
5392-10-9: 01587
5392-12-1: 16597
5392-19-8: 23605
5392-35-8: 07079
5392-86-9: 03072
5393-22-6: 24632
5393-41-9: 09173
5393-49-7: 02196
5393-75-9: 00447
5394-18-3: 15368
5395-04-0: 20142
5395-20-0: 03740
5395-43-7: 24246
5395-47-1: 19423
5395-75-5: 11749
5395-85-7: 01912
5395-94-8: 24850
5396-24-7: 00521
5396-38-3: 03230
5396-58-7: 08229
5396-64-5: 21840
5396-91-8: 18583
5396-94-1: 23572
5397-37-5: 12748
5398-08-3: 20853
5398-10-7: 20842
5398-24-3: 07835
5398-25-4: 18211
5398-27-6: 02173
5398-44-7: 22866
5398-69-6: 06420
5398-75-4: 18933
5399-02-0: 13414
5399-06-4: 16111
5399-68-8: 01645
5400-80-6: 15074
5400-81-7: 05912
5400-88-4: 09773
5401-94-5: 14978
5402-20-0: 22938
5402-73-3: 05502
5404-88-6: 22150
5405-13-0: 04939
5405-34-5: 08860
5405-41-4: 08431
5405-58-3: 14051
5405-79-8: 14350
5405-96-9: 07433
5406-10-0: 13685
5406-12-2: 05169
5406-18-8: 05272
5406-79-1: 22962
5407-58-9: 07047
5407-68-1: 13049
5407-87-4: 22543
5407-98-7: 15906
5408-55-9: 11133
5408-57-1: 17765
5408-76-4: 08920
5408-86-6: 07949
5409-15-4: 16668
5409-32-5: 16802
5409-55-2: 17018
5409-99-4: 15951
5410-26-4: 01428
5410-29-7: 01437
5410-37-7: 01435
5410-89-9: 01398
5410-97-9: 11135
5411-50-7: 03024
5411-53-0: 02675
5411-56-3: 04996
5411-70-1: 04629
5412-43-1: 07575
5412-64-6: 04889
5412-92-0: 17574
5413-05-8: 04119
5413-58-1: 04202
5413-65-0: 01304
5414-19-7: 11859
5414-21-1: 18365
5415-44-1: 22177
5416-12-6: 00225
5416-30-8: 15561
5416-43-3: 25073
5416-55-7: 10098
5416-70-6: 22024
5416-93-3: 03559
5416-94-4: 24639
5417-32-3: 11309
5417-50-5: 00368
5417-52-7: 00209
5417-63-0: 16960
5417-68-5: 24949
5417-84-5: 22154
5417-86-7: 22161
5418-10-0: 22492
5418-51-9: 22994
5418-86-0: 15716
5419-55-6: 07746
5419-82-9: 00338
5420-17-7: 17494
5421-04-5: 04200
5421-45-4: 06530
5422-53-7: 00102
5422-88-8: 15909
5423-27-8: 22064
5424-19-1: 16036
5424-21-5: 23060
5424-77-1: 00248
5425-50-3: 05255
5425-83-2: 16726
5426-52-8: 20404
5426-65-3: 22471

5427-08-7: 19199
5427-99-6: 13333
5428-32-0: 11928
5428-37-5: 12947
5428-54-6: 19497
5428-89-7: 23021
5428-95-5: 00003
5429-01-6: 23802
5429-28-7: 06161
5429-85-6: 17479
5431-44-7: 22886
5432-09-7: 23540
5432-17-7: 06262
5432-61-1: 09777
5432-85-9: 10257
5433-01-2: 02678
5434-03-7: 04957
5434-20-8: 04529
5434-27-5: 20551
5435-83-6: 00562
5436-21-5: 08644
5437-12-7: 05443
5437-38-7: 06496
5437-98-9: 07858
5438-13-1: 16322
5438-19-7: 06570
5438-36-8: 01708
5438-54-0: 03195
5439-01-0: 22788
5439-25-8: 19330
5439-39-4: 20787
5440-04-0: 01429
5440-71-1: 00457
5441-16-7: 03118
5441-51-0: 10243
5441-52-1: 10129
5443-49-2: 21614
5445-24-9: 10198
5445-29-4: 17672
5445-30-7: 10112
5445-31-8: 17284
5445-51-2: 09556
5445-85-2: 00391
5447-02-9: 01583
5447-37-0: 24571
5447-93-8: 00544
5447-96-1: 20514
5447-99-4: 18518
5448-43-1: 23864
5448-47-5: 15181
5449-69-4: 16000
5451-39-8: 22765
5451-58-1: 20932
5451-65-0: 24400
5451-80-9: 18426
5452-35-7: 09626
5452-37-9: 10451
5452-75-5: 09988
5453-44-1: 08919
5453-67-8: 22899
5453-80-5: 07374
5454-28-4: 13616
5454-83-1: 18406
5455-24-3: 17654
5456-94-0: 03015
5457-44-3: 14151
5458-48-0: 02980
5458-54-8: 16521
5458-59-3: 17690
5459-10-9: 20437
5459-40-5: 03347
5459-58-5: 00550
5459-85-8: 04467
5459-93-8: 09776
5459-95-0: 20757
5460-29-7: 15371
5460-31-1: 19495
5460-32-2: 03528
5461-23-4: 01108
5461-32-5: 05243
5461-51-8: 00306
5462-27-1: 05587
5462-97-5: 24576
5463-10-5: 22149
5463-59-2: 22958
5463-64-9: 05444
5464-22-2: 12048
5464-79-9: 07040
5464-87-9: 19675
5465-09-8: 10237
5465-13-4: 03973
5465-65-6: 12323
5466-23-9: 01700
5466-84-2: 15322
5467-04-9: 01954
5467-13-0: 25157
5467-64-1: 00059
5467-74-3: 07756
5467-78-7: 24423
5468-23-5: 20792
5468-97-3: 04834
5469-19-2: 02732
5469-45-4: 21774
5469-69-2: 22520
5470-18-8: 22606
5470-22-4: 22832
5470-28-0: 20109
5470-44-0: 13024
5470-47-3: 04115
5470-65-5: 19122
5470-66-6: 22688
5470-91-7: 22025
5470-96-2: 23731
5471-26-1: 16592
5471-30-7: 17061
5471-51-2: 08661
5471-90-9: 05373
5471-96-5: 21999
5472-01-5: 21447
5472-05-9: 21448
5472-13-9: 05127
5472-38-8: 08163
5472-99-1: 02862
5487-14-9: 16174
5488-16-4: 04415
5488-50-6: 25452
5496-04-8: 22482
5500-21-0: 10789
5502-88-5: 10340
5503-12-8: 10365
5507-44-8: 24054
5509-08-0: 05461
5509-65-9: 02166
5509-82-0: 21518
5510-78-1: 19790
5510-81-6: 19788
5510-99-6: 19090
5515-64-0: 10063
5518-76-3: 23107
5519-42-6: 22427
5519-50-6: 20294
5521-60-8: 22369
5522-95-2: 24267
5524-05-0: 10258
5524-60-7: 19909
5524-62-9: 19908
5529-38-4: 03188
5531-65-7: 17448
5532-90-1: 09110
5538-41-0: 19147
5540-60-3: 00170
5541-67-3: 23819
5541-68-4: 23821
5544-00-3: 23457
5544-06-9: 23500
5544-07-0: 23497

5551-11-1: 01620
5556-07-0: 12442
5557-31-3: 17481
5557-88-0: 18758
5558-29-2: 04339
5558-31-6: 04340
5558-42-9: 04361
5558-66-7: 04270
5558-78-1: 04347
5560-69-0: 16875
5563-78-0: 10170
5568-33-2: 01617
5570-71-8: 22081
5570-72-9: 21482
5581-35-1: 04810
5581-75-9: 04876
5582-82-1: 13678
5586-09-4: 19954
5586-15-2: 11418
5587-63-3: 10731
5587-79-1: 10757
5588-20-5: 14858
5589-96-8: 00491
5594-20-7: 20096
5597-27-3: 07344
5597-81-9: 16283
5597-89-7: 18133
5599-71-3: 09164
5600-21-5: 23022
5600-62-4: 04372
5601-60-5: 10973
5606-39-3: 02087
5609-09-6: 13802
5611-17-6: 08343
5612-44-2: 04473
5614-32-4: 07793
5614-64-2: 22211
5615-81-6: 23404
5617-41-4: 09874
5617-42-5: 10539
5617-74-3: 17884
5617-92-5: 10831
5619-07-8: 19721
5621-02-3: 23026
5621-13-6: 15140
5621-14-7: 15142
5621-16-9: 15136
5622-83-3: 22996
5623-45-0: 15872
5630-56-8: 15866
5631-75-4: 06891
5632-91-7: 15528
5633-55-6: 19584
5634-22-0: 03295
5634-30-0: 10524
5634-54-8: 21708
5636-65-7: 14475
5637-83-2: 24988
5637-96-7: 03833
5638-26-6: 13449
5649-36-5: 15139
5650-10-2: 02370
5650-40-8: 21498
5650-41-9: 21499
5651-14-9: 16610
5651-47-8: 06489
5652-38-0: 01005
5653-21-4: 00045
5653-46-3: 06341
5653-62-3: 06696
5653-67-8: 05033
5653-80-5: 13710
5655-61-8: 07317
5656-83-7: 16616
5657-41-0: 05308
5660-37-7: 23908
5661-71-2: 01544
5663-56-9: 04108
5663-96-7: 17852
5666-17-1: 21659
5666-21-7: 21668
5668-93-9: 07507
5669-09-0: 20604
5669-14-7: 05215
5673-07-4: 03172
5675-22-9: 13436
5675-51-4: 11505
5676-58-4: 07178
5676-61-9: 21944
5676-82-4: 02436
5680-61-5: 20819
5680-80-8: 23962
5680-83-1: 13341
5682-00-8: 24725
5682-41-7: 15563
5683-33-0: 22544
5683-44-3: 18344
5685-06-3: 24537
5685-47-2: 10781
5685-82-5: 14765
5687-80-9: 10834
5687-92-3: 24620
5689-04-3: 15104
5690-24-4: 06752
5690-26-6: 16895
5692-26-2: 01855
5692-27-3: 01854
5693-09-4: 08156
5697-01-8: 17069
5697-02-9: 16929
5697-04-1: 16608
5697-55-2: 20738
5697-85-8: 02464
5699-17-2: 18869
5699-58-1: 18422
5699-74-1: 18367
5699-78-5: 14371
5700-29-8: 04711
5700-67-4: 04739
5701-87-1: 05913
5702-69-2: 22450
5702-75-0: 00213
5703-21-9: 04085
5703-26-4: 04089
5703-52-6: 05257
5704-20-1: 18584
5704-66-5: 21608
5705-15-7: 14653
5706-12-7: 15097
5707-44-8: 07508
5711-41-1: 08897
5715-23-1: 10128
5716-20-1: 05001
5718-88-7: 24547
5720-05-8: 07762
5722-93-0: 06342
5724-56-1: 06701
5728-52-9: 07554
5732-47-8: 11729
5736-03-8: 11323
5736-70-9: 09479
5736-88-9: 01598
5736-89-0: 12311
5736-94-7: 01691
5738-24-9: 05669
5742-19-8: 00589
5743-97-5: 18892
5744-03-6: 12722
5744-51-4: 22437
5748-36-7: 07582
5748-43-6: 07596
5750-00-5: 07933
5752-08-9: 22145
5756-43-4: 14040
5756-56-9: 08985
5763-48-4: 10615

5763-61-1: 04904
5765-44-6: 15496
5766-03-0: 15569
5766-40-5: 04206
5768-10-5: 10111
5769-35-7: 11086
5773-56-8: 04866
5773-66-0: 16690
5773-67-1: 16691
5773-87-5: 16687
5773-98-8: 16656
5773-99-9: 17066
5774-26-5: 21542
5774-32-3: 22131
5775-82-6: 22388
5775-86-0: 22389
5776-50-1: 15295
5776-51-2: 15294
5779-51-1: 04791
5779-72-6: 01797
5779-94-2: 01660
5779-95-3: 01662
5780-08-5: 01780
5780-13-2: 07950
5781-02-2: 19308
5781-53-3: 00540
5785-66-0: 05021
5787-28-0: 03768
5787-29-1: 03767
5787-31-5: 07910
5787-33-7: 07911
5789-30-0: 03008
5792-36-9: 21469
5794-03-6: 07273
5794-04-7: 07274
5794-13-8: 01458
5794-24-1: 01456
5794-88-7: 05890
5796-07-6: 09103
5796-17-8: 25241
5796-23-6: 04771
5796-31-6: 01487
5796-78-1: 12290
5798-75-4: 05996
5798-79-8: 04307
5798-81-2: 08571
5798-88-9: 08340
5798-94-7: 01825
5802-60-8: 14593
5802-82-4: 13848
5805-58-3: 05639
5805-76-5: 05617
5805-95-8: 02295
5807-64-7: 07627
5807-65-8: 07626
5807-76-1: 02736
5809-07-4: 12895
5811-48-3: 07345
5813-64-9: 20447
5813-86-5: 01922
5814-42-6: 21339
5814-85-7: 03661
5817-26-5: 20371
5820-22-4: 03760
5824-40-8: 04910
5831-22-1: 12252
5833-18-1: 04174
5833-19-2: 00480
5834-25-3: 00195
5835-47-2: 24549
5836-06-6: 00966
5836-28-2: 11352
5836-66-8: 20553
5837-45-6: 13454
5837-78-5: 08903
5842-00-2: 12066
5846-43-5: 10200
5847-59-6: 19118
5848-56-6: 03393
5851-43-4: 05623
5853-00-9: 14569
5855-52-7: 23529
5855-72-1: 07611
5856-62-2: 08527
5856-63-3: 08526
5856-77-9: 08683
5856-82-6: 08675
5857-36-3: 18591
5857-86-3: 10199
5858-18-4: 05547
5859-45-0: 16288
5860-70-8: 22839
5862-40-8: 12188
5862-62-4: 20592
5865-15-6: 20591
5867-45-8: 00366
5870-61-1: 12547
5870-68-8: 18446
5870-93-9: 08422
5872-08-2: 07301
5874-90-8: 01019
5875-24-1: 20410
5876-87-9: 11469
5876-92-6: 04671
5877-51-0: 02055
5877-53-2: 05391
5877-55-4: 02407
5877-58-7: 02408
5878-19-3: 21562
5878-61-5: 22196
5881-17-4: 17616
5882-44-0: 02192
5882-45-1: 02154
5883-17-0: 21649
5888-29-9: 11718
5891-06-5: 12068
5891-21-4: 18545
5893-05-0: 13350
5893-91-4: 16067
5896-38-8: 13115
5897-76-7: 04985
5906-73-0: 24163
5907-43-7: 09420
5908-87-2: 11456
5909-24-0: 23106
5910-87-2: 17181
5910-89-4: 22323
5911-04-6: 17212
5911-08-0: 10769
5912-58-3: 12115
5912-86-7: 19429
5912-87-8: 19430
5913-06-4: 12016
5913-85-9: 20211
5913-87-1: 20212
5913-89-3: 20213
5913-94-0: 20217
5916-96-1: 12924
5917-47-5: 20161
5921-54-0: 08667
5921-73-3: 17332
5921-80-2: 07961
5921-82-4: 13653
5921-84-6: 13680
5922-60-1: 06664
5922-69-0: 12820
5928-63-2: 12412
5928-66-5: 12410
5928-67-6: 12411
5928-72-3: 05483
5930-28-9: 19006
5931-54-4: 02243
5932-68-3: 19428
5933-35-7: 06532
5934-25-8: 16708
5934-29-2: 14573

5934-50-9: 04221
5934-56-5: 14746
5936-28-7: 15342
5937-53-1: 09089
5937-83-7: 18398
5941-05-9: 00648
5943-30-6: 11402
5943-31-7: 25210
5943-35-1: 24410
5944-41-2: 22632
5949-05-3: 17792
5949-07-5: 09422
5950-19-6: 08635
5951-67-7: 10316
5953-69-5: 01058
5953-79-7: 11339
5957-05-1: 25004
5957-68-6: 07308
5959-52-4: 16655
5960-69-0: 12496
5961-35-3: 19471
5961-36-4: 19365
5961-55-7: 16586
5961-59-1: 02324
5962-00-5: 07141
5962-19-6: 09418
5963-77-9: 18783
5964-64-7: 11336
5965-66-2: 13205
5965-92-4: 10867
5966-51-8: 21360
5967-84-0: 22200
5971-96-0: 11364
5973-11-5: 07919
5973-71-7: 01661
5974-11-8: 16065
5976-47-6: 21984
5978-55-2: 17579
5978-70-1: 17720
5978-81-4: 04267
5978-95-0: 14745
5980-96-1: 04061
5981-10-2: 25200
5984-58-7: 13474
5985-23-9: 13249
5985-91-1: 25036
5985-99-9: 20215
5986-55-0: 15815
5987-98-4: 12413
5989-27-5: 10334
5989-54-8: 10335
5989-81-1: 13195
5990-63-6: 08147
5991-76-4: 13298
5992-90-5: 25264
6000-43-7: 13320
6001-64-5: 21428
6001-87-2: 21105
6003-11-8: 01295
6004-44-0: 21994
6004-60-0: 12335
6004-79-1: 15804
6005-18-1: 08451
6006-06-0: 25176
6006-33-3: 09967
6006-34-4: 10578
6006-65-1: 15609
6006-90-2: 18048
6006-95-7: 09946
6007-87-0: 24815
6008-27-1: 17971
6009-50-3: 11356
6011-11-6: 14767
6011-64-9: 16721
6012-97-1: 24768
6018-27-5: 01364
6020-39-9: 10880
6020-51-5: 14878
6020-67-3: 04225
6020-87-7: 13283
6027-89-0: 13386
6031-02-3: 03725
6032-29-7: 18477
6033-23-4: 13652
6033-24-5: 13651
6033-61-0: 13979
6035-40-1: 15338
6035-49-0: 06865
6036-46-0: 18980
6040-62-6: 17052
6044-68-4: 21707
6045-53-0: 11668
6050-13-1: 11235
6050-26-6: 20430
6050-34-6: 10652
6051-00-9: 09987
6051-40-7: 24310
6051-43-0: 05894
6051-52-1: 05058
6051-53-2: 08960
6051-87-2: 17087
6051-98-5: 05772
6052-68-2: 23005
6055-06-7: 16068
6055-52-3: 14105
6057-60-9: 07585
6057-94-9: 20327
6061-06-9: 07789
6061-10-5: 17684
6064-27-3: 11544
6064-76-2: 00791
6064-90-0: 13393
6064-96-6: 22044
6065-01-6: 17302
6065-32-3: 08863
6065-59-4: 20871
6065-82-3: 00595
6065-90-3: 18221
6065-93-6: 21702
6066-82-6: 23394
6068-69-5: 02415
6068-72-0: 07203
6069-71-2: 07341
6069-97-2: 07287
6069-98-3: 09911
6072-57-7: 15090
6077-72-1: 10833
6078-26-8: 00952
6078-62-2: 05922
6079-73-8: 06335
6080-79-1: 18771
6088-94-4: 17187
6089-09-4: 18784
6090-16-0: 13447
6091-44-7: 20174
6091-50-5: 10427
6091-52-7: 10425
6091-58-3: 08227
6091-64-1: 05989
6092-71-3: 09086
6093-03-4: 24321
6093-67-0: 06846
6093-68-1: 06847
6094-02-6: 14408
6099-04-3: 21880
6099-90-7: 05590
6100-34-1: 20268
6100-74-9: 12418
6106-81-6: 04227
6108-61-8: 13947
6109-22-4: 01152
6111-61-1: 08599
6111-78-0: 01988
6111-88-2: 18205
6111-99-5: 21124
6112-71-6: 09498

6114-18-7: 17500
6117-80-2: 08841
6117-91-5: 08927
6118-13-4: 08939
6118-14-5: 08938
6119-47-7: 09435
6120-13-4: 07270
6120-95-2: 10817
6124-64-7: 17234
6125-21-9: 20587
6125-24-2: 03798
6125-79-7: 05660
6127-92-0: 08878
6127-93-1: 08879
6129-13-1: 20915
6130-75-2: 04017
6131-24-4: 14065
6131-25-5: 13542
6134-66-3: 08377
6135-29-1: 00779
6135-36-0: 25380
6136-66-9: 15999
6136-67-0: 16003
6137-03-7: 18552
6137-08-2: 17756
6137-11-7: 13716
6137-15-1: 17763
6137-26-4: 11542
6138-41-6: 22960
6138-90-5: 17533
6139-84-0: 07812
6140-17-6: 03789
6140-61-0: 18366
6140-65-4: 10728
6140-80-3: 21724
6141-57-7: 12976
6141-58-8: 13006
6141-60-2: 13020
6141-68-0: 12889
6142-53-6: 01508
6146-52-7: 15163
6147-14-4: 15584
6147-43-9: 13818
6147-92-8: 17764
6148-33-0: 12999
6148-34-1: 13002
6148-37-4: 12914
6149-41-3: 21198
6149-46-8: 18429
6153-16-8: 10329
6153-56-6: 11983
6155-35-7: 15579
6155-36-8: 15580
6156-18-9: 20497
6156-25-8: 15697
6158-45-8: 16443
6158-54-9: 05969
6159-25-7: 20950
6159-55-3: 23445
6159-56-4: 23444
6159-66-6: 11147
6161-50-8: 07480
6161-83-7: 12127
6161-86-0: 12203
6162-79-4: 09083
6163-58-2: 19755
6163-64-0: 20661
6163-66-2: 20680
6163-75-3: 19825
6164-78-9: 22365
6164-98-3: 15772
6165-44-2: 09809
6166-86-5: 10695
6166-87-6: 10296
6168-72-5: 21297
6168-83-8: 08429
6169-06-8: 17721
6170-08-7: 20328
6170-23-6: 21298
6174-95-4: 12614
6175-23-1: 17232
6175-49-1: 11540
6180-61-6: 21346
6180-80-9: 15830
6185-14-4: 11231
6190-36-9: 04542
6191-58-8: 04214
6192-52-5: 05468
6195-92-2: 10671
6196-58-3: 12229
6196-60-7: 18639
6200-63-1: 14752
6201-65-6: 04687
6203-08-3: 07378
6203-18-5: 21623
6210-51-1: 14305
6211-15-0: 16070
6213-85-0: 05420
6213-90-7: 08876
6214-18-2: 05470
6214-25-1: 08881
6214-28-4: 08875
6214-44-4: 05060
6214-45-5: 04999
6221-55-2: 07414
6222-00-0: 00806
6222-02-2: 00840
6222-03-3: 00812
6222-35-1: 21122
6223-78-5: 14028
6223-83-2: 12769
6224-63-1: 19756
6225-06-5: 18153
6231-18-1: 22624
6231-36-3: 25204
6232-88-8: 06016
6236-09-5: 08172
6236-88-0: 09868
6237-59-8: 04875
6237-95-2: 23058
6237-96-3: 23042
6237-97-4: 23053
6240-11-5: 25122
6240-55-7: 16341
6245-13-2: 06751
6245-93-8: 15197
6248-20-0: 01649
6250-72-2: 17171
6257-64-3: 02019
6258-06-6: 01335
6258-66-8: 04961
6259-19-4: 11227
6259-42-3: 02102
6261-22-9: 18790
6262-27-7: 01963
6263-62-3: 03446
6264-93-3: 09660
6265-03-8: 19035
6265-06-1: 19033
6265-07-2: 19034
6265-13-0: 19473
6265-26-5: 13114
6267-02-3: 00965
6267-24-9: 11855
6267-99-8: 16168
6268-00-4: 04360
6268-14-0: 13371
6268-37-7: 02937
6269-89-2: 20085
6269-92-7: 17590
6270-16-2: 08595
6270-34-4: 12979
6270-46-8: 23124
6270-55-9: 21234
6270-56-0: 12896
6270-89-9: 03681

6271-08-5: 00148
6271-78-9: 22805
6272-36-2: 05481
6272-38-4: 19573
6272-40-8: 15837
6272-52-2: 07567
6274-16-4: 09082
6274-73-3: 06872
6275-81-6: 15505
6276-03-5: 12760
6276-05-7: 12786
6276-48-8: 05710
6276-54-6: 20442
6276-62-6: 09728
6277-38-9: 12449
6278-69-9: 07054
6279-05-6: 15896
6279-48-7: 02531
6279-86-3: 15769
6279-88-5: 16673
6279-98-7: 13996
6280-00-8: 14175
6280-01-9: 18348
6280-03-1: 17248
6280-48-4: 06052
6280-57-5: 12821
6280-88-2: 06110
6280-96-2: 19591
6280-98-4: 03291
6281-02-3: 19245
6282-65-1: 02743
6283-25-6: 02105
6283-26-7: 06401
6283-85-8: 07289
6284-08-8: 11474
6284-40-8: 13185
6284-79-3: 15922
6284-82-8: 17789
6284-84-0: 20057
6285-05-8: 21458
6285-06-9: 18796
6285-34-3: 15912
6285-64-9: 20405
6286-32-4: 24691
6286-46-0: 06007
6287-12-3: 05437
6287-38-3: 01631
6287-43-0: 02155
6287-47-4: 19074
6289-73-2: 13042
6290-03-5: 08210
6290-24-0: 06260
6290-49-9: 00675
6290-86-4: 07604
6290-94-4: 16796
6291-23-2: 19529
6291-88-9: 21281
6292-62-2: 06703
6292-82-6: 20221
6292-91-7: 05751
6293-11-4: 13122
6293-20-5: 13280
6294-31-1: 14088
6294-34-4: 19810
6294-39-9: 09805
6294-40-2: 09807
6294-89-9: 14708
6295-94-9: 05375
6297-03-6: 17384
6297-48-9: 18442
6298-19-7: 22561
6298-37-9: 23852
6298-65-3: 10492
6299-08-7: 01432
6299-25-8: 23061
6299-67-8: 02174
6300-80-7: 01666
6301-12-8: 24580
6302-94-9: 11434
6303-58-8: 08507
6304-04-7: 06122
6304-18-3: 22684
6304-23-0: 22698
6304-33-2: 04318
6304-46-7: 16943
6304-50-3: 11497
6304-89-8: 05860
6305-43-7: 08239
6305-52-8: 16270
6306-07-6: 00026
6306-30-5: 14349
6306-52-1: 25538
6306-60-1: 04315
6307-13-7: 15385
6307-83-1: 06022
6307-92-2: 04076
6308-96-9: 25290
6308-98-1: 04813
6309-50-8: 11515
6310-09-4: 12330
6310-21-0: 02197
6310-44-7: 08639
6310-64-1: 19068
6311-22-4: 12781
6311-27-9: 14739
6311-43-9: 07500
6311-60-0: 03026
6313-17-3: 06422
6313-33-3: 15777
6313-37-7: 02179
6313-44-6: 01349
6314-28-9: 07127
6314-97-2: 03135
6315-03-3: 06492
6315-19-1: 16646
6315-52-2: 11992
6315-89-5: 02178
6315-96-4: 21520
6316-09-2: 23111
6317-49-3: 13585
6317-78-8: 16032
6318-13-4: 13367
6318-36-1: 24562
6318-37-2: 24563
6318-64-5: 12259
6319-21-7: 11067
6319-40-0: 06024
6320-03-2: 05544
6320-15-6: 23049
6320-40-7: 03995
6320-51-0: 24550
6320-52-1: 09077
6320-96-3: 20407
6322-01-6: 21356
6322-49-2: 08633
6324-01-2: 04128
6324-13-6: 09112
6324-65-8: 21531
6325-62-8: 01112
6325-91-3: 05647
6325-93-5: 05387
6326-44-9: 20837
6326-56-3: 08612
6326-83-6: 00502
6327-52-2: 09108
6327-59-9: 01469
6328-44-5: 06538
6329-16-4: 04523
6330-51-4: 12830
6333-07-9: 15983
6333-15-9: 01944
6334-18-5: 01626
6334-37-8: 16285
6335-07-5: 23958
6335-34-8: 00103
6336-01-2: 24860

6336-70-5: 05949
6336-79-4: 16739
6337-19-5: 01195
6338-19-8: 08804
6338-45-0: 14770
6338-83-6: 13270
6339-75-9: 15732
6339-79-3: 24575
6339-87-3: 24821
6341-60-2: 20865
6341-92-0: 15207
6341-97-5: 19222
6341-98-6: 19172
6342-60-5: 06096
6342-77-4: 05236
6343-54-0: 12836
6343-76-6: 12839
6344-28-1: 12317
6344-60-1: 12792
6344-61-2: 12774
6344-62-3: 12784
6344-63-4: 12710
6344-66-7: 12712
6344-72-5: 23883
6345-02-4: 09247
6345-63-7: 15739
6346-09-4: 07872
6351-10-6: 15075
6358-07-2: 18997
6358-08-3: 18998
6358-09-4: 18999
6358-15-2: 19052
6358-64-1: 02073
6358-77-6: 02038
6361-21-3: 01618
6361-22-4: 01619
6361-23-5: 01628
6363-53-7: 13238
6366-18-3: 01998
6366-20-7: 01985
6366-23-0: 02002
6366-24-1: 01978
6370-23-6: 05411
6370-24-7: 05404
6370-47-4: 19533
6372-01-6: 08880
6373-11-1: 00001
6373-50-8: 02126
6374-70-5: 11110
6374-84-1: 01223
6374-92-1: 15210
6375-37-7: 01197
6377-82-8: 16103
6378-11-6: 09333
6378-51-4: 25421
6378-65-0: 14238
6378-90-1: 00227
6380-09-2: 24938
6380-10-5: 05361
6380-21-8: 19586
6380-23-0: 03341
6380-24-1: 03192
6380-28-5: 19477
6380-34-3: 01449
6382-06-5: 21193
6382-07-6: 12150
6382-13-4: 17446
6383-11-5: 07417
6386-63-6: 00532
6393-01-7: 04483
6397-09-7: 06992
6397-10-0: 06993
6397-29-1: 07000
6397-44-0: 06995
6397-45-1: 06996
6398-98-7: 19195
6400-14-2: 04450
6402-36-4: 11556
6406-96-8: 01134
6407-29-0: 15914
6407-30-3: 15915
6407-31-4: 15509
6407-55-2: 01273
6410-09-9: 16999
6410-10-2: 17000
6414-58-0: 20874
6414-69-3: 21209
6414-72-8: 24262
6416-39-3: 15020
6416-55-3: 01171
6416-59-7: 04468
6418-46-8: 11583
6423-02-5: 14016
6423-29-6: 16710
6436-90-4: 13347
6437-03-2: 01350
6437-04-3: 01355
6437-09-8: 14927
6437-10-1: 06170
6441-22-1: 09037
6443-91-0: 12645
6443-92-1: 13745
6448-90-4: 01270
6452-49-9: 02921
6452-61-5: 24223
6453-93-6: 16029
6453-99-2: 16041
6454-03-1: 15838
6456-01-5: 14853
6461-99-0: 06735
6468-47-9: 06854
6478-86-0: 05643
6480-68-8: 23739
6482-24-2: 11769
6482-26-4: 00566
6482-34-4: 09190
6483-86-9: 01337
6485-40-1: 10420
6485-83-2: 24156
6486-93-7: 01316
6493-73-8: 25208
6493-77-2: 10376
6493-83-0: 21507
6496-48-6: 23371
6497-83-2: 22768
6498-47-1: 08941
6502-38-1: 22010
6508-04-9: 06728
6509-38-2: 11741
6510-65-2: 09167
6511-74-6: 01354
6511-75-7: 06169
6512-83-0: 06255
6513-13-9: 25515
6515-36-2: 06889
6517-15-3: 06789
6526-72-3: 03672
6529-53-9: 02646
6530-11-6: 01475
6531-31-3: 14887
6531-35-7: 01279
6531-86-8: 07407
6533-53-5: 13883
6536-33-0: 09423
6538-02-9: 11619
6549-19-5: 16244
6552-46-1: 05020
6553-48-6: 09528
6554-98-9: 19567
6556-12-3: 13242
6556-35-0: 11557
6556-37-2: 16722
6560-83-4: 23658
6561-63-3: 21170
6565-55-5: 04081
6565-70-4: 04724

6568-58-7: 10447
6569-51-3: 07737
6573-52-0: 10446
6575-07-1: 06689
6575-09-3: 06685
6575-10-6: 06682
6579-55-1: 21401
6583-06-8: 06989
6583-71-7: 12832
6587-24-2: 06125
6592-85-4: 11341
6596-35-6: 01162
6596-57-2: 08080
6596-66-3: 04290
6596-69-6: 05480
6600-40-4: 17352
6602-02-4: 24655
6602-28-4: 22975
6602-32-0: 22970
6602-54-6: 22774
6602-56-8: 22939
6606-18-4: 01555
6606-59-3: 21900
6607-45-0: 03061
6607-66-5: 08076
6609-39-8: 05610
6609-56-9: 06715
6609-57-0: 06704
6610-29-3: 14670
6613-44-1: 07209
6617-04-5: 09993
6621-60-9: 14219
6622-55-5: 06561
6622-76-0: 08904
6622-92-0: 23259
6623-29-6: 17193
6623-30-9: 09249
6623-59-2: 03075
6623-81-0: 23140
6623-91-2: 23669
6624-48-2: 14958
6624-53-9: 03814
6624-76-6: 17162
6624-79-9: 11571
6626-15-9: 04684
6626-23-9: 16290
6626-32-0: 19342
6626-33-1: 10903
6626-84-2: 20880
6627-55-0: 19108
6627-61-8: 09502
6627-72-1: 07321
6627-74-3: 10391
6627-76-5: 17285
6628-00-8: 09796
6628-04-2: 23528
6628-06-4: 19519
6628-18-8: 11752
6628-21-3: 21145
6628-31-5: 17231
6628-77-9: 22566
6628-79-1: 18455
6628-80-4: 10114
6628-84-8: 05954
6628-86-0: 01622
6628-97-3: 16819
6629-04-5: 09057
6629-80-7: 07173
6629-96-5: 11277
6630-11-1: 12015
6630-33-7: 01584
6630-86-0: 12023
6631-89-6: 22488
6632-68-4: 23118
6635-20-7: 01718
6635-48-9: 14699
6635-57-0: 18282
6635-86-5: 22553
6636-23-3: 20299
6636-78-8: 22971
6639-19-6: 01964
6639-30-1: 04023
6639-32-3: 09066
6639-36-7: 11217
6639-82-3: 23877
6640-25-1: 15891
6640-27-3: 19169
6640-34-2: 07166
6640-77-3: 06465
6641-04-9: 19149
6641-83-4: 18454
6642-26-8: 22213
6646-78-2: 14755
6655-90-9: 16755
6660-65-7: 04595
6665-86-7: 06928
6665-97-0: 03184
6669-13-2: 03221
6670-92-4: 09207
6672-30-6: 10672
6672-64-6: 11204
6676-90-0: 11071
6676-95-5: 19509
6682-06-0: 15033
6682-69-5: 15088
6682-71-9: 15014
6683-92-7: 18573
6688-11-5: 10456
6689-22-1: 11378
6693-29-4: 09792
6693-30-7: 09791
6693-31-8: 09790
6694-75-3: 19181
6696-41-9: 13161
6697-12-7: 13914
6697-15-0: 18047
6698-74-4: 00879
6700-28-3: 11180
6701-26-4: 12405
6702-21-2: 15520
6703-78-2: 13395
6703-79-3: 10510
6703-80-6: 13391
6703-81-7: 10512
6703-82-8: 10538
6703-99-7: 09825
6704-00-3: 09824
6705-31-3: 22371
6705-89-1: 07425
6706-59-8: 13183
6709-22-4: 10734
6709-39-3: 13441
6711-46-2: 11944
6712-35-2: 05132
6712-78-3: 07386
6714-38-1: 09440
6719-35-3: 05695
6723-07-5: 15873
6728-26-3: 14383
6734-98-1: 14415
6735-88-2: 06168
6738-23-4: 03575
6739-34-0: 10025
6740-88-1: 10225
6742-54-7: 04082
6745-75-1: 01746
6748-95-4: 01370
6749-92-4: 07537
6750-03-4: 17179
6750-34-1: 11539
6751-75-3: 04683
6753-98-6: 10860
6755-54-0: 12604
6759-98-4: 22810
6759-99-5: 22799
6760-00-5: 22820

6760-01-6: 22798
6760-02-7: 01891
6760-03-8: 01889
6760-04-9: 01888
6760-05-0: 01887
6760-06-1: 01886
6760-07-2: 01885
6760-08-3: 01884
6760-09-4: 01883
6760-14-1: 22840
6765-39-5: 13429
6776-19-8: 08894
6778-87-6: 19840
6779-09-5: 19824
6780-13-8: 11957
6780-38-7: 15410
6780-49-0: 15778
6781-42-6: 12504
6781-98-2: 02795
6784-18-5: 10677
6785-23-5: 10587
6786-36-3: 17458
6789-80-6: 14388
6789-88-4: 06291
6789-94-2: 20128
6790-27-8: 11567
6790-47-2: 09986
6795-75-1: 12631
6802-75-1: 20858
6803-95-8: 25024
6811-63-8: 17868
6812-38-0: 09882
6812-39-1: 10541
6812-78-8: 17836
6818-07-1: 14258
6824-60-8: 10361
6826-03-5: 14875
6829-40-9: 20793
6829-41-0: 20845
6829-82-9: 05332
6830-82-6: 04113
6831-89-6: 22408
6831-91-0: 22423
6832-93-5: 09023
6832-98-0: 13856
6833-51-8: 00023
6836-19-7: 17013
6836-38-0: 11538
6842-78-0: 07455
6843-49-8: 14866
6843-66-9: 24044
6843-89-6: 16674
6846-11-3: 14865
6848-13-1: 02076
6848-84-6: 18235
6849-18-9: 18709
6850-36-8: 09775
6850-57-3: 04924
6851-44-1: 04010
6851-46-3: 04026
6852-54-6: 11733
6853-14-1: 05051
6854-06-4: 01890
6854-84-8: 00613
6860-63-5: 07613
6861-64-9: 00016
6868-97-9: 08657
6873-15-0: 23501
6874-06-2: 17804
6874-07-3: 17524
6874-09-5: 17774
6874-10-8: 17780
6874-29-9: 17796
6874-31-3: 17627
6874-32-4: 17805
6874-33-5: 13443
6874-34-6: 17526
6874-35-7: 17529
6874-36-8: 17525
6874-41-5: 17799
6874-44-8: 17781
6874-47-1: 13729
6874-76-6: 09424
6874-98-2: 23934
6876-10-4: 09897
6876-23-9: 09840
6876-66-0: 01932
6883-66-5: 07538
6886-16-4: 14274
6887-59-8: 22953
6891-45-8: 19134
6893-26-1: 13244
6894-69-5: 10733
6897-81-0: 16425
6897-88-7: 16560
6901-13-9: 00091
6902-91-6: 09593
6906-32-7: 21658
6907-59-1: 15461
6908-52-7: 07555
6909-81-5: 22000
6911-45-1: 24699
6912-05-6: 10708
6912-07-8: 13860
6913-13-9: 24445
6914-76-7: 10814
6917-76-6: 23994
6920-22-5: 14152
6921-05-7: 20450
6921-17-1: 14847
6921-26-2: 16374
6921-27-3: 22111
6921-28-4: 22099
6921-35-3: 17976
6921-64-8: 12423
6923-20-2: 21696
6924-67-0: 23891
6925-65-1: 20949
6928-51-4: 07138
6928-52-5: 07142
6928-55-8: 06869
6928-56-9: 06870
6928-85-4: 20025
6931-19-7: 23664
6931-71-1: 14170
6932-05-4: 09548
6933-10-4: 02045
6933-57-9: 00343
6935-06-4: 09070
6935-27-9: 22557
6935-65-5: 01965
6935-80-4: 09046
6936-40-9: 03947
6937-74-2: 01217
6938-94-9: 14119
6939-11-3: 23252
6939-30-6: 14814
6939-33-9: 16935
6939-34-0: 16179
6939-75-9: 18469
6939-89-5: 06580
6939-93-1: 04592
6939-97-5: 12725
6940-09-6: 17452
6940-50-7: 04139
6940-76-7: 20525
6940-78-9: 07912
6941-12-4: 21169
6941-69-1: 00509
6942-59-2: 12226
6942-98-9: 23602
6942-99-0: 03034
6943-00-6: 23388
6943-08-4: 23791
6943-20-0: 24863
6943-58-4: 21375

6946-29-8: 05923
6946-35-6: 06444
6947-94-0: 13004
6948-30-7: 01588
6948-59-0: 18950
6949-73-1: 12793
6949-81-1: 22529
6949-98-0: 13903
6950-30-7: 14811
6950-43-2: 06025
6950-79-4: 02376
6950-84-1: 25476
6951-08-2: 05748
6952-59-6: 06672
6952-93-8: 10811
6953-83-9: 12258
6954-48-9: 16753
6954-55-8: 12762
6956-39-4: 06896
6956-82-7: 00496
6957-17-1: 14365
6957-27-3: 15446
6957-50-2: 00073
6957-89-7: 00579
6959-06-4: 08585
6961-41-7: 14818
6961-82-6: 05347
6962-09-0: 04240
6963-03-7: 05718
6963-44-6: 18336
6964-04-1: 13591
6965-67-9: 22668
6965-69-1: 00630
6966-10-5: 05045
6967-12-0: 14963
6967-43-7: 08584
6968-22-5: 05942
6968-28-1: 04544
6969-16-0: 25177
6969-28-4: 22198
6969-49-9: 06315
6969-50-2: 22759
6969-52-4: 22583
6969-90-0: 09069
6970-57-6: 22252
6970-83-8: 09114
6971-38-6: 19504
6971-40-0: 18610
6971-44-4: 22923
6971-51-3: 05093
6972-05-0: 24871
6972-27-6: 23130
6972-37-8: 22845
6972-47-0: 19648
6972-61-8: 21837
6972-71-0: 02213
6973-60-0: 23329
6974-05-6: 00507
6974-10-3: 15408
6974-46-5: 03110
6974-77-2: 20506
6975-60-6: 21553
6975-92-4: 14397
6975-98-0: 10923
6975-99-1: 11568
6976-00-7: 25315
6976-72-3: 14235
6982-25-8: 08219
6986-48-7: 11860
6994-25-8: 22440
6994-40-7: 06762
6994-63-4: 19318
6994-82-7: 14106
6995-20-6: 16426
6995-30-8: 16424
6995-79-5: 10065
6996-44-7: 22821
6996-45-8: 22809
6996-92-5: 05280
6998-82-9: 10041
7000-52-4: 14907
7003-48-7: 17329
7005-47-2: 21326
7005-72-3: 02922
7006-52-2: 02097
7011-98-5: 15352
7012-16-0: 12730
7013-05-0: 08310
7013-11-8: 08729
7018-84-0: 03742
7019-01-4: 02479
7020-84-0: 11933
7021-09-2: 04243
7024-33-1: 11705
7025-19-6: 24554
7028-22-0: 10589
7029-06-3: 17657
7035-03-2: 04332
7036-98-8: 17308
7040-43-9: 12888
7044-91-9: 01180
7045-71-8: 25279
7045-89-8: 25181
7058-01-7: 09924
7065-46-5: 08684
7065-92-1: 23866
7069-38-7: 21695
7073-36-1: 07201
7073-94-1: 02677
7076-23-5: 22740
7081-40-5: 20191
7081-78-9: 11778
7085-93-0: 06271
7086-07-9: 20700
7087-68-5: 20476
7094-26-0: 09970
7094-27-1: 09972
7094-34-0: 15850
7098-07-9: 14775
7098-21-7: 25214
7098-22-8: 24411
7099-25-4: 09518
7099-42-5: 05694
7101-96-4: 23547
7111-77-5: 07593
7116-86-1: 14403
7116-94-1: 09574
7116-95-2: 07527
7119-89-3: 15669
7119-92-8: 11720
7119-94-0: 20192
7132-64-1: 18068
7132-70-9: 00974
7133-36-0: 10560
7137-11-3: 15260
7137-12-4: 15254
7137-13-5: 15261
7137-14-6: 15257
7137-15-7: 15253
7137-16-8: 15255
7137-17-9: 15258
7137-55-5: 02765
7137-56-6: 02979
7138-28-5: 12007
7143-16-0: 15733
7144-05-0: 20264
7144-65-2: 17988
7144-68-5: 12777
7145-20-2: 14417
7145-35-9: 13297
7145-38-2: 00857
7145-49-5: 19418
7145-99-5: 05735
7146-60-3: 17602
7147-89-9: 19182
7148-03-0: 07594

7148-07-4: 23353
7148-74-5: 21597
7149-10-2: 19030
7149-32-8: 12989
7149-63-5: 20107
7149-70-4: 02690
7149-75-9: 02099
7149-79-3: 00162
7151-64-6: 12734
7151-68-0: 06451
7151-75-9: 14916
7152-15-0: 18458
7152-19-4: 09503
7152-42-3: 19685
7152-51-4: 01059
7152-80-9: 24633
7152-88-7: 06367
7153-14-2: 10221
7154-16-7: 13348
7154-66-7: 07193
7154-73-6: 23408
7154-75-8: 18779
7154-79-2: 18269
7154-80-5: 13571
7162-59-6: 01690
7163-15-7: 09302
7163-25-9: 16679
7164-98-9: 14788
7168-56-1: 07721
7169-34-8: 05821
7170-38-9: 21274
7170-44-7: 25298
7187-01-1: 21762
7188-38-7: 20635
7190-64-9: 08494
7191-16-4: 11293
7193-78-4: 12457
7197-93-5: 12951
7200-25-1: 01378
7204-34-4: 20048
7205-90-5: 02785
7207-49-0: 17761
7208-05-1: 17933
7208-47-1: 13186
7208-92-6: 11973
7209-38-3: 20113
7212-09-1: 15256
7212-48-8: 17432
7214-50-8: 10418
7216-27-5: 10026
7216-46-8: 05177
7216-56-0: 17784
7217-83-6: 17428
7218-02-2: 04484
7220-26-0: 14045
7220-56-6: 19691
7220-81-7: 10495
7223-30-5: 22094
7223-38-3: 22097
7225-64-1: 13412
7225-65-2: 01976
7225-66-3: 25150
7225-67-4: 13531
7225-68-5: 10567
7225-69-6: 09940
7225-70-9: 03832
7225-71-0: 25281
7227-96-5: 17982
7231-33-6: 22414
7235-40-7: 09271
7236-47-7: 08704
7239-24-9: 20471
7241-98-7: 13134
7242-17-3: 08816
7244-67-9: 01022
7244-77-1: 03819
7250-53-5: 23741
7250-84-2: 24487
7250-85-3: 08780
7250-95-5: 16158
7251-06-1: 20105
7251-36-7: 23887
7251-61-8: 23881
7252-83-7: 11762
7254-21-9: 20230
7255-28-9: 10080
7255-80-3: 06804
7258-49-3: 10689
7259-71-4: 03287
7283-69-4: 08797
7283-70-7: 08796
7283-96-7: 24794
7285-83-8: 24995
7287-81-2: 05041
7287-82-3: 05040
7290-03-1: 11646
7290-05-3: 11644
7291-33-0: 00390
7291-41-0: 11652
7296-55-1: 01369
7296-56-2: 01371
7297-25-8: 08262
7297-52-1: 02433
7299-89-0: 04535
7300-03-0: 13761
7300-34-7: 20441
7301-19-1: 10489
7305-03-5: 09065
7305-71-7: 24495
7306-96-9: 08522
7307-02-0: 13593
7307-55-3: 25269
7309-43-5: 08021
7310-67-0: 15412
7310-68-1: 15411
7310-94-3: 01608
7311-34-4: 01657
7311-63-9: 24803
7314-06-9: 02253
7314-37-6: 24365
7314-44-5: 05034
7314-85-4: 25113
7315-32-4: 05733
7315-96-0: 16713
7318-00-5: 08854
7318-67-4: 13940
7318-82-3: 07759
7318-94-7: 11239
7319-00-8: 13939
7321-48-4: 17812
7321-55-3: 20773
7325-46-4: 04414
7326-19-4: 05225
7327-60-8: 00454
7327-66-4: 14197
7327-98-2: 10943
7328-17-8: 21846
7328-33-8: 10898
7332-93-6: 18145
7335-02-6: 20195
7335-06-0: 23357
7335-07-1: 23367
7335-12-8: 10195
7335-25-3: 06050
7335-26-4: 06437
7335-27-5: 06062
7336-54-1: 00112
7341-16-4: 14192
7341-17-5: 14193
7341-22-2: 10952
7341-63-1: 24892
7343-34-2: 25052
7344-34-5: 15049
7345-79-1: 21791
7345-82-6: 21836
7350-88-1: 18819

7351-74-8: 16325
7355-22-8: 06005
7357-22-4: 11913
7357-93-9: 08734
7359-72-0: 04020
7361-31-1: 25246
7361-80-0: 17472
7364-19-4: 03229
7364-25-2: 14983
7367-38-6: 17301
7368-62-9: 12215
7369-50-8: 03427
7371-64-4: 09571
7371-67-7: 09560
7372-85-2: 07488
7372-86-3: 16401
7372-87-4: 18848
7372-88-5: 11216
7372-92-1: 15052
7373-13-9: 17456
7375-15-7: 23421
7376-03-6: 16033
7378-37-2: 07277
7378-94-1: 17511
7379-12-6: 14361
7379-18-2: 07584
7379-30-8: 11751
7380-40-7: 13129
7383-11-1: 19032
7383-14-4: 00471
7383-20-2: 17330
7383-61-1: 00671
7383-90-6: 07491
7385-10-6: 24069
7385-78-6: 18634
7385-82-2: 14429
7389-14-2: 23095
7391-72-2: 07497
7396-21-6: 12270
7396-28-3: 22116
7396-92-1: 15918
7397-46-8: 07747
7398-82-5: 02560
7399-49-7: 03690
7399-50-0: 22661
7400-08-0: 21869
7400-39-7: 09250
7402-93-9: 19843
7403-42-1: 18571
7403-66-9: 00127
7403-75-0: 00271
7409-09-8: 06272
7411-49-6: 07683
7413-09-4: 19814
7415-31-8: 08770
7417-18-7: 04850
7417-19-8: 04837
7417-48-3: 18092
7418-32-8: 12833
7422-78-8: 14654
7423-69-0: 14400
7424-54-6: 13595
7425-27-6: 08439
7425-45-8: 08347
7425-48-1: 08351
7425-60-7: 17242
7428-85-5: 15073
7428-91-3: 11205
7431-95-0: 17501
7432-10-2: 01490
7433-56-9: 11004
7433-78-5: 11005
7434-96-0: 12352
7436-02-4: 10697
7436-07-9: 15022
7436-90-0: 03012
7439-33-0: 09569
7442-07-1: 07070
7442-22-0: 22903
7446-43-7: 20457
7450-61-5: 09113
7454-47-9: 05351
7455-09-6: 21137
7455-58-5: 21399
7459-46-3: 12080
7460-30-2: 25035
7460-54-0: 17931
7460-90-4: 07532
7461-41-8: 01841
7461-79-2: 12813
7461-93-0: 21231
7462-74-0: 20408
7463-35-6: 00161
7463-51-6: 19103
7464-48-4: 06445
7466-14-0: 23160
7466-38-8: 11063
7467-35-8: 05641
7468-01-1: 15606
7469-77-4: 16928
7469-83-2: 10018
7470-14-6: 18909
7473-93-0: 22031
7474-10-4: 10432
7474-78-4: 05433
7474-83-1: 16751
7474-84-2: 16765
7474-86-4: 16742
7474-90-0: 16820
7474-97-7: 16635
7475-11-8: 01322
7475-29-8: 01181
7475-39-0: 20847
7475-56-1: 04159
7477-57-8: 16764
7477-73-8: 24428
7479-05-2: 25231
7480-80-0: 15050
7488-99-5: 09275
7493-58-5: 18350
7493-69-8: 08410
7493-71-2: 08906
7493-82-5: 13636
7494-67-9: 07440
7494-95-3: 04140
7495-46-7: 07546
7496-02-8: 09410
7496-10-8: 01979
7498-52-4: 12532
7498-54-6: 06227
7498-57-9: 16576
7498-66-0: 15890
7498-76-2: 10726
7498-87-5: 17016
7499-06-1: 06102
7499-07-2: 06100
7499-09-4: 17855
7499-32-3: 06943
7499-61-8: 07429
7499-65-2: 16259
7499-66-3: 16204
7499-80-1: 18955
7500-45-0: 04117
7500-53-0: 04412
7500-55-2: 10386
7500-78-9: 01818
7501-02-2: 12102
7501-11-3: 02778
7501-30-6: 24465
7501-32-8: 12809
7504-86-1: 09102
7505-02-4: 13053
7505-62-6: 02004
7506-37-8: 15387
7506-82-3: 19808
7508-74-9: 03283

7509-22-0: 08949
7509-96-8: 14812
7510-33-0: 02238
7510-49-8: 05972
7511-32-2: 06526
7515-29-9: 09578
7521-41-7: 22791
7521-80-4: 23982
7522-79-4: 05797
7523-15-1: 09199
7523-44-6: 08934
7524-50-7: 19726
7524-63-2: 16513
7525-62-4: 03349
7526-26-3: 19836
7529-63-7: 20203
7533-40-6: 18482
7535-15-1: 07618
7539-25-5: 01117
7540-51-4: 17832
7548-68-7: 23213
7550-39-2: 17925
7552-07-0: 24444
7554-12-3: 08169
7554-65-6: 22417
7555-72-8: 15911
7556-79-8: 00804
7556-83-4: 00839
7558-47-6: 05455
7559-04-8: 09732
7560-65-8: 10304
7561-04-8: 00573
7562-61-0: 11146
7564-63-8: 03348
7566-42-9: 07652
7567-22-8: 10462
7568-92-5: 04820
7568-93-6: 04976
7570-26-5: 11837
7570-45-8: 09179
7570-96-9: 16492
7570-97-0: 16489
7570-98-1: 16491
7572-29-4: 12640
7576-88-7: 23855
7578-63-4: 12689
7578-64-5: 12690
7578-68-9: 06868
7578-71-4: 24658
7578-72-5: 24656
7580-85-0: 12145
7595-53-1: 12808
7597-18-4: 14979
7597-91-3: 00461
7598-80-3: 07661
7611-43-0: 01467
7611-86-1: 08715
7611-87-2: 08714
7612-96-6: 11065
7613-08-3: 00967
7613-09-4: 15484
7613-10-7: 01153
7614-93-9: 03750
7615-57-8: 25029
7616-22-0: 06815
7619-08-1: 17365
7620-45-3: 15485
7623-01-0: 10848
7623-11-2: 08679
7623-13-4: 21604
7626-68-8: 14594
7634-42-6: 08225
7634-96-0: 21215
7635-11-2: 01033
7635-51-0: 16095
7642-04-8: 17801
7642-09-3: 14433
7642-10-6: 13756
7642-15-1: 17811
7643-75-6: 01365
7647-95-2: 00836
7648-01-3: 24568
7649-92-5: 06277
7651-83-4: 15474
7651-86-7: 18945
7657-09-2: 03033
7658-08-4: 13236
7659-45-2: 20973
7659-86-1: 00664
7661-39-4: 23573
7661-47-4: 23649
7661-55-4: 23675
7664-80-4: 00672
7665-72-7: 18003
7665-85-2: 09369
7667-55-2: 09976
7667-58-5: 09979
7667-59-6: 09978
7667-60-9: 09980
7667-80-3: 13843
7669-54-7: 05285
7672-76-6: 20110
7674-36-4: 22190
7681-84-7: 12955
7686-77-3: 10730
7688-21-3: 14414
7693-46-1: 09233
7693-52-9: 19123
7693-77-8: 17967
7694-45-3: 10374
7696-51-7: 23282
7697-25-8: 06095
7697-29-2: 06101
7697-66-7: 15435
7698-91-1: 23941
7698-94-4: 23942
7699-00-5: 21185
7705-14-8: 10332
7711-32-2: 18894
7712-60-9: 18664
7712-66-5: 24909
7713-69-1: 07893
7720-42-5: 16625
7724-48-3: 23434
7731-28-4: 10154
7731-29-5: 10155
7734-05-6: 12514
7735-33-3: 18413
7738-93-4: 01044
7745-40-6: 01096
7745-91-7: 02043
7745-93-9: 02697
7747-84-4: 22102
7755-92-2: 20099
7756-96-9: 05865
7761-80-0: 09754
7766-23-6: 19110
7768-28-7: 04847
7770-16-3: 18839
7771-44-0: 11598
7773-60-6: 19367
7774-74-5: 24826
7778-87-2: 13637
7779-27-3: 24990
7779-75-1: 08493
7779-77-3: 05870
7779-80-8: 13633
7781-98-8: 06325
7782-21-0: 04294
7782-24-3: 04257
7782-25-4: 04295
7782-26-5: 04256
7785-26-4: 07370
7786-44-9: 17185
7786-67-6: 10165
7787-20-4: 07348

7787-72-6: 16312
7789-57-3: 24093
7789-89-1: 20716
7789-90-4: 20402
7789-92-6: 20728
7799-56-6: 20564
7803-51-2: 19731
7803-62-5: 23973
8003-05-2: 15597
10008-73-8: 13089
10016-52-1: 11132
10017-37-5: 20481
10020-96-9: 04327
10021-92-8: 10042
10024-74-5: 04941
10024-89-2: 16104
10025-78-2: 24097
10026-78-5: 12202
10029-09-1: 10049
10030-62-3: 13159
10030-80-5: 15583
10030-92-9: 22814
10031-82-0: 01672
10031-87-5: 00620
10033-99-5: 07783
10038-40-1: 05139
10039-63-1: 12058
10042-29-2: 00822
10042-59-8: 13649
10047-18-4: 12353
10047-28-6: 00657
10048-32-5: 22275
10059-13-9: 25284
10060-33-0: 16724
10061-01-5: 21698
10061-02-6: 21697
10071-60-0: 18659
10075-38-4: 08771
10075-50-0: 15122
10075-63-5: 16352
10075-73-7: 16249
10088-95-6: 22296
10089-43-7: 19767
10098-89-2: 15549
10099-71-5: 08815
10108-56-2: 09767
10124-73-9: 14391
10126-69-9: 11591
10130-87-7: 05512
10130-89-9: 06124
10130-91-3: 10211
10133-34-3: 07117
10136-52-4: 13432
10138-10-0: 08498
10138-17-7: 11276
10138-89-3: 08075
10140-89-3: 20386
10141-22-7: 20387
10143-32-5: 21397
10143-54-1: 20984
10146-44-8: 01230
10146-67-5: 01813
10147-36-1: 21006
10151-95-8: 00346
10152-76-8: 21726
10157-76-3: 05462
10160-99-3: 13852
10166-05-9: 15463
10166-08-2: 04091
10169-09-2: 09755
10172-89-1: 19718
10173-84-9: 05830
10179-91-6: 22158
10182-48-6: 22909
10193-98-3: 00668
10193-99-4: 00656
10203-08-4: 01632
10203-28-8: 11536
10203-30-2: 11537
10203-48-2: 07085
10203-58-4: 20861
10203-59-5: 20831
10207-58-6: 11258
10210-32-9: 12271
10210-64-7: 07251
10212-74-5: 09094
10215-25-5: 24899
10220-20-9: 20159
10220-33-4: 04390
10220-83-4: 10076
10222-95-4: 04060
10227-50-6: 14888
10229-10-4: 18791
10230-68-9: 21572
10233-13-3: 11530
10236-47-2: 06878
10240-08-1: 16931
10240-10-5: 16939
10250-45-0: 13684
10253-28-8: 01182
10253-83-5: 24464
10254-28-1: 11165
10254-29-2: 23581
10254-30-5: 23762
10257-92-8: 15161
10259-22-0: 06441
10264-05-8: 18160
10268-65-2: 06092
10268-69-6: 05871
10268-70-9: 05884
10268-71-0: 06439
10268-82-3: 09558
10269-01-9: 04884
10271-73-5: 00147
10273-87-7: 07553
10273-90-2: 22697
10276-09-2: 08921
10276-30-9: 25011
10283-15-5: 19146
10287-53-3: 06220
10289-68-6: 25164
10290-37-6: 25413
10290-61-6: 16055
10292-98-5: 07358
10293-00-2: 07336
10293-06-8: 07335
10297-05-9: 07936
10297-09-3: 17856
10297-38-8: 24286
10298-80-3: 02863
10302-94-0: 08825
10303-64-7: 18432
10303-88-5: 21161
10303-96-5: 12814
10307-14-9: 07897
10307-15-0: 08085
10309-50-9: 07346
10311-74-7: 03325
10312-55-7: 04613
10312-83-1: 00042
10313-82-3: 15443
10315-89-6: 20189
10316-00-4: 16118
10316-79-7: 10636
10317-10-9: 21312
10321-71-8: 18702
10323-39-4: 19106
10323-40-7: 08335
10324-58-0: 20197
10324-64-8: 14386
10325-40-3: 08574
10325-41-4: 08575
10326-41-7: 21179
10327-01-2: 05168
10327-08-9: 21074
10328-92-4: 07172

10330-21-9: 24278
10330-31-1: 17751
10334-13-1: 07325
10334-26-6: 07297
10339-67-0: 13816
10340-91-7: 03566
10342-50-4: 25207
10342-83-3: 21444
10343-99-4: 23606
10344-62-4: 01478
10349-57-2: 23742
10352-60-0: 23785
10353-44-3: 07011
10353-86-3: 21402
10354-00-4: 11108
10355-53-0: 24319
10359-65-6: 20658
10366-35-5: 22804
10368-43-1: 04828
10373-81-6: 07343
10374-74-0: 24391
10375-96-9: 03199
10388-10-0: 02777
10394-35-1: 05614
10394-64-6: 02307
10394-96-4: 20775
10395-45-6: 10428
10403-18-6: 13204
10403-47-1: 02048
10403-60-8: 08070
10403-73-3: 02577
10403-82-4: 24280
10403-85-7: 24281
10404-51-0: 17082
10404-52-1: 17083
10405-51-3: 24923
10405-52-4: 24922
10405-54-6: 00754
10405-85-3: 17300
10408-16-9: 07398
10410-35-2: 05791
10412-98-3: 14188
10413-81-7: 12086
10420-90-3: 13968
10420-91-4: 13972
10424-29-0: 08196
10424-38-1: 20748
10425-05-5: 15949
10425-83-9: 15094
10425-90-8: 04058
10428-64-5: 07956
10435-55-9: 06265
10435-81-1: 17664
10435-93-5: 17665
10436-75-6: 02374
10439-23-3: 05299
10443-65-9: 21783
10447-93-5: 14771
10460-33-0: 19613
10463-48-6: 08727
10468-17-4: 02085
10468-59-4: 17020
10468-60-7: 17021
10468-61-8: 17022
10470-04-9: 02967
10471-14-4: 11838
10473-14-0: 08943
10473-16-2: 04406
10476-95-6: 21753
10477-40-4: 02816
10477-99-3: 06240
10478-00-9: 06245
10478-01-0: 06241
10478-02-1: 06236
10478-03-2: 06242
10478-07-6: 06244
10478-12-3: 18697
10479-25-1: 04917
10479-30-8: 00099
10484-09-0: 06322
10486-19-8: 25141
10486-58-5: 06972
10486-61-0: 24746
10487-71-5: 08979
10487-96-4: 10653
10488-25-2: 10158
10488-94-5: 17691
10489-97-1: 09827
10493-44-4: 08713
10495-45-1: 20688
10498-35-8: 09834
10500-57-9: 23704
10503-96-5: 07934
10504-35-5: 01451
10508-09-5: 18808
10512-65-9: 07010
10513-96-9: 15405
10519-06-9: 02755
10519-20-7: 17857
10519-33-2: 11014
10519-87-6: 24029
10521-91-2: 05164
10522-26-6: 25303
10523-68-9: 25110
10524-08-0: 18627
10527-16-9: 01166
10528-67-3: 10097
10537-10-7: 16166
10539-14-7: 19001
10541-56-7: 12495
10541-83-0: 06480
10542-47-9: 21665
10544-63-5: 08892
10546-24-4: 16214
10548-77-3: 21472
10556-98-6: 24986
10557-71-8: 24005
10558-21-1: 24279
10560-13-1: 04610
10563-23-2: 20762
10567-22-3: 20914
10568-38-4: 03419
10573-35-0: 13518
10574-17-1: 03761
10574-36-4: 14426
10574-37-5: 18666
10575-43-6: 13452
10575-87-8: 13503
10577-44-3: 03618
10588-10-0: 18459
10588-31-5: 00926
10590-73-5: 15201
10595-51-4: 02401
10595-72-9: 21287
10601-19-1: 15191
10602-34-3: 08779
10604-59-8: 15146
10606-38-9: 03684
10606-72-1: 04193
10606-73-2: 04149
11018-89-6: 09260
11018-93-2: 24438
11024-24-1: 13162
11028-42-5: 09291
11040-28-1: 24252
12078-25-0: 09461
12082-08-5: 09393
12091-58-6: 12682
12093-10-6: 12676
12114-61-3: 12663
12122-67-7: 25583
12129-67-8: 09394
12180-80-2: 12669
12243-60-6: 12656
12277-09-7: 12685
12277-39-3: 12684

12288-75-4: 12672
12288-76-5: 12649
12289-46-2: 12657
12320-47-7: 12654
12320-65-9: 12651
12321-03-8: 12650
12321-04-9: 12664
12321-37-8: 12652
12552-53-3: 24915
13012-59-4: 20534
13014-18-1: 03121
13014-24-9: 03120
13017-60-2: 15567
13019-20-0: 13715
13019-22-2: 11013
13020-57-0: 15993
13021-21-1: 14922
13021-24-4: 14921
13021-27-7: 14928
13021-29-9: 13304
13021-30-2: 13305
13021-31-3: 23458
13022-02-1: 22226
13023-00-2: 08371
13023-07-9: 08326
13024-46-9: 23456
13024-49-2: 02466
13024-80-1: 03503
13027-43-5: 03372
13029-08-8: 07467
13029-09-9: 07465
13029-23-7: 16664
13029-24-8: 16167
13030-26-7: 06782
13031-32-8: 15723
13031-43-1: 12256
13036-81-2: 24985
13036-91-4: 09137
13037-56-4: 24586
13037-71-3: 19126
13037-86-0: 19377
13038-21-6: 17307
13041-70-8: 02716
13041-79-7: 06732
13045-99-3: 08782
13046-50-9: 12595
13047-04-6: 04643
13047-06-8: 15876
13047-12-6: 22461
13048-33-4: 21859
13051-89-3: 22370
13053-81-1: 00379
13054-87-0: 08525
13057-17-5: 15635
13057-19-7: 15680
13057-72-2: 06929
13058-16-7: 25228
13063-43-9: 20772
13065-07-1: 16514
13067-94-2: 23597
13070-25-2: 09701
13073-25-1: 19120
13073-29-5: 19498
13073-35-3: 02566
13076-17-0: 11283
13079-20-4: 18150
13080-75-6: 11096
13080-90-5: 07387
13081-17-9: 19144
13081-67-9: 24035
13082-21-8: 23464
13084-22-5: 24540
13086-84-5: 19802
13091-23-1: 22608
13091-80-0: 07149
13095-47-1: 18309
13095-48-2: 18310
13096-31-6: 20245
13096-96-3: 14967
13097-10-4: 24582
13099-50-8: 02935
13102-31-3: 19562
13104-55-7: 21259
13105-10-7: 12542
13105-52-7: 09048
13107-13-6: 24056
13108-19-5: 10956
13112-64-6: 18612
13113-71-8: 04213
13114-62-0: 25477
13114-90-4: 25376
13115-28-1: 16224
13116-53-5: 20698
13116-60-4: 20696
13122-33-3: 04405
13122-67-3: 04877
13123-92-7: 14605
13125-68-3: 02708
13129-23-2: 13003
13129-69-6: 15218
13133-20-5: 07713
13133-29-4: 21333
13134-55-9: 22341
13136-33-9: 25069
13138-48-2: 07437
13138-51-7: 08054
13140-83-5: 25462
13142-86-4: 25439
13145-46-5: 24413
13147-06-3: 24395
13148-24-8: 16621
13148-65-7: 17899
13150-76-0: 24283
13151-04-7: 13741
13151-05-8: 13740
13151-17-2: 14430
13151-34-3: 10924
13151-35-4: 10925
13151-55-8: 09629
13151-92-3: 25145
13152-94-8: 16013
13152-95-9: 15954
13158-31-1: 04525
13161-30-3: 22968
13163-73-0: 10265
13165-77-0: 05293
13165-79-2: 05295
13167-36-7: 21823
13171-00-1: 00845
13171-63-6: 08113
13172-91-3: 13763
13176-46-0: 08342
13183-68-1: 20736
13190-77-7: 10622
13190-97-1: 14375
13194-68-8: 02305
13194-69-9: 02302
13195-79-4: 12339
13195-80-7: 21688
13201-16-6: 21257
13205-48-6: 06509
13205-58-8: 17196
13206-46-7: 20379
13206-60-5: 19571
13206-67-2: 25390
13207-43-7: 14675
13207-45-9: 14665
13207-47-1: 14666
13207-66-4: 23805
13209-15-9: 02557
13209-38-6: 04982
13211-15-9: 07340
13214-66-9: 04358
13214-70-5: 16328
13220-80-9: 00659
13222-26-9: 21601

13222-85-0: 06470
13228-40-5: 15170
13228-41-6: 15169
13231-00-0: 22212
13236-39-0: 00858
13238-84-1: 22321
13241-33-3: 06876
13243-65-7: 16782
13246-14-5: 12403
13246-52-1: 14222
13248-54-9: 09223
13249-58-6: 02633
13250-12-9: 07887
13250-46-9: 00945
13250-95-8: 20863
13252-27-2: 09151
13253-91-3: 01463
13254-34-7: 13657
13256-21-8: 24670
13257-48-2: 18358
13257-95-9: 05296
13258-63-4: 22913
13268-79-6: 11641
13269-19-7: 22569
13269-35-7: 04218
13269-52-8: 14432
13275-68-8: 24447
13278-67-6: 14674
13279-87-3: 08067
13281-50-0: 13138
13284-42-9: 18691
13286-73-2: 25148
13286-90-3: 08026
13286-92-5: 08024
13286-95-8: 03277
13289-29-7: 09149
13290-74-9: 02893
13290-96-5: 04611
13292-05-2: 18927
13296-65-6: 02625
13296-94-1: 02047
13299-20-2: 13196
13302-98-2: 19488
13306-60-0: 16121
13307-61-4: 03780
13307-67-0: 15196
13308-61-7: 17512
13311-76-7: 13778
13312-42-0: 06950
13313-91-2: 00777
13314-96-0: 15399
13323-74-5: 24221
13323-81-4: 05105
13324-66-8: 04898
13325-10-5: 08529
13326-06-2: 05259
13327-31-6: 23661
13329-40-3: 12446
13329-78-7: 10947
13331-19-6: 04731
13331-27-6: 07764
13336-52-2: 19782
13343-98-1: 08013
13344-00-8: 07979
13345-26-1: 06953
13347-42-7: 19145
13351-61-6: 05267
13351-73-0: 07153
13360-52-6: 13206
13360-57-1: 24275
13360-61-7: 18079
13360-63-9: 07875
13360-64-0: 22339
13360-65-1: 22337
13361-35-8: 00518
13361-36-9: 00513
13361-63-2: 18774
13362-36-2: 09703
13362-80-6: 23640
13364-16-4: 18167
13364-95-9: 00990
13367-26-5: 07419
13367-29-8: 07418
13368-65-5: 10253
13372-74-2: 10906
13379-22-1: 16821
13388-93-7: 12459
13389-42-9: 17800
13389-87-2: 13140
13389-98-5: 12995
13390-02-8: 12958
13393-60-7: 10994
13395-85-2: 19160
13398-94-2: 04848
13401-56-4: 21310
13402-32-9: 03025
13402-35-2: 10465
13402-36-3: 10467
13402-37-4: 10463
13403-37-7: 17996
13404-51-8: 25492
13408-09-8: 18323
13410-61-2: 19876
13414-95-4: 07088
13417-43-1: 08762
13418-77-4: 23027
13420-87-6: 11196
13420-88-7: 11193
13423-15-9: 12930
13426-55-6: 07388
13429-07-7: 21408
13432-25-2: 14310
13436-48-1: 14972
13438-26-1: 04456
13440-22-7: 05357
13440-24-9: 03009
13440-25-0: 03278
13442-89-2: 18238
13444-02-5: 11080
13444-24-1: 20279
13450-77-6: 13296
13465-77-5: 11360
13466-38-1: 22979
13466-40-5: 18086
13466-41-6: 22992
13466-98-3: 15390
13469-52-8: 19715
13471-42-6: 14304
13472-80-5: 23007
13475-75-7: 18063
13475-76-8: 11446
13475-77-9: 11584
13475-78-0: 13529
13475-79-1: 18234
13475-81-5: 14077
13475-82-6: 13554
13477-53-7: 08308
13478-91-6: 19036
13481-70-4: 19236
13482-62-7: 19754
13483-03-9: 24491
13487-99-5: 09583
13491-79-7: 10131
13492-22-3: 09358
13493-47-5: 00252
13494-10-5: 12359
13501-73-0: 14275
13505-34-5: 13594
13505-39-0: 08616
13506-76-8: 06495
13508-96-8: 22687
13511-38-1: 21111
13511-83-6: 18941
13512-31-7: 25262
13513-82-1: 05098
13518-40-6: 25536

13519-74-9: 02081
13524-04-4: 05012
13529-27-6: 12878
13531-52-7: 20752
13532-18-8: 21256
13532-52-0: 20432
13534-97-9: 22559
13534-99-1: 22534
13545-04-5: 08153
13545-65-8: 16607
13552-21-1: 08568
13552-31-3: 20893
13556-55-3: 16510
13556-58-6: 16405
13556-81-5: 11038
13557-98-7: 21291
13562-76-0: 08492
13565-36-1: 03500
13577-19-0: 16638
13602-12-5: 22875
13602-13-6: 08769
13602-84-1: 22901
13603-04-8: 23355
13603-21-9: 20140
13604-13-2: 07058
13605-19-1: 12204
13615-35-5: 00313
13615-40-2: 16930
13618-83-2: 09832
13618-93-4: 15239
13620-57-0: 11037
13629-73-7: 03623
13629-82-8: 07576
13631-45-3: 05430
13632-93-4: 03145
13632-94-5: 03149
13632-95-6: 03147
13636-53-8: 14617
13639-16-2: 12059
13640-28-3: 13078
13643-20-4: 20469
13658-56-5: 16264
13659-98-8: 04778
13665-04-8: 12343
13665-98-0: 02445
13669-10-8: 24820
13669-42-6: 23732
13669-51-7: 23764
13672-18-9: 16233
13673-30-8: 22580
13673-53-5: 15422
13673-92-2: 04651
13677-32-2: 12861
13678-54-1: 24735
13678-59-6: 12912
13678-68-7: 12061
13679-41-9: 12921
13679-46-4: 12906
13679-70-4: 24796
13679-74-8: 12480
13679-75-9: 21527
13688-75-0: 24110
13693-59-9: 19622
13698-16-3: 09059
13702-35-7: 04103
13706-86-0: 14177
13708-12-8: 23882
13711-39-2: 16156
13720-01-9: 18881
13720-06-4: 16329
13720-94-0: 23750
13721-54-5: 14525
13722-96-8: 23912
13728-58-0: 02001
13730-55-7: 06209
13732-80-4: 03146
13744-68-8: 05322
13744-81-5: 02235
13748-90-8: 18434
13749-37-6: 20515
13750-81-7: 14807
13750-95-3: 24941
13750-96-4: 17034
13752-83-5: 01366
13754-19-3: 23109
13755-14-1: 03978
13757-91-0: 13700
13794-14-4: 08505
13799-90-1: 04621
13805-40-8: 25083
13807-91-5: 21419
13808-64-5: 22391
13810-83-8: 04580
13811-01-3: 09682
13811-71-7: 08144
13820-09-2: 18275
13821-77-7: 10437
13828-34-7: 09914
13831-30-6: 00448
13835-81-9: 24109
13837-48-4: 18872
13847-93-3: 17036
13851-11-1: 07311
13854-85-8: 07357
13855-77-1: 09099
13856-85-4: 05009
13857-12-0: 14942
13858-67-8: 18360
13860-69-0: 14943
13865-19-5: 08499
13865-48-0: 03745
13877-91-3: 17775
13878-44-9: 09467
13886-04-9: 00221
13887-55-3: 00533
13893-04-4: 24846
13893-05-5: 24845
13894-60-5: 14472
13898-58-3: 05975
13898-73-2: 10336
13900-80-6: 09584
13905-48-1: 09806
13907-59-0: 25486
13907-60-3: 25498
13907-61-4: 05981
13907-92-1: 25364
13907-99-8: 25487
13908-00-4: 25397
13908-05-9: 13259
13908-16-2: 25399
13908-34-4: 25383
13908-36-6: 25356
13908-44-6: 06076
13908-45-7: 06078
13908-56-0: 25396
13908-57-1: 25389
13908-59-3: 25395
13908-62-8: 25408
13908-63-9: 25409
13908-65-1: 25485
13908-71-9: 25461
13908-79-7: 25382
13909-02-9: 25387
13909-05-2: 10595
13909-25-6: 06077
13909-28-9: 01840
13909-29-0: 00720
13909-73-4: 12530
13916-39-7: 00181
13917-06-1: 17110
13920-94-0: 11396
13921-90-9: 05203
13922-41-3: 07763
13925-00-3: 22336
13936-99-7: 19465
13937-08-1: 20844

13943-70-9: 15092
13949-77-4: 09699
13956-29-1: 04715
13963-35-4: 03744
13974-80-6: 12716
13979-28-7: 08889
13988-38-0: 10761
13992-75-1: 23381
13993-65-2: 19690
13997-70-1: 02939
13997-73-4: 19194
14002-51-8: 07588
14003-34-0: 23906
14003-66-8: 14782
14007-64-8: 04178
14008-30-1: 21355
14010-23-2: 13416
14020-52-1: 19459
14021-92-2: 10945
14024-17-0: 15300
14024-18-1: 15303
14024-48-7: 09460
14024-56-7: 15554
14024-63-6: 25581
14026-63-2: 01539
14026-69-8: 01502
14027-78-2: 14138
14035-94-0: 18314
14039-65-7: 16800
14041-01-1: 12308
14045-23-9: 22941
14045-26-2: 20172
14062-18-1: 04248
14062-19-2: 04264
14062-78-3: 01966
14064-10-9: 20805
14064-35-8: 06731
14064-41-6: 03614
14064-43-8: 02926
14064-45-0: 02715
14064-67-6: 02927
14064-68-7: 06733
14064-72-3: 02219
14064-82-5: 02462
14068-53-2: 24446
14070-37-2: 12799
14070-68-9: 10799
14072-86-7: 10313
14073-97-3: 10270
14077-58-8: 00936
14085-33-7: 10494
14085-34-8: 05764
14090-77-8: 12022
14090-87-0: 17651
14090-88-1: 17235
14092-20-7: 14528
14108-88-4: 16518
14116-69-9: 13234
14121-36-9: 22744
14121-97-2: 01916
14122-00-0: 19216
14122-18-0: 13168
14123-52-5: 23369
14131-07-8: 12217
14131-63-6: 13203
14135-28-5: 17035
14140-02-4: 16798
14140-03-5: 16760
14142-16-6: 20188
14152-56-8: 23889
14155-25-0: 13169
14159-48-9: 21005
14160-70-4: 23930
14161-11-6: 22525
14163-82-7: 19886
14166-21-3: 15318
14167-59-0: 24415
14168-83-3: 01115
14178-30-4: 16035
14179-94-3: 18772
14180-63-3: 21735
14188-79-5: 16317
14191-92-5: 01085
14191-95-8: 04331
14192-12-2: 06186
14200-76-1: 13235
14205-39-1: 08856
14210-25-4: 24426
14210-97-0: 04747
14214-31-4: 21682
14216-34-3: 06450
14222-60-7: 22784
14227-95-3: 23352
14228-16-1: 20244
14230-00-3: 24604
14236-12-5: 11278
14248-22-7: 05216
14250-79-4: 13641
14252-23-4: 23698
14252-42-7: 11748
14252-45-0: 20495
14255-23-3: 14418
14255-44-8: 14073
14257-69-3: 13202
14268-54-3: 21755
14268-66-7: 05725
14272-54-9: 18755
14273-06-4: 22105
14273-85-9: 08440
14274-07-8: 08914
14276-95-0: 15030
14282-76-9: 24680
14284-93-6: 23924
14289-73-7: 03905
14290-23-4: 16138
14290-92-7: 20608
14295-64-8: 21249
14296-80-1: 09521
14301-11-2: 02224
14301-34-9: 00392
14303-70-9: 24381
14304-27-9: 18757
14304-30-4: 11002
14315-11-8: 07109
14315-14-1: 07110
14316-61-1: 00725
14320-58-2: 04272
14321-25-6: 00893
14321-27-8: 04915
14331-54-5: 23065
14337-82-7: 19775
14338-36-4: 04124
14348-38-0: 06014
14348-41-5: 06012
14352-61-5: 09992
14362-31-3: 20044
14371-10-9: 21634
14376-79-5: 10283
14379-80-7: 13319
14381-66-9: 01135
14397-64-9: 15069
14398-36-8: 08978
14400-94-3: 19603
14401-72-0: 12348
14401-73-1: 12342
14401-75-3: 12385
14405-49-3: 15302
14411-56-4: 03228
14411-75-7: 03247
14415-44-2: 06823
14426-20-1: 12141
14428-98-9: 04948
14436-32-9: 11010
14442-48-9: 18695
14444-77-0: 03136
14446-67-4: 20210

14447-18-8: 00554
14452-30-3: 12445
14452-34-7: 19315
14453-59-9: 16922
14459-29-1: 20310
14464-29-0: 23383
14472-14-1: 19114
14473-55-3: 24369
14473-90-6: 21809
14473-91-7: 21792
14474-22-7: 06551
14476-01-8: 16460
14476-30-3: 20712
14476-37-0: 25312
14478-14-9: 21979
14490-79-0: 24341
14492-62-7: 15591
14495-51-3: 02599
14526-22-8: 15304
14527-41-4: 24530
14529-53-4: 22646
14538-09-1: 22040
14544-96-8: 12749
14545-08-5: 12644
14547-69-4: 22818
14547-70-7: 22816
14547-71-8: 22803
14547-74-1: 01936
14547-75-2: 01924
14547-76-3: 01849
14547-77-4: 01827
14547-79-6: 01848
14548-30-2: 16022
14548-31-3: 14527
14548-42-6: 22023
14548-44-8: 15903
14548-45-9: 15879
14548-46-0: 16037
14548-47-1: 16005
14548-48-2: 15904
14548-49-3: 15880
14548-50-6: 23832
14548-51-7: 09256
14548-52-8: 09257
14548-56-2: 22825
14554-09-7: 16793
14558-14-6: 16541
14561-19-4: 19779
14570-44-6: 24638
14572-81-7: 03332
14575-59-8: 25056
14575-61-2: 23389
14579-91-0: 23966
14590-54-6: 10830
14593-43-2: 03902
14595-35-8: 08806
14596-92-0: 13937
14602-62-1: 20711
14609-79-1: 18531
14620-52-1: 11492
14620-53-2: 24356
14621-02-4: 22824
14621-03-5: 22806
14623-58-6: 23164
14625-61-7: 05625
14630-40-1: 24068
14631-45-9: 20979
14633-82-0: 19770
14636-23-8: 24637
14641-93-1: 13194
14642-47-8: 18383
14642-48-9: 18386
14660-50-5: 22944
14660-52-7: 18405
14667-55-1: 22358
14679-13-1: 04059
14685-29-1: 10846
14686-13-6: 13744
14686-14-7: 13755
14689-97-5: 11840
14691-35-1: 19341
14697-46-2: 18384
14698-30-7: 25079
14703-48-1: 21423
14703-76-5: 19283
14703-77-6: 19284
14703-80-1: 19286
14703-86-7: 19446
14705-46-5: 19865
14711-84-3: 00753
14712-23-3: 20263
14714-50-2: 04329
14716-89-3: 15499
14718-07-1: 18951
14719-83-6: 06114
14720-74-2: 13559
14731-25-0: 24462
14737-89-4: 21839
14737-91-8: 21878
14737-94-1: 21805
14739-72-1: 13390
14746-98-6: 14842
14746-99-7: 14838
14752-60-4: 18770
14752-61-5: 18777
14752-75-1: 03494
14755-02-3: 21868
14755-16-9: 19285
14758-47-5: 03572
14759-74-1: 16917
14770-96-8: 15977
14779-16-9: 07080
14779-17-0: 07043
14783-68-7: 15553
14804-32-1: 03418
14804-59-2: 21616
14805-64-2: 16609
14811-92-8: 17404
14811-95-1: 11575
14812-59-0: 11291
14813-68-4: 17555
14814-75-6: 04460
14845-35-3: 10412
14845-55-7: 10674
14847-51-9: 19109
14850-22-7: 17808
14850-23-8: 17810
14852-31-4: 13905
14869-39-7: 12389
14898-79-4: 08567
14898-86-3: 05120
14905-79-4: 03779
14906-37-7: 22870
14916-79-1: 13839
14916-80-4: 17860
14919-01-8: 17807
14920-89-9: 12884
14924-53-9: 09553
14927-67-4: 22503
14938-35-3: 19553
14944-13-9: 15818
14944-26-4: 17017
14949-01-0: 00085
14963-40-7: 10755
14965-34-5: 12183
14969-33-6: 07720
14976-57-9: 23351
14979-39-6: 13679
14996-78-2: 09651
15009-91-3: 22705
15011-25-3: 05382
15012-61-0: 16803
15013-37-3: 08023
15014-25-2: 20799
15015-46-0: 05082
15018-62-9: 23096

15024-10-9: 06065
15028-10-1: 15057
15038-67-2: 19601
15045-43-9: 12932
15058-72-7: 12998
15068-90-3: 07333
15069-48-4: 14960
15070-88-9: 05268
15071-36-0: 08542
15076-00-3: 10012
15103-48-7: 22936
15104-61-7: 20682
15111-56-5: 10375
15115-58-9: 05208
15116-17-3: 04712
15119-34-3: 06792
15121-84-3: 04861
15128-82-2: 22974
15131-55-2: 10719
15132-01-1: 22407
15148-80-8: 21371
15150-39-7: 00516
15164-44-0: 01739
15173-56-5: 01052
15174-02-4: 19420
15174-69-3: 01723
15181-11-0: 02563
15181-13-2: 03395
15192-80-0: 17782
15196-83-5: 06090
15206-03-8: 11195
15206-55-0: 04283
15219-34-8: 12025
15222-64-7: 21780
15224-25-6: 15997
15231-91-1: 16968
15232-76-5: 17845
15232-96-9: 02969
15233-65-5: 04751
15240-91-2: 04890
15247-96-8: 05770
15250-22-3: 17705
15250-29-0: 21931
15254-64-5: 18851
15258-44-3: 02082
15258-73-8: 05024
15268-49-2: 20665
15286-48-3: 02410
15286-50-7: 02123
15286-88-1: 09570
15290-29-6: 06638
15301-40-3: 23767
15320-72-6: 08884
15328-87-7: 24587
15329-10-9: 10414
15331-16-5: 20714
15333-22-9: 08883
15336-72-8: 07691
15336-73-9: 19826
15356-60-2: 10183
15356-62-4: 00938
15356-70-4: 10176
15358-48-2: 15785
15361-99-6: 03967
15364-55-3: 05599
15375-48-1: 19686
15394-30-6: 04524
15394-33-9: 07908
15396-38-0: 06598
15408-59-0: 02201
15409-60-6: 10240
15411-43-5: 02255
15424-06-3: 08867
15424-14-3: 04381
15430-98-5: 18080
15440-82-1: 00323
15440-88-7: 09118
15440-98-9: 03524
15441-54-0: 24617
15451-44-2: 16801
15457-50-8: 01918
15460-06-7: 19101
15462-91-6: 18549
15462-94-9: 17075
15463-91-9: 05206
15466-93-0: 10120
15466-94-1: 10119
15469-77-9: 11007
15474-90-5: 14570
15480-00-9: 12126
15485-80-0: 06926
15501-33-4: 20626
15504-46-8: 06571
15506-54-4: 09575
15507-52-5: 09384
15516-60-6: 06403
15519-73-0: 07641
15521-18-3: 21324
15529-46-1: 09294
15532-75-9: 20093
15540-90-6: 15211
15541-26-1: 04853
15547-89-4: 10248
15563-53-8: 01956
15567-98-3: 07030
15567-99-4: 07027
15568-00-0: 07028
15568-02-2: 07029
15570-12-4: 05552
15570-45-3: 02982
15572-56-2: 20477
15577-04-5: 10986
15580-32-2: 21266
15588-91-7: 23511
15600-01-8: 20620
15600-08-5: 09367
15601-87-3: 07032
15601-89-5: 07026
15606-95-8: 01390
15625-89-5: 21851
15638-06-9: 18874
15638-08-1: 18840
15638-09-2: 18841
15644-80-1: 23787
15644-82-3: 23786
15647-08-2: 19988
15647-11-7: 10108
15647-47-9: 18401
15649-56-6: 18689
15657-96-2: 20963
15667-21-7: 13071
15667-99-9: 19542
15672-96-5: 18152
15679-01-3: 09530
15679-02-4: 09531
15679-12-6: 24522
15679-24-0: 22504
15680-28-1: 09470
15690-25-2: 21974
15708-30-2: 09292
15721-05-8: 10843
15726-15-5: 13725
15727-65-8: 16409
15727-77-2: 21250
15733-49-0: 09913
15753-50-1: 10051
15760-35-7: 09546
15760-36-8: 09555
15761-83-8: 23147
15764-04-2: 17028
15764-16-6: 01659
15764-52-0: 19547
15775-81-2: 22730
15778-30-0: 15263
15778-31-1: 15244
15778-32-2: 15243

15778-33-3: 15262
15778-36-6: 15259
15780-57-1: 00860
15780-61-7: 08265
15780-63-9: 08263
15780-64-0: 00862
15780-65-1: 00861
15784-82-4: 09513
15786-26-2: 09053
15788-02-0: 08973
15788-78-0: 06428
15789-03-4: 01905
15793-48-3: 02080
15793-77-8: 23841
15802-97-8: 22426
15802-99-0: 22424
15804-19-0: 23899
15813-24-8: 21788
15814-45-6: 05256
15821-13-3: 23737
15826-80-9: 07396
15827-34-6: 19067
15834-24-9: 13822
15834-78-3: 22312
15840-60-5: 14445
15846-25-0: 08700
15848-22-3: 18483
15852-73-0: 04990
15854-58-7: 21879
15861-24-2: 15188
15862-34-7: 22980
15869-80-4: 13521
15869-85-9: 17216
15869-86-0: 17617
15869-87-1: 17601
15869-89-3: 17604
15869-92-8: 17608
15869-93-9: 17609
15869-94-0: 17610
15869-95-1: 17611
15869-96-2: 17612
15870-10-7: 13738
15872-48-7: 06290
15874-03-0: 13865
15877-57-3: 18143
15878-00-9: 22393
15890-36-5: 10161
15890-40-1: 10583
15892-23-6: 08566
15894-30-1: 21787
15901-49-2: 10856
15904-30-0: 18585
15909-76-9: 21160
15910-23-3: 09370
15914-84-8: 16850
15918-08-8: 13545
15922-53-9: 25042
15925-05-0: 01403
15925-47-0: 08264
15932-80-6: 10271
15933-07-0: 08484
15945-07-0: 05521
15950-66-0: 19638
15953-45-4: 22397
15953-73-8: 22395
15963-69-6: 09816
15967-72-3: 20749
15968-56-6: 19288
15969-08-1: 19416
15969-09-2: 19218
15971-29-6: 16598
15971-88-7: 09608
15979-79-0: 02071
15979-87-0: 02428
15980-15-1: 17908
15988-11-1: 25072
15989-02-3: 16135
15998-93-3: 14893
16000-39-8: 16585
16003-14-8: 12922
16005-15-5: 08890
16020-17-0: 16538
16020-34-1: 16924
16021-20-8: 03438
16026-71-4: 23797
16029-28-0: 07714
16034-46-1: 22443
16056-11-4: 02528
16064-21-4: 23499
16066-09-4: 24409
16067-01-9: 17760
16078-25-4: 20181
16079-88-2: 14846
16080-75-4: 10228
16085-05-5: 07644
16090-33-8: 04763
16094-35-2: 06411
16096-32-5: 15156
16106-59-5: 14402
16108-50-2: 12407
16109-68-5: 11050
16110-09-1: 22617
16110-99-9: 04552
16116-32-8: 14885
16116-38-4: 14876
16116-39-5: 14834
16116-40-8: 14873
16116-43-1: 14833
16133-26-9: 22945
16133-49-6: 02336
16135-43-6: 15468
16136-58-6: 15200
16136-84-8: 21678
16136-85-9: 21677
16158-75-1: 05562
16170-33-5: 07066
16174-92-8: 11199
16174-97-3: 11198
16177-46-1: 09527
16179-36-5: 18431
16183-46-3: 08615
16187-90-9: 05192
16188-57-1: 19396
16189-06-3: 20438
16197-80-1: 21464
16197-90-3: 08882
16200-52-5: 23286
16204-57-2: 15024
16204-58-3: 14957
16204-72-1: 15028
16205-90-6: 14554
16214-27-0: 15055
16220-46-5: 24609
16230-71-0: 17984
16239-18-2: 16976
16245-79-7: 02442
16246-07-4: 09079
16262-72-9: 19677
16264-67-8: 15171
16273-71-5: 22623
16273-79-3: 22579
16275-44-8: 00215
16277-67-1: 03622
16290-94-1: 23064
16292-86-7: 19045
16292-88-9: 02332
16292-90-3: 19041
16292-95-8: 19417
16292-99-2: 19445
16299-05-1: 25549
16302-35-5: 22236
16306-39-1: 18829
16308-92-2: 05043
16320-20-0: 18382
16323-43-6: 21970
16326-97-9: 10627

16326-98-0: 10628
16330-02-2: 24931
16330-03-3: 01098
16332-06-2: 00291
16335-92-5: 22883
16346-16-0: 18083
16353-02-9: 23493
16357-59-8: 23746
16369-05-4: 08530
16369-17-8: 18481
16369-21-4: 12240
16383-89-4: 02263
16386-93-9: 18713
16387-70-5: 17554
16387-71-6: 10902
16387-72-7: 24345
16400-69-4: 22276
16409-44-2: 17545
16416-32-3: 24336
16416-34-5: 14374
16419-60-6: 07760
16420-13-6: 09142
16426-64-5: 06019
16429-03-1: 10664
16432-53-4: 14154
16433-43-5: 04371
16433-88-8: 12729
16434-59-6: 01980
16434-97-2: 07668
16435-50-0: 18663
16440-97-4: 15087
16442-14-1: 03793
16452-01-0: 02322
16452-11-2: 16206
16464-38-3: 14562
16466-24-3: 13093
16478-84-5: 18154
16479-17-7: 16706
16483-73-1: 13439
16484-17-6: 15105
16485-23-7: 20487
16488-98-5: 18800
16488-99-6: 18799
16490-19-0: 23100
16491-15-9: 10706
16491-62-6: 08885
16503-25-6: 07782
16503-45-0: 05207
16503-53-0: 05261
16509-46-9: 17541
16513-94-3: 19563
16518-62-0: 02033
16520-62-0: 02772
16527-88-1: 22710
16532-02-8: 23976
16532-79-9: 04309
16533-50-9: 19592
16538-89-9: 10455
16538-93-5: 10453
16548-06-4: 12597
16554-45-3: 02330
16554-47-5: 02331
16560-43-3: 23659
16567-18-3: 23567
16580-23-7: 09908
16580-24-8: 09909
16582-38-0: 02896
16583-73-6: 02948
16587-32-9: 07122
16587-33-0: 11220
16587-34-1: 17105
16587-35-2: 17101
16587-37-4: 17104
16587-38-5: 17103
16587-39-6: 07125
16587-40-9: 07095
16587-41-0: 07098
16587-42-1: 07101
16587-43-2: 07104
16587-44-3: 07105
16587-45-4: 07092
16587-46-5: 07106
16587-47-6: 07111
16587-48-7: 07094
16587-50-1: 07097
16587-51-2: 07103
16587-52-3: 11215
16587-53-4: 11221
16587-54-5: 07100
16587-58-9: 05676
16587-59-0: 17102
16587-65-8: 07123
16587-71-6: 10242
16588-16-2: 06113
16588-24-2: 02605
16588-34-4: 01621
16600-92-3: 05592
16605-36-0: 09905
16605-91-7: 07468
16611-15-7: 02112
16617-46-2: 22430
16618-72-7: 15096
16619-12-8: 15095
16619-38-8: 22029
16619-55-9: 20956
16629-19-9: 24823
16631-63-3: 10554
16638-61-2: 09278
16650-36-5: 10766
16650-52-5: 16627
16650-53-6: 16628
16650-55-8: 16622
16650-61-6: 17059
16650-65-0: 17060
16661-99-7: 09831
16664-31-6: 17335
16664-45-2: 18843
16667-01-9: 17073
16668-45-4: 25116
16669-99-1: 21997
16671-89-9: 24645
16674-04-7: 21115
16688-98-5: 20988
16694-18-1: 24802
16694-52-3: 20538
16696-65-4: 25274
16704-90-8: 01039
16704-91-9: 01040
16709-86-7: 23998
16713-66-9: 10607
16714-21-9: 16245
16714-77-5: 08640
16715-80-3: 07840
16716-13-5: 23909
16722-39-7: 06797
16725-71-6: 07326
16726-67-3: 16623
16732-66-4: 14595
16736-42-8: 17534
16737-44-3: 03683
16739-20-1: 23928
16744-89-1: 14502
16745-94-1: 14399
16746-02-4: 14067
16746-85-3: 14406
16746-86-4: 14395
16746-87-5: 14396
16747-25-4: 14089
16747-26-5: 14090
16747-28-7: 14092
16747-30-1: 14095
16747-31-2: 14096
16747-32-3: 18240
16747-33-4: 18241
16747-38-9: 18272
16747-42-5: 14081

16747-44-7: 18265
16747-45-8: 18266
16747-50-5: 10533
16749-11-4: 09836
16750-67-7: 18990
16751-59-0: 13468
16754-49-7: 20655
16757-80-5: 06946
16765-63-2: 24615
16778-10-2: 08565
16778-11-3: 21352
16778-14-6: 07185
16778-16-8: 05603
16789-46-1: 14052
16790-23-1: 02261
16790-49-1: 07833
16791-94-9: 07726
16803-92-2: 05311
16803-95-5: 05328
16803-96-6: 05327
16803-97-7: 05304
16805-76-8: 19897
16805-77-9: 19957
16805-78-0: 19969
16806-29-4: 05338
16813-18-6: 13421
16813-28-8: 07725
16813-30-2: 07724
16813-33-5: 14840
16813-34-6: 14839
16813-36-8: 23134
16813-41-5: 07723
16813-42-6: 07722
16824-02-5: 15344
16827-42-2: 17062
16830-40-3: 11517
16832-35-2: 12706
16834-57-4: 19844
16837-43-7: 05171
16840-84-9: 17174
16841-48-8: 21736
16844-08-9: 17586
16849-88-0: 20774
16863-34-6: 09310
16863-37-9: 09316
16863-41-5: 09311
16863-48-2: 09324
16863-49-3: 09326
16867-03-1: 22969
16867-04-2: 22987
16867-28-0: 23017
16867-53-1: 23016
16870-28-3: 06353
16872-50-7: 14841
16872-65-4: 01192
16873-50-0: 08090
16874-33-2: 12994
16879-02-0: 22982
16883-48-0: 10584
16889-72-8: 21228
16898-52-5: 20208
16906-54-0: 03279
16909-09-4: 21838
16924-34-8: 11241
16927-27-8: 11126
16932-48-2: 15857
16932-49-3: 06699
16934-07-9: 09426
16934-08-0: 09425
16939-57-4: 02739
16943-12-7: 24608
16943-20-7: 14664
16943-33-2: 24601
16943-34-3: 24602
16943-36-5: 25010
16943-78-5: 23446
16947-63-0: 02209
16956-42-6: 14700
16957-70-3: 18696
16958-25-1: 18444
16958-78-4: 00775
16960-08-0: 19568
16968-19-7: 03308
16980-60-2: 10121
16982-97-1: 17097
16982-98-2: 17096
16982-99-3: 17094
16983-04-3: 17093
16983-05-4: 17092
16985-17-4: 09477
16994-13-1: 12265
16995-35-0: 21584
16995-36-1: 21536
17001-28-4: 17426
17005-59-3: 24002
17012-47-4: 23671
17015-11-1: 14303
17024-18-9: 18885
17024-19-0: 18886
17026-49-2: 04636
17040-62-9: 05249
17041-81-5: 05491
17042-16-9: 14362
17042-21-6: 18577
17043-56-0: 18803
17044-50-7: 11111
17044-70-1: 12344
17046-24-1: 08174
17047-46-0: 10863
17055-13-9: 21440
17055-17-3: 12289
17055-18-4: 21441
17056-93-8: 12429
17056-99-4: 23901
17057-82-8: 15010
17057-88-4: 07493
17057-91-9: 16555
17057-92-0: 16400
17057-93-1: 16399
17059-45-9: 03264
17059-48-2: 15012
17059-52-8: 05801
17059-53-9: 16402
17061-90-4: 03119
17061-92-6: 02951
17064-77-6: 02435
17065-03-1: 19386
17066-67-0: 16311
17071-01-1: 22063
17071-47-5: 08030
17071-52-2: 07924
17074-78-1: 21070
17074-82-7: 21069
17074-84-9: 21156
17074-86-1: 21071
17075-03-5: 22495
17075-14-8: 22704
17082-09-6: 22038
17082-12-1: 11058
17085-91-5: 16365
17088-22-1: 22505
17091-40-6: 12212
17094-21-2: 08476
17094-34-7: 18421
17094-36-9: 20790
17095-46-4: 21364
17096-37-6: 08845
17102-64-6: 13965
17112-85-5: 04120
17113-33-6: 12920
17119-73-2: 23051
17122-21-3: 03807
17123-20-5: 09713
17123-21-6: 09712
17123-23-8: 06966
17135-74-9: 16327

17135-78-3: 01128
17136-36-6: 13346
17164-77-1: 07113
17170-82-0: 14780
17174-06-0: 06017
17174-09-3: 06507
17174-11-7: 06163
17174-67-3: 20934
17174-69-5: 20947
17174-98-0: 01850
17176-77-1: 19985
17177-98-9: 02562
17181-49-6: 21361
17182-95-5: 23752
17199-29-0: 04191
17201-43-3: 06676
17202-19-6: 17785
17202-20-9: 17786
17206-52-9: 10245
17207-42-0: 08324
17213-57-9: 07207
17213-58-0: 01873
17216-08-9: 17007
17216-62-5: 20808
17224-72-5: 09563
17224-73-6: 10612
17226-28-7: 08036
17226-34-5: 14435
17228-23-8: 22042
17228-64-7: 22602
17228-71-6: 22976
17236-58-7: 24818
17240-33-4: 14910
17241-09-7: 03687
17241-13-3: 03688
17249-75-1: 24701
17249-76-2: 24703
17249-77-3: 24779
17249-79-5: 24700
17249-82-0: 24693
17252-51-6: 22734
17253-11-1: 05719
17256-39-2: 20472
17259-74-4: 12130
17260-71-8: 05348
17269-94-2: 03226
17269-95-3: 19136
17269-96-4: 19135
17275-82-0: 06198
17280-60-3: 20256
17283-45-3: 07318
17286-26-9: 16862
17293-45-7: 09794
17293-48-0: 07916
17293-49-1: 02702
17293-55-9: 03747
17297-04-0: 04379
17299-41-1: 10431
17301-23-4: 25276
17301-94-9: 17214
17302-01-1: 13523
17302-02-2: 14030
17302-04-4: 13528
17312-50-4: 10917
17312-63-9: 17201
17312-74-2: 10920
17312-80-0: 25275
17324-04-8: 16988
17334-50-8: 06939
17336-90-2: 23834
17341-93-4: 09239
17345-77-6: 04646
17347-45-4: 19911
17348-59-3: 20657
17348-76-4: 12693
17351-07-4: 10033
17356-19-3: 10646
17357-74-3: 12477
17360-11-1: 25252
17361-75-0: 11950
17362-04-8: 12070
17373-93-2: 05110
17374-19-5: 17924
17374-23-1: 24443
17376-04-4: 03536
17380-83-5: 10116
17384-76-8: 16416
17392-83-5: 21189
17397-37-4: 04348
17397-53-4: 13104
17397-57-8: 13105
17397-58-9: 13103
17398-57-1: 11625
17402-80-1: 07116
17402-86-7: 07118
17406-45-0: 13164
17407-56-6: 08437
17408-14-9: 12525
17420-30-3: 06669
17422-32-1: 15124
17423-48-2: 18825
17430-12-5: 11550
17432-37-0: 01781
17437-39-7: 09525
17447-00-6: 18701
17447-01-7: 18700
17448-96-3: 20838
17449-03-5: 20848
17451-00-2: 06821
17474-44-1: 02363
17479-64-0: 06387
17484-36-5: 03600
17492-65-8: 23861
17492-68-1: 23879
17496-14-9: 15089
17507-11-8: 24189
17508-50-8: 02120
17510-66-6: 15270
17512-08-2: 22386
17512-11-7: 22387
17518-43-3: 11738
17521-25-4: 24561
17521-29-8: 24579
17523-06-7: 24913
17526-35-1: 02010
17531-68-9: 24819
17531-72-5: 24785
17534-14-4: 01339
17534-15-5: 04782
17534-33-7: 11441
17534-44-0: 17049
17550-03-7: 17348
17556-48-8: 21430
17564-61-3: 09150
17570-26-2: 21881
17576-35-1: 21748
17576-53-3: 23689
17577-93-4: 17838
17579-70-3: 05180
17579-79-2: 16990
17584-12-2: 24980
17587-22-3: 17658
17587-33-6: 17182
17596-61-1: 04819
17608-41-2: 09363
17608-76-3: 11616
17609-31-3: 17771
17609-47-1: 25537
17609-80-2: 18996
17612-91-8: 12054
17618-76-7: 13759
17624-02-1: 25556
17624-08-7: 22940
17626-40-3: 06692
17627-77-9: 15481
17635-26-6: 23893

17636-18-9: 01173
17639-76-8: 21219
17639-93-9: 21095
17640-02-7: 21140
17640-07-2: 21135
17640-08-3: 21142
17640-26-5: 00617
17642-12-5: 24939
17642-74-9: 17910
17648-05-4: 03971
17655-74-2: 03373
17658-32-1: 18504
17670-89-2: 17433
17672-88-7: 05731
17679-92-4: 17841
17684-33-2: 05191
17686-87-2: 16222
17687-22-8: 23104
17694-11-0: 12253
17694-16-5: 11249
17694-63-2: 00427
17696-11-6: 17673
17696-37-6: 19296
17696-53-6: 19426
17696-62-7: 06332
17696-64-9: 00517
17699-05-7: 07369
17700-09-3: 04028
17702-41-9: 10889
17704-30-2: 00576
17705-40-7: 01860
17708-39-3: 02057
17715-00-3: 09956
17723-04-5: 00332
17723-05-6: 00335
17730-53-9: 23492
17738-09-9: 05176
17738-24-8: 00922
17738-25-9: 00923
17738-26-0: 00441
17738-27-1: 00439
17738-28-2: 00440
17738-30-6: 00407
17738-31-7: 00409
17738-32-8: 00412
17738-33-9: 00410
17738-34-0: 00411
17738-35-1: 00431
17738-36-2: 00432
17738-37-3: 00425
17738-39-5: 00430
17738-40-8: 00428
17738-41-9: 00429
17738-48-6: 00917
17738-49-7: 00921
17738-50-0: 00919
17738-51-1: 00920
17738-52-2: 00442
17738-53-3: 00443
17738-54-4: 00438
17738-55-5: 00437
17738-56-6: 00918
17739-97-8: 19490
17743-62-3: 00178
17745-45-8: 07769
17746-05-3: 25317
17746-08-6: 18078
17750-24-2: 22808
17750-93-5: 06948
17751-20-1: 00202
17751-22-3: 00203
17754-68-6: 04488
17754-90-4: 01642
17759-88-5: 20508
17764-91-9: 22076
17764-92-0: 22077
17764-93-1: 21443
17768-28-4: 25121
17772-19-9: 00413
17772-20-2: 00336
17773-30-7: 17415
17773-65-8: 08764
17785-58-9: 23462
17789-14-9: 11313
17804-59-0: 18303
17831-71-9: 21953
17837-41-1: 24621
17849-38-6: 05003
17851-27-3: 03451
17852-28-7: 15823
17854-07-8: 03316
17865-85-9: 18054
17874-76-9: 22898
17878-02-3: 05716
17878-44-3: 23978
17880-86-3: 21605
17881-65-1: 24001
17890-41-4: 24178
17892-51-2: 06798
17892-52-3: 06800
17892-53-4: 06799
17911-94-3: 19564
17912-17-3: 17592
17933-03-8: 07761
17937-68-7: 24030
17940-63-5: 10694
17951-84-7: 13150
17952-11-3: 18237
17954-23-3: 23745
17964-77-1: 23995
17966-13-1: 25336
17966-60-8: 01012
17972-08-6: 16059
17973-92-1: 19016
17980-09-5: 18854
17988-31-7: 24060
17994-17-1: 15268
17998-02-6: 08223
18002-79-4: 23965
18011-32-0: 22599
18017-73-7: 04410
18023-33-1: 24059
18023-36-4: 24177
18030-61-0: 23974
18036-81-2: 24150
18036-82-3: 24149
18052-90-9: 22332
18054-10-9: 06590
18055-70-4: 11383
18056-96-7: 24012
18065-05-9: 12400
18066-89-2: 02586
18068-06-9: 10147
18069-17-5: 18312
18076-09-0: 09024
18076-11-4: 08703
18084-26-9: 01221
18084-97-4: 02814
18087-10-0: 19323
18092-96-1: 05428
18099-96-2: 16752
18099-99-5: 16754
18102-38-0: 03580
18106-46-2: 11368
18112-99-7: 19944
18113-04-7: 19167
18123-20-1: 00975
18143-31-2: 24051
18143-83-4: 11430
18144-47-3: 05879
18144-52-0: 16115
18144-54-2: 16114
18156-74-6: 14796
18162-48-6: 23988
18166-01-3: 16060
18169-57-8: 24112

18171-59-0: 23987
18178-59-1: 24053
18185-81-4: 17822
18187-24-1: 11377
18189-05-4: 05867
18217-12-4: 13705
18231-53-3: 18232
18233-24-4: 19583
18236-57-2: 24025
18236-89-0: 24024
18239-10-6: 15892
18240-50-1: 13038
18243-41-9: 23977
18250-08-3: 25041
18256-48-9: 12415
18261-92-2: 17457
18262-85-6: 03450
18263-25-7: 13880
18264-71-6: 02426
18264-75-0: 14704
18268-81-0: 09747
18271-22-2: 01940
18274-34-5: 16720
18276-53-4: 23321
18278-34-7: 01714
18281-05-5: 11590
18281-94-2: 11747
18282-40-1: 03416
18282-59-2: 02613
18283-93-7: 24327
18294-93-4: 25287
18295-59-5: 18147
18300-91-9: 18066
18309-32-5: 07392
18320-91-7: 23311
18338-40-4: 08068
18356-20-2: 07464
18358-63-9: 06481
18362-36-2: 01721
18362-64-6: 13599
18362-97-5: 18453
18365-42-9: 21618
18368-61-1: 22690
18368-63-3: 22605
18368-71-3: 22694
18368-73-5: 22691
18368-76-8: 22603
18371-52-3: 25501
18373-31-4: 21039
18381-45-8: 08250
18383-34-1: 07368
18386-50-0: 20943
18388-45-9: 24038
18395-92-1: 24013
18397-07-4: 09496
18398-36-2: 16758
18398-37-3: 16756
18402-22-7: 24120
18402-76-1: 11375
18419-53-9: 23993
18420-46-7: 10935
18423-69-3: 13055
18424-96-9: 16136
18425-34-8: 16137
18427-44-6: 17461
18428-05-2: 07102
18428-76-7: 01697
18430-74-5: 21465
18433-97-1: 22333
18433-98-2: 22327
18440-30-7: 13383
18441-43-5: 05800
18441-55-9: 19363
18441-61-7: 23717
18444-66-1: 17342
18444-79-6: 24193
18450-16-3: 18971
18450-73-2: 13643
18450-74-3: 13642
18450-78-7: 13617
18453-16-2: 01930
18453-17-3: 01847
18458-44-1: 13775
18472-06-5: 23806
18474-60-7: 15203
18476-57-8: 17535
18479-51-1: 17835
18479-57-7: 17725
18490-38-5: 17120
18490-59-0: 17440
18492-37-0: 07347
18495-23-3: 17848
18495-26-6: 13831
18495-30-2: 20697
18503-90-7: 15601
18505-77-6: 02083
18506-47-3: 00198
18507-23-8: 01931
18508-00-4: 02005
18510-29-7: 24096
18514-46-0: 13351
18521-07-8: 17827
18523-22-3: 12291
18523-48-3: 00475
18524-86-2: 13618
18530-30-8: 07295
18545-44-3: 08372
18545-83-0: 18313
18589-29-2: 13534
18592-13-7: 23131
18595-90-9: 09153
18596-83-3: 09152
18614-11-4: 12972
18619-97-1: 22864
18631-82-8: 24315
18636-52-7: 15035
18636-54-9: 15036
18640-74-9: 24527
18643-86-2: 04619
18645-22-2: 11372
18649-64-4: 12900
18651-57-5: 17585
18653-98-0: 19344
18664-94-3: 20969
18666-24-5: 24067
18668-68-3: 07775
18674-50-5: 07353
18685-70-6: 15287
18698-97-0: 04136
18699-48-4: 04616
18699-87-1: 22686
18704-37-5: 23772
18707-60-3: 08898
18708-70-8: 04032
18719-43-2: 20807
18720-42-8: 24070
18729-48-1: 10651
18733-07-8: 15989
18746-23-1: 25354
18747-42-7: 21569
18748-27-1: 24295
18751-10-5: 24155
18752-86-8: 24151
18752-92-6: 24082
18755-28-7: 22242
18755-31-2: 08217
18755-32-3: 12901
18755-33-4: 20809
18755-34-5: 18320
18755-35-6: 18340
18755-76-5: 22231
18755-77-6: 22232
18755-84-5: 15622
18755-87-8: 18484
18758-34-4: 10849
18761-07-4: 11025

18765-09-8: 24152
18773-38-1: 07674
18773-77-8: 09510
18775-73-0: 01527
18787-63-8: 13912
18787-64-9: 13913
18787-66-1: 18076
18791-49-6: 21206
18794-00-8: 15413
18794-01-9: 15416
18794-77-9: 24743
18794-78-0: 24740
18794-84-8: 11549
18794-96-2: 08404
18799-85-4: 24799
18800-00-5: 24787
18800-02-7: 24789
18803-19-5: 15985
18803-25-3: 15839
18811-30-8: 14828
18812-62-9: 14492
18815-75-3: 13292
18819-63-1: 21808
18819-66-4: 21810
18826-95-4: 08209
18826-96-5: 24969
18829-56-6: 17293
18839-90-2: 19368
18840-04-5: 24071
18840-45-4: 25529
18841-79-7: 13167
18851-33-7: 18963
18855-13-5: 07635
18866-73-4: 08767
18869-72-2: 13411
18894-98-9: 24788
18899-86-0: 24136
18904-54-6: 09473
18905-91-4: 14358
18908-07-1: 03558
18908-66-2: 13485
18912-37-3: 14707
18913-04-7: 12222
18913-31-0: 07804
18917-91-4: 01036
18920-70-2: 15991
18925-42-3: 18766
18931-71-0: 09529
18932-14-4: 17632
18936-17-9: 08253
18937-76-3: 25352
18937-79-6: 14555
18952-87-9: 22418
18952-94-8: 07743
18957-52-3: 23167
18961-95-0: 19968
18961-96-1: 19967
18963-00-3: 20721
18963-01-4: 20701
18967-27-6: 21919
18971-59-0: 18254
18978-78-4: 23544
18979-50-5: 19593
18979-53-8: 19554
18979-60-7: 04732
18979-61-8: 04686
18979-62-9: 04716
18982-44-0: 03111
18982-54-2: 04989
18984-16-2: 03109
18991-98-5: 00483
18992-39-7: 21422
18992-64-8: 09156
18992-80-8: 12371
18992-86-4: 09154
18996-77-5: 04503
18998-48-6: 25039
18999-28-5: 13767
19020-26-9: 23647
19023-68-8: 23853
19031-42-6: 23854
19031-43-7: 23859
19039-21-5: 04792
19039-22-6: 04793
19041-15-7: 13087
19041-66-8: 01045
19043-03-9: 10151
19047-85-9: 19819
19056-41-8: 02037
19064-67-6: 22530
19072-79-8: 19133
19074-59-0: 19267
19084-86-7: 07164
19086-26-1: 18053
19089-47-5: 21327
19089-92-0: 08896
19093-48-2: 24895
19094-75-8: 00535
19112-42-6: 03317
19113-77-0: 14315
19113-78-1: 14312
19125-84-9: 16326
19128-84-8: 08506
19132-06-0: 08222
19134-89-5: 22003
19139-31-2: 08802
19145-60-9: 01491
19145-96-1: 10045
19155-24-9: 15275
19163-05-4: 07748
19177-04-9: 03693
19177-84-5: 09215
19184-67-9: 18304
19200-24-9: 13210
19200-26-1: 13209
19200-31-8: 25569
19200-33-0: 25568
19201-34-4: 21267
19202-04-1: 16088
19221-08-0: 19539
19224-27-2: 08216
19230-27-4: 03002
19241-35-1: 02855
19241-37-3: 02856
19244-91-8: 20037
19245-41-1: 19343
19246-38-9: 09828
19246-39-0: 08570
19249-04-8: 08108
19255-35-7: 14622
19255-37-9: 14621
19255-90-4: 08293
19256-80-5: 16941
19264-68-7: 09162
19265-24-8: 18544
19269-28-4: 13978
19270-93-0: 18490
19287-45-7: 11238
19294-04-3: 02551
19301-82-7: 09414
19303-88-9: 19612
19310-59-9: 21643
19310-63-5: 21641
19317-11-4: 11548
19319-33-6: 04280
19319-35-8: 04279
19329-89-6: 21191
19335-11-6: 14962
19338-43-3: 17960
19351-48-5: 20024
19351-52-1: 20074
19354-27-9: 12928
19360-67-9: 06042
19361-59-2: 01782
19361-91-2: 05429
19370-34-4: 03774

19374-46-0: 10581
19383-04-1: 17948
19386-06-2: 25126
19393-45-4: 08826
19393-87-4: 16991
19393-92-1: 02611
19395-58-5: 23494
19396-83-9: 07293
19398-47-1: 08579
19398-53-9: 18212
19398-61-9: 03092
19398-75-5: 10516
19398-77-7: 14031
19402-87-0: 09176
19406-51-0: 02358
19410-27-6: 22182
19410-28-7: 22184
19410-29-8: 22183
19410-30-1: 22180
19410-32-3: 22171
19410-33-4: 22169
19410-36-7: 22168
19410-37-8: 22170
19410-66-3: 22266
19410-90-3: 25009
19410-92-5: 13023
19410-93-6: 24976
19411-56-4: 16665
19413-49-1: 15301
19420-29-2: 16483
19434-30-1: 15990
19434-42-5: 07667
19434-65-2: 20381
19437-26-4: 15965
19438-10-9: 06326
19438-61-0: 15320
19442-87-6: 15568
19447-29-1: 17314
19463-48-0: 01615
19472-74-3: 04308
19476-96-1: 05359
19481-82-4: 20966
19482-57-6: 17850
19483-78-4: 08849
19484-57-2: 06826
19485-03-1: 21933
19486-60-3: 07526
19489-10-2: 09865
19493-31-3: 07446
19493-44-8: 15440
19493-45-9: 15441
19510-15-7: 04825
19519-75-6: 02707
19522-67-9: 11955
19523-57-0: 16594
19534-08-8: 10059
19542-89-3: 09465
19543-81-8: 16122
19549-70-3: 13666
19549-71-4: 13667
19549-72-5: 13668
19549-73-6: 13669
19549-74-7: 13670
19549-77-0: 13682
19549-79-2: 13686
19549-83-8: 13707
19549-84-9: 13719
19549-97-4: 13835
19550-03-9: 14293
19550-07-3: 14311
19550-09-5: 14316
19550-10-8: 14341
19550-14-2: 14353
19550-40-4: 10312
19550-48-2: 10705
19550-90-4: 14504
19575-07-6: 23749
19577-87-8: 00420
19578-19-9: 00421
19589-46-9: 01352
19589-48-1: 01356
19594-02-6: 03374
19598-31-3: 04824
19601-07-1: 22181
19601-08-2: 22185
19611-22-4: 06807
19613-76-4: 02622
19614-16-5: 02703
19617-44-8: 14940
19635-52-0: 01768
19641-57-7: 05068
19655-56-2: 23650
19655-60-8: 23648
19670-51-0: 13885
19671-18-2: 11756
19674-60-3: 23136
19678-58-1: 20187
19686-73-8: 21363
19689-18-0: 10999
19690-59-6: 14601
19692-35-4: 00834
19694-02-1: 22513
19698-38-5: 12543
19700-96-0: 13954
19700-97-1: 13953
19715-19-6: 05982
19718-45-7: 16496
19727-83-4: 15131
19733-77-8: 07165
19735-89-8: 22479
19752-55-7: 02610
19768-02-6: 13119
19780-01-9: 20154
19780-10-0: 11543
19780-40-6: 13659
19780-56-4: 10712
19780-63-3: 18508
19780-65-5: 18553
19780-66-6: 18638
19780-67-7: 18675
19780-85-9: 20275
19780-90-6: 14158
19780-96-2: 17277
19781-10-3: 17731
19781-11-4: 17745
19781-13-6: 17747
19781-27-2: 17736
19781-43-2: 13907
19781-47-6: 09844
19781-68-1: 10777
19781-69-2: 10778
19781-72-7: 13394
19781-73-8: 09873
19781-77-2: 13779
19788-35-3: 15497
19788-48-8: 00669
19798-80-2: 22538
19798-81-3: 22536
19799-42-9: 24814
19806-17-8: 04413
19811-05-3: 15921
19812-63-6: 24361
19814-71-2: 03844
19818-37-2: 25074
19818-55-4: 14584
19819-95-5: 04830
19819-98-8: 04856
19821-84-2: 13828
19823-40-6: 24263
19824-16-9: 20797
19829-56-2: 03302
19829-72-2: 04624
19832-87-2: 09704
19832-98-5: 17015
19837-85-5: 05310
19837-91-3: 00322

19837-92-4: 00177
19841-72-6: 17730
19846-01-6: 04126
19846-06-1: 15566
19850-35-2: 15209
19851-61-7: 04617
19856-91-8: 24929
19856-92-9: 24930
19856-95-2: 15752
19860-56-1: 21164
19866-50-3: 18698
19866-51-4: 18699
19879-11-9: 10117
19879-12-0: 10118
19880-19-4: 06997
19880-20-7: 06998
19889-37-3: 08411
19891-74-8: 09283
19894-97-4: 07385
19915-11-8: 18136
19922-78-2: 16499
19942-78-0: 24647
19943-90-9: 11200
19947-22-9: 03436
19950-83-5: 24478
19952-47-7: 07035
19957-45-0: 05425
19957-52-9: 11707
19958-58-8: 01801
19961-27-4: 20475
19967-24-9: 04378
19967-57-8: 20378
19968-85-5: 10107
19975-63-4: 24557
19975-64-5: 24558
19983-44-9: 07121
19987-13-4: 20535
19992-45-1: 09795
19995-26-7: 07089
20001-64-3: 02779
20002-32-8: 11993
20002-34-0: 12002
20009-26-1: 16495
20020-02-4: 16500
20023-47-6: 10202
20024-90-2: 03440
20024-91-3: 03439
20027-77-4: 16516
20029-25-8: 24971
20030-32-4: 10355
20034-02-0: 05642
20035-32-9: 01593
20047-41-0: 21077
20052-93-1: 18393
20053-22-9: 22380
20053-34-3: 18391
20053-37-6: 18392
20054-75-5: 07013
20054-76-6: 07012
20057-83-4: 06257
20063-92-7: 17298
20074-79-7: 19801
20075-68-7: 07019
20089-07-0: 08270
20098-48-0: 04029
20103-09-7: 04477
20103-10-0: 04750
20126-73-2: 18052
20126-74-3: 11572
20133-93-1: 21372
20139-55-3: 07841
20173-04-0: 22922
20176-11-8: 16659
20177-02-0: 18665
20182-99-4: 08176
20184-89-8: 17315
20184-91-2: 17316
20185-55-1: 06491
20188-28-7: 19604
20191-75-7: 16201
20191-76-8: 16159
20193-20-8: 20453
20198-49-6: 14547
20200-74-2: 24186
20210-97-3: 06302
20213-30-3: 16437
20220-77-3: 02310
20231-81-6: 09265
20232-11-5: 15051
20235-19-2: 01375
20237-34-7: 13936
20241-76-3: 00467
20244-71-7: 12556
20246-26-8: 24904
20248-64-0: 04453
20258-95-1: 16593
20265-97-8: 02315
20266-00-6: 23350
20266-65-3: 20912
20269-19-6: 20262
20270-50-2: 14123
20273-27-2: 02547
20277-92-3: 13366
20278-84-6: 13567
20278-85-7: 13564
20278-87-9: 13570
20278-88-0: 13572
20278-89-1: 13573
20279-21-4: 12705
20279-53-2: 12980
20280-10-8: 15727
20280-93-7: 09474
20281-39-4: 07543
20281-44-1: 07544
20281-83-8: 18494
20281-85-0: 08541
20281-86-1: 14289
20281-91-8: 08581
20282-28-4: 07545
20291-60-5: 14071
20291-61-6: 13547
20291-72-9: 18842
20291-73-0: 18849
20291-74-1: 18846
20291-75-2: 18896
20291-77-4: 18899
20291-79-6: 18898
20291-91-2: 14043
20291-95-6: 13560
20294-76-2: 17390
20296-50-8: 07371
20309-48-2: 24306
20312-36-1: 05226
20312-37-2: 18433
20316-12-5: 02000
20316-32-9: 14471
20320-55-2: 20005
20327-68-8: 01300
20329-98-0: 21977
20330-45-4: 00223
20332-10-9: 24226
20333-41-9: 11400
20336-15-6: 22758
20338-26-5: 08034
20347-65-3: 07316
20348-74-7: 09845
20357-25-9: 01658
20357-79-3: 15006
20359-55-1: 18538
20367-35-5: 03035
20378-06-7: 00961
20389-49-5: 17358
20395-25-9: 20718
20396-10-5: 09026
20407-74-3: 08274
20416-04-0: 13116

20426-48-6: 22005
20427-22-9: 22166
20427-81-0: 04737
20427-93-4: 19827
20430-82-4: 22186
20432-26-2: 04181
20442-79-9: 07515
20443-90-7: 22229
20459-86-3: 16186
20459-90-9: 16187
20460-82-6: 18390
20474-93-5: 08918
20484-34-8: 19246
20485-41-0: 24531
20487-40-5: 21168
20490-22-6: 19673
20493-83-8: 11968
20493-84-9: 20769
20493-85-0: 11967
20493-86-1: 20768
20493-87-2: 11969
20493-88-3: 20770
20494-02-4: 12078
20494-03-5: 12079
20494-04-6: 12076
20494-05-7: 12077
20503-92-8: 17949
20511-03-9: 08961
20511-12-0: 22547
20513-25-1: 14872
20520-64-3: 10774
20524-86-1: 15687
20531-84-4: 08403
20549-47-7: 10163
20561-15-3: 18605
20570-25-6: 05713
20582-55-2: 24532
20589-63-3: 18818
20589-85-9: 21737
20599-21-7: 17849
20599-27-3: 18654
20615-29-6: 17188
20615-75-2: 02541
20618-04-6: 22035
20621-70-9: 05212
20627-73-0: 03180
20629-35-0: 08862
20634-92-8: 20454
20636-41-3: 23138
20637-49-4: 20731
20638-52-2: 22243
20640-10-2: 12073
20651-75-6: 02767
20651-76-7: 02766
20653-41-2: 07408
20653-90-1: 17646
20657-84-5: 15250
20657-85-6: 15246
20657-87-8: 15251
20657-88-9: 15245
20657-89-0: 15248
20657-90-3: 15247
20661-61-4: 00892
20668-26-2: 23635
20668-28-4: 23636
20668-29-5: 23637
20680-10-8: 20600
20698-91-3: 04195
20704-99-8: 09209
20712-05-4: 05658
20712-06-5: 05659
20712-12-3: 04970
20712-13-4: 07020
20712-14-5: 07025
20717-73-1: 20732
20717-74-2: 20733
20717-79-7: 16662
20731-18-4: 17328
20731-48-0: 06008
20731-74-2: 24758
20731-93-5: 05070
20734-67-2: 04681
20734-71-8: 19412
20735-08-4: 07144
20739-58-6: 17859
20743-50-4: 24429
20747-49-3: 10182
20748-86-1: 13590
20751-75-1: 07001
20751-76-2: 07002
20751-77-3: 06999
20752-34-5: 10181
20754-85-2: 12687
20759-15-3: 08101
20759-62-0: 09475
20760-12-7: 05271
20762-71-4: 22964
20764-63-0: 13208
20767-04-8: 09449
20769-85-1: 21593
20776-48-1: 05891
20776-50-5: 05889
20776-51-6: 05888
20776-60-7: 05907
20776-61-8: 05904
20776-62-9: 05901
20783-52-2: 15249
20795-08-8: 14581
20795-09-9: 14580
20805-20-3: 17966
20805-23-6: 17968
20805-27-0: 17961
20805-28-1: 17962
20805-30-5: 17957
20805-32-7: 17965
20809-46-5: 17767
20810-18-8: 17331
20816-76-6: 16914
20820-82-0: 22748
20824-52-6: 21155
20826-04-4: 22841
20830-75-5: 09261
20834-61-1: 19094
20843-07-6: 12887
20850-43-5: 05730
20859-25-0: 08125
20862-47-9: 09478
20864-49-7: 12261
20865-97-8: 23489
20865-98-9: 23488
20865-99-0: 23487
20875-63-2: 17019
20876-30-6: 04252
20876-99-7: 01906
20881-09-8: 13218
20884-70-2: 01919
20884-71-3: 01864
20884-72-4: 01863
20884-74-6: 01907
20884-75-7: 01961
20884-76-8: 20207
20884-77-9: 01920
20884-78-0: 01866
20884-85-9: 06429
20884-95-1: 11338
20884-99-5: 06430
20887-40-5: 23491
20892-46-0: 15231
20902-45-8: 25530
20902-95-8: 06810
20911-93-7: 15544
20913-25-1: 10816
20917-46-8: 22227
20922-76-3: 25426
20922-98-9: 24607
20928-05-6: 11223

20937-86-4: 16243
20954-32-9: 08383
20954-34-1: 08397
20954-35-2: 08362
20970-32-5: 22528
20972-43-4: 11060
20984-22-9: 06975
20984-33-2: 23820
20985-15-3: 09036
21017-50-5: 15229
21017-87-8: 24528
21017-89-0: 24529
21020-18-8: 08533
21020-19-9: 18977
21020-24-6: 08991
21020-26-8: 18775
21020-27-9: 18781
21023-42-7: 07156
21023-43-8: 11192
21030-43-3: 01481
21035-59-6: 22919
21039-95-2: 09459
21040-45-9: 21992
21050-95-3: 07147
21057-13-6: 20046
21057-14-7: 20066
21057-26-1: 20054
21057-29-4: 20052
21057-31-8: 20050
21057-32-9: 20051
21057-35-2: 20043
21057-36-3: 20032
21057-37-4: 20031
21057-40-9: 20027
21057-47-6: 20081
21057-48-7: 20047
21057-50-1: 20053
21057-52-3: 20073
21057-54-5: 20067
21057-56-7: 20092
21062-20-4: 00939
21064-19-7: 09616
21066-72-8: 08143
21072-37-7: 02965
21072-38-8: 02966
21078-47-7: 09489
21078-87-5: 13908
21081-94-7: 19786
21081-95-8: 19787
21081-96-9: 19789
21083-47-6: 14896
21083-50-1: 14892
21086-87-3: 06474
21091-83-8: 20042
21091-85-0: 20041
21091-87-2: 20072
21091-90-7: 20030
21091-92-9: 20029
21091-93-0: 20028
21091-98-5: 20080
21092-01-3: 20075
21092-02-4: 20063
21099-82-1: 17959
21101-60-0: 19543
21102-13-6: 14302
21120-79-6: 02822
21120-80-9: 11780
21122-64-5: 01968
21122-69-0: 01973
21122-70-3: 01876
21122-71-4: 01957
21122-72-5: 01969
21122-74-7: 01970
21122-75-8: 01861
21122-76-9: 01972
21124-33-4: 09133
21128-00-7: 08179
21129-09-9: 24363
21129-27-1: 10169
21131-99-7: 23496
21132-74-1: 07703
21134-25-8: 20017
21139-21-9: 24190
21147-77-3: 23490
21149-38-2: 00397
21160-03-2: 03281
21161-11-5: 04607
21168-46-7: 23766
21179-48-6: 21129
21184-21-4: 09508
21185-61-5: 00903
21185-66-0: 07242
21185-68-2: 07241
21185-94-4: 07245
21203-68-9: 22689
21204-86-4: 05967
21206-60-0: 01482
21209-30-3: 17491
21210-43-5: 04196
21211-65-4: 23310
21222-05-9: 15870
21229-95-8: 02477
21233-55-6: 20071
21236-74-8: 19516
21239-22-5: 18301
21239-99-6: 01971
21244-24-6: 15388
21248-02-2: 06949
21255-16-3: 16212
21262-52-2: 20411
21263-59-2: 23495
21266-88-6: 13500
21266-90-0: 13501
21290-60-8: 00902
21293-29-8: 18112
21296-92-4: 15174
21299-26-3: 10224
21299-27-4: 10223
21303-50-4: 07076
21304-38-1: 19727
21304-39-2: 12337
21306-21-8: 09752
21315-20-8: 19520
21325-49-5: 23231
21326-63-6: 08578
21327-86-6: 06097
21351-21-3: 18903
21352-22-7: 23521
21368-68-3: 07350
21370-71-8: 17023
21386-21-0: 07658
21387-93-9: 03805
21391-98-0: 10367
21400-25-9: 21740
21400-41-9: 07800
21402-19-7: 24538
21411-35-8: 03144
21423-81-4: 06687
21423-84-7: 06684
21432-98-4: 13260
21436-44-2: 06464
21436-75-9: 14671
21436-93-1: 20101
21436-96-4: 02183
21443-96-9: 14964
21461-09-6: 06400
21462-92-0: 19487
21472-86-6: 04021
21489-32-7: 05374
21490-63-1: 18002
21492-30-8: 20098
21504-43-8: 21705
21507-41-5: 01077
21508-19-0: 12946
21512-82-3: 00568
21520-88-7: 17893

21520-91-2: 17896
21520-97-8: 17892
21521-20-0: 25066
21531-91-9: 14153
21532-73-0: 18884
21535-45-5: 11100
21535-97-7: 05799
21536-60-7: 25076
21536-61-8: 25077
21538-06-7: 05597
21540-57-8: 03654
21545-31-3: 16786
21551-47-3: 22001
21553-89-9: 24836
21553-91-3: 24835
21554-70-1: 06884
21564-91-0: 16509
21564-92-1: 16511
21570-35-4: 13690
21571-91-5: 18220
21577-50-4: 08887
21579-78-2: 20122
21586-21-0: 18291
21598-06-1: 15198
21598-82-3: 01111
21615-34-9: 07221
21617-12-9: 23618
21617-18-5: 23615
21620-60-0: 21162
21626-41-5: 22556
21633-79-4: 20869
21639-41-8: 05510
21650-54-4: 11045
21655-48-1: 20059
21660-38-8: 25067
21660-39-9: 25064
21661-50-7: 18904
21668-09-7: 23932
21674-96-4: 00088
21679-46-9: 09462
21690-30-2: 13081
21690-31-3: 13082
21690-34-6: 13094
21690-36-8: 13062
21690-37-9: 13061
21690-38-0: 13063
21690-63-1: 12926
21693-54-9: 16517
21694-48-4: 12956
21700-31-2: 20681
21702-79-4: 06143
21711-65-9: 16467
21717-29-3: 22546
21720-48-9: 21981
21720-68-3: 17024
21721-78-8: 01832
21732-47-8: 24323
21732-93-4: 04428
21739-92-4: 06002
21739-93-5: 05999
21752-80-7: 24088
21762-12-9: 04132
21762-25-4: 04127
21777-84-4: 03698
21784-30-5: 16094
21788-49-8: 08832
21810-40-2: 10031
21849-92-3: 07666
21849-97-8: 16747
21856-89-3: 14344
21862-63-5: 10134
21877-54-3: 23069
21884-26-4: 18467
21889-88-3: 13797
21890-35-7: 07780
21892-80-8: 07184
21894-94-0: 03874
21900-27-6: 07196
21900-42-5: 07208
21900-59-4: 07195
21911-67-1: 13332
21911-76-2: 00321
21915-53-7: 18039
21917-20-4: 10613
21917-23-7: 10043
21928-43-8: 20173
21929-69-1: 13343
21938-47-6: 14604
21948-47-0: 17598
21964-23-8: 18656
21964-44-3: 17306
21964-48-7: 25138
21968-08-1: 11009
21969-11-9: 02912
21972-99-6: 23325
21974-48-1: 20225
21981-48-6: 18517
21984-93-0: 13008
21987-21-3: 25165
21991-00-4: 07630
21994-77-4: 14468
21994-78-5: 14467
22001-17-8: 02545
22013-33-8: 05720
22021-59-6: 16385
22029-76-1: 08944
22031-64-7: 21654
22037-28-1: 12865
22037-73-6: 08709
22038-56-8: 08869
22038-57-9: 08874
22042-71-3: 00632
22056-82-2: 00039
22057-88-1: 12633
22065-49-2: 01749
22070-40-2: 25484
22077-55-0: 03915
22077-56-1: 03914
22080-96-2: 01706
22083-74-5: 22699
22085-98-9: 01528
22100-66-9: 03982
22104-78-5: 17821
22115-41-9: 06674
22118-09-8: 00929
22118-12-3: 08678
22128-62-7: 09219
22128-63-8: 09224
22134-75-4: 05313
22137-85-5: 24271
22139-13-5: 03338
22139-77-1: 04729
22144-60-1: 05149
22151-32-2: 06801
22152-33-6: 19839
22156-92-9: 08004
22157-31-9: 07927
22160-08-3: 20167
22160-12-9: 14245
22160-37-8: 23607
22160-38-9: 23609
22160-39-0: 18438
22160-40-3: 18443
22160-41-4: 14251
22180-30-9: 20794
22181-43-7: 00548
22191-19-1: 07004
22191-22-6: 07003
22191-26-0: 07017
22191-27-1: 07009
22191-28-2: 07015
22191-29-3: 07016
22191-30-6: 07014
22191-31-7: 07018
22191-32-8: 07005
22191-33-9: 07006

22191-36-2: 07008
22191-37-3: 07007
22192-81-0: 01835
22192-82-1: 01836
22193-20-0: 25407
22194-12-3: 12195
22198-72-7: 22470
22209-47-8: 21061
22214-31-9: 08953
22219-02-9: 06592
22227-75-4: 21703
22232-54-8: 14815
22241-18-5: 01698
22245-48-3: 22514
22253-11-8: 15023
22254-01-9: 23008
22254-04-2: 23011
22278-81-5: 24963
22280-11-1: 11246
22287-02-1: 17287
22291-52-7: 24912
22303-26-0: 00139
22303-30-6: 00388
22304-71-8: 05670
22307-03-5: 20083
22307-05-7: 20086
22308-32-3: 22089
22319-28-4: 13809
22319-31-9: 13810
22327-39-5: 10419
22331-38-0: 11579
22339-51-1: 16856
22347-82-6: 07151
22349-03-7: 09933
22349-59-3: 18844
22362-65-8: 21454
22362-76-1: 22074
22362-84-1: 22088
22364-68-7: 04335
22371-68-2: 11174
22374-89-6: 05187
22381-86-8: 25338
22388-07-4: 07065
22395-84-2: 25510
22397-44-0: 14785
22398-09-0: 20219
22399-98-0: 11454
22401-25-8: 19854
22409-37-6: 20433
22409-91-2: 21275
22418-49-1: 21722
22421-97-2: 21114
22424-58-4: 03719
22425-07-6: 10100
22430-23-5: 15578
22433-68-7: 23038
22433-76-7: 00371
22438-36-4: 08022
22438-39-7: 10927
22438-40-0: 24622
22439-48-1: 11142
22439-58-3: 12338
22439-61-8: 11208
22439-99-2: 20078
22440-38-6: 16672
22445-41-6: 03531
22446-40-8: 04110
22452-74-0: 06583
22469-15-4: 21776
22483-09-6: 11702
22487-59-8: 23052
22487-87-2: 13753
22496-14-6: 14904
22502-99-4: 05370
22509-74-6: 15414
22514-98-3: 22167
22520-40-7: 17649
22524-32-9: 00492
22531-20-0: 16408
22532-24-7: 09996
22532-60-1: 01589
22532-62-3: 01592
22539-65-7: 03245
22540-60-9: 22761
22545-12-6: 04845
22545-13-7: 04846
22554-27-4: 10251
22554-29-6: 10250
22564-83-6: 15589
22564-99-4: 17539
22567-21-1: 10164
22591-15-7: 10276
22592-18-3: 13841
22610-61-3: 17405
22610-63-5: 17403
22618-23-1: 19271
22620-27-5: 22843
22621-74-5: 07328
22625-57-6: 17956
22632-10-6: 23040
22646-79-3: 10384
22663-64-5: 17264
22672-74-8: 11187
22686-81-3: 23510
22692-70-2: 24257
22701-60-6: 08672
22701-62-8: 25472
22701-63-9: 25446
22704-30-9: 25482
22710-11-8: 21099
22715-27-1: 23108
22717-57-3: 06378
22721-08-0: 00422
22721-09-1: 05394
22726-00-7: 01823
22734-04-9: 10727
22742-59-2: 18284
22743-00-6: 16694
22750-93-2: 18802
22760-85-6: 02337
22760-87-8: 02114
22760-88-9: 02400
22760-89-0: 02434
22764-04-1: 20100
22764-81-4: 09063
22767-77-7: 16493
22767-95-9: 05242
22767-96-0: 05253
22768-19-0: 13750
22768-20-3: 13749
22769-84-2: 13794
22771-66-0: 25410
22771-67-1: 25358
22771-99-9: 16090
22772-75-4: 22762
22772-76-5: 22763
22772-77-6: 22764
22772-82-3: 22673
22772-87-8: 22733
22772-91-4: 22665
22786-85-2: 00408
22795-31-9: 00414
22795-32-0: 00418
22795-59-1: 00086
22795-62-6: 00435
22796-67-4: 14776
22800-68-6: 05178
22805-21-6: 15559
22805-42-1: 05066
22809-37-6: 14367
22811-02-5: 00665
22812-33-5: 09450
22817-17-0: 00973
22819-27-8: 09044
22820-60-6: 21359
22821-80-3: 00327

22825-14-5: 17005
22839-47-0: 19723
22856-77-5: 02438
22859-35-4: 12627
22868-76-4: 23066
22870-43-5: 22254
22877-76-5: 14830
22900-08-9: 10160
22922-67-4: 14390
22924-15-8: 01671
22927-07-7: 03304
22927-13-5: 01680
22948-02-3: 05540
22948-12-5: 07904
22948-59-0: 08284
22948-60-3: 08290
22948-63-6: 08289
22948-67-0: 08288
22949-08-2: 15120
22951-47-9: 08287
22951-48-0: 08292
22951-49-1: 08285
22951-50-4: 08286
22951-52-6: 22381
22951-55-9: 22382
22951-60-6: 22379
22954-83-2: 22277
22955-19-7: 06824
22956-94-1: 08481
22961-02-0: 00416
22962-97-6: 24605
22975-58-2: 03441
22996-18-5: 05016
23000-25-1: 12727
23003-22-7: 22950
23010-00-6: 08721
23010-04-0: 07973
23010-42-6: 00083
23012-11-5: 17934
23014-13-3: 00556
23014-19-9: 17047
23030-43-5: 04757
23048-13-7: 10750
23050-90-0: 06162
23050-91-1: 06223
23050-96-6: 06229
23056-29-3: 20130
23063-36-7: 03098
23067-32-5: 13317
23069-99-0: 12835
23071-04-7: 23361
23074-36-4: 08708
23074-42-2: 25117
23079-65-4: 02754
23081-86-9: 20297
23097-98-5: 13836
23107-77-9: 08291
23107-79-1: 22384
23132-29-8: 04357
23132-52-7: 03240
23140-52-5: 22127
23141-61-9: 23684
23144-68-5: 00384
23149-36-2: 04772
23153-09-5: 02419
23156-75-4: 04766
23159-07-1: 23412
23183-11-1: 07394
23187-02-2: 14871
23187-03-3: 14579
23197-67-3: 20849
23197-68-4: 15565
23197-69-5: 20828
23197-70-8: 16842
23197-72-0: 20862
23204-80-0: 02934
23213-34-5: 23145
23218-93-1: 05945
23228-74-2: 24659
23253-96-5: 13379
23255-20-1: 22826
23256-55-5: 08593
23265-69-2: 19872
23271-77-4: 17964
23280-38-8: 00199
23283-97-8: 10185
23284-84-6: 13338
23287-26-5: 06374
23308-82-9: 05099
23316-35-0: 06919
23346-57-8: 08096
23350-58-5: 08698
23365-29-9: 25245
23378-11-2: 08720
23378-88-3: 05504
23399-81-7: 00766
23408-16-4: 19012
23410-40-4: 11938
23418-37-3: 17749
23418-38-4: 17274
23418-82-8: 10457
23421-81-0: 17031
23429-04-1: 17944
23429-05-2: 17943
23431-48-3: 03759
23436-19-3: 21413
23438-77-9: 10434
23444-65-7: 16790
23454-01-5: 07813
23468-31-7: 05729
23470-12-4: 14559
23495-98-9: 09429
23496-33-5: 13284
23504-03-2: 04713
23511-70-8: 08509
23516-74-7: 05367
23518-57-2: 22133
23537-25-9: 01027
23545-35-9: 15375
23545-36-0: 20019
23545-40-6: 20018
23546-12-5: 13427
23563-47-5: 24926
23569-45-1: 22429
23571-23-5: 14753
23571-25-7: 14827
23571-42-8: 14760
23571-47-3: 24269
23571-54-2: 14805
23574-01-8: 01007
23576-81-0: 14929
23576-87-6: 14935
23578-35-0: 14108
23580-51-0: 17861
23580-52-1: 22653
23585-79-7: 14757
23588-63-8: 07031
23588-64-9: 07021
23588-65-0: 07022
23588-67-2: 07024
23594-30-1: 17942
23598-72-3: 15498
23617-71-2: 06205
23619-59-2: 19587
23649-33-4: 14826
23650-12-6: 07023
23658-96-0: 19669
23666-11-7: 24304
23673-40-7: 17095
23674-14-8: 01157
23674-65-9: 07072
23676-09-7: 06264
23679-20-1: 19821
23689-34-1: 19210
23699-41-4: 00901
23705-39-7: 05385

23708-56-7: 25301
23711-41-3: 22475
23711-45-7: 22453
23711-48-0: 22489
23713-56-6: 10326
23727-16-4: 07377
23730-10-1: 25420
23731-06-8: 01616
23732-40-3: 13072
23749-65-7: 04025
23764-31-0: 20765
23769-39-3: 25217
23783-42-8: 24405
23786-14-3: 04249
23804-34-4: 10027
23804-41-3: 10028
23806-24-8: 24807
23809-12-3: 07878
23822-15-3: 24244
23834-14-2: 23584
23843-52-9: 17387
23852-97-3: 05886
23853-48-7: 17117
23853-49-8: 16209
23853-53-4: 01535
23853-55-6: 20220
23853-56-7: 16120
23853-57-8: 16096
23853-58-9: 20082
23853-59-0: 16236
23853-67-0: 01514
23853-68-1: 16533
23865-95-4: 07285
23867-85-8: 24977
23867-87-0: 21335
23867-88-1: 24974
23867-89-2: 14747
23867-90-5: 23275
23867-92-7: 23274
23867-93-8: 25013
23867-95-0: 14748
23867-96-1: 24979
23867-97-2: 24975
23868-26-0: 24978
23875-36-7: 25012
23876-13-3: 05123
23877-94-3: 23679
23896-95-9: 19813
23900-71-2: 24482
23900-73-4: 24476
23900-77-8: 24477
23905-46-6: 05494
23910-22-7: 10839
23932-27-6: 23010
23932-84-5: 15347
23937-68-0: 00456
23938-42-3: 25125
23939-20-0: 00602
23942-28-1: 13151
23942-33-8: 22445
23947-37-7: 23835
23956-39-0: 13141
23975-31-7: 00907
23982-57-2: 21568
23982-94-7: 17902
23983-01-9: 17903
24004-45-3: 12468
24010-03-5: 23723
24010-26-2: 23451
24010-29-5: 23624
24010-34-2: 23450
24018-44-8: 16660
24019-66-7: 01735
24033-03-2: 02906
24043-80-9: 09476
24044-52-8: 24551
24050-49-5: 17042
24070-68-6: 10599
24070-69-7: 12461
24078-21-5: 06498
24083-13-4: 01765
24083-16-7: 01624
24085-65-2: 12620
24092-46-4: 15507
24102-52-1: 11148
24102-61-2: 04186
24106-05-6: 00106
24123-89-5: 01495
24127-62-6: 21790
24133-90-2: 14909
24133-93-5: 14802
24133-96-8: 14803
24133-97-9: 14804
24133-99-1: 14798
24134-06-3: 14837
24138-83-8: 24544
24139-56-8: 18111
24142-77-6: 03763
24142-78-7: 12624
24152-39-4: 20141
24154-37-8: 02516
24160-53-0: 21976
24162-64-9: 03010
24164-44-1: 01009
24168-56-7: 16728
24190-29-2: 08977
24195-02-6: 22894
24195-03-7: 22896
24195-07-1: 22893
24197-40-8: 19279
24221-86-1: 05118
24229-53-6: 11923
24234-56-8: 22900
24234-61-5: 12381
24234-64-8: 22884
24243-71-8: 21002
24247-53-8: 08690
24251-86-3: 11491
24253-30-3: 14511
24262-63-3: 06327
24264-08-2: 08947
24271-85-0: 17793
24272-67-1: 00462
24272-68-2: 00871
24280-93-1: 14470
24281-25-2: 00262
24297-12-9: 22124
24305-36-0: 10891
24305-56-4: 04700
24306-18-1: 13431
24313-88-0: 02518
24317-94-0: 13611
24317-95-1: 10954
24323-23-7: 10968
24323-24-8: 10970
24324-17-2: 12771
24327-08-0: 12087
24332-22-7: 13796
24332-24-9: 10146
24332-28-3: 09901
24337-52-8: 10085
24341-62-6: 13303
24342-04-9: 18788
24342-44-7: 04097
24342-47-0: 12284
24342-50-5: 12500
24346-19-8: 04084
24346-22-3: 12482
24346-56-3: 05222
24347-58-8: 08220
24356-00-1: 18626
24358-62-1: 04887
24358-65-4: 05186
24359-85-1: 14903
24364-02-1: 21775
24388-84-9: 18151

24393-61-1: 21948
24393-70-2: 07324
24398-88-7: 05993
24402-72-0: 00340
24410-50-2: 05792
24410-51-3: 05789
24410-52-4: 05794
24413-59-0: 10102
24415-29-0: 24934
24415-66-5: 25089
24423-11-8: 18834
24425-40-9: 14989
24425-97-6: 20618
24426-16-2: 11922
24426-40-2: 00885
24438-17-3: 15274
24443-15-0: 07957
24446-63-7: 04292
24454-82-8: 01639
24459-40-3: 09611
24463-19-2: 01130
24463-41-0: 10021
24465-45-0: 20250
24473-00-5: 01786
24480-08-8: 22273
24480-28-2: 17906
24480-38-4: 09285
24482-35-7: 17907
24483-72-5: 06593
24490-31-1: 21495
24503-62-6: 21602
24517-64-4: 22773
24528-50-5: 04476
24535-04-4: 25295
24535-08-8: 14719
24535-12-4: 17681
24539-56-8: 04574
24544-04-5: 02021
24545-81-1: 07405
24547-70-4: 09445
24547-71-5: 09442
24547-72-6: 09443
24547-73-7: 09444
24547-75-9: 00044
24547-76-0: 00034
24547-78-2: 00048
24547-79-3: 00043
24549-06-2: 02273
24550-41-2: 12621
24550-42-3: 20490
24552-04-3: 17566
24557-10-6: 21820
24560-98-3: 18040
24567-08-6: 22409
24579-70-2: 09060
24580-48-1: 10188
24580-52-7: 10190
24581-93-9: 00581
24589-96-6: 00645
24596-19-8: 02034
24598-45-6: 21268
24599-58-4: 03174
24603-65-4: 09075
24606-87-9: 17911
24606-89-1: 17913
24606-90-4: 17915
24606-91-5: 17912
24606-92-6: 17916
24606-94-8: 17914
24612-70-2: 09963
24612-75-7: 09961
24612-99-5: 00628
24618-31-3: 09802
24621-61-2: 08211
24622-72-8: 23360
24623-24-3: 15093
24626-94-6: 00333
24626-99-1: 00334
24641-56-3: 09025
24644-78-8: 15091
24664-49-1: 22476
24664-57-1: 22463
24664-63-9: 22467
24665-75-6: 22466
24665-76-7: 22465
24665-83-6: 22454
24673-56-1: 05813
24677-78-9: 01643
24686-78-0: 20296
24716-09-4: 04142
24723-77-1: 00989
24724-07-0: 25294
24734-68-7: 05198
24743-21-3: 07435
24743-62-2: 21050
24743-64-4: 21051
24743-69-9: 21056
24743-71-3: 21053
24748-78-5: 18389
24748-79-6: 18387
24756-04-5: 18394
24756-07-8: 18388
24764-40-7: 23294
24764-47-4: 00866
24764-93-0: 19347
24764-97-4: 07809
24769-39-9: 24694
24774-58-1: 14017
24775-57-3: 24888
24776-55-4: 19287
24796-87-0: 10109
24799-18-6: 04981
24815-24-5: 25572
24821-22-5: 05985
24823-81-2: 20730
24824-25-7: 16377
24824-26-8: 16380
24824-27-9: 16381
24824-28-0: 16697
24828-59-9: 10744
24829-12-7: 19631
24833-14-5: 01359
24840-05-9: 21766
24844-28-8: 10197
24870-11-9: 08911
24885-45-8: 19397
24890-63-9: 18974
24894-63-1: 12377
24894-78-8: 16933
24910-63-2: 18668
24910-69-8: 03113
24914-08-7: 13356
24914-09-8: 13355
24918-02-3: 14933
24918-04-5: 14932
24918-13-6: 14939
24918-20-5: 14938
24918-22-7: 14937
24918-25-0: 14931
24918-28-3: 14934
24932-48-7: 03005
24932-49-8: 03004
24934-47-2: 16372
24948-66-1: 11017
24949-35-7: 20496
24951-10-8: 14930
24964-01-0: 06652
24964-02-1: 06650
24964-03-2: 06651
24964-07-6: 06657
24964-09-8: 06656
24964-12-3: 06654
24964-16-7: 06653
24964-64-5: 06708
24965-84-2: 10249
24965-87-5: 10254

24965-92-2: 10152
24965-94-4: 10153
24966-88-9: 23166
24973-49-7: 07586
25006-60-4: 00578
25016-01-7: 01596
25016-08-4: 11047
25016-20-0: 22438
25016-24-4: 22444
25017-68-9: 12779
25023-01-2: 12731
25032-74-0: 15847
25044-01-3: 18740
25056-62-6: 24859
25056-70-6: 00780
25059-43-2: 04341
25081-39-4: 06216
25084-93-9: 03068
25109-57-3: 08638
25117-74-2: 06705
25118-28-9: 14013
25129-13-9: 01010
25129-65-1: 21856
25141-25-7: 00609
25144-04-1: 10650
25144-05-2: 10649
25148-87-2: 08777
25152-83-4: 10890
25158-44-5: 16851
25160-69-4: 24858
25162-00-9: 22700
25167-70-8: 18620
25168-24-5: 00569
25172-42-3: 10655
25172-43-4: 10656
25177-45-1: 18901
25182-84-7: 21343
25183-32-8: 09251
25183-33-9: 09248
25185-95-9: 01630
25186-47-4: 03091
25187-00-2: 15998
25187-01-3: 15846
25192-03-4: 09123
25201-40-5: 13451
25229-42-9: 08886
25234-83-7: 18457
25237-06-3: 08000
25238-52-2: 20761
25244-30-8: 02722
25245-33-4: 03527
25263-44-9: 04330
25265-71-8: 21295
25287-19-8: 02119
25287-21-2: 01105
25287-23-4: 01100
25287-24-5: 02326
25287-25-6: 01109
25287-26-7: 17124
25287-27-8: 17116
25287-28-9: 02343
25287-29-0: 24946
25287-31-4: 17121
25287-32-5: 17111
25287-38-1: 17113
25287-39-2: 01107
25287-40-5: 24947
25287-41-6: 01099
25287-42-7: 01113
25287-44-9: 17123
25287-45-0: 17114
25287-46-1: 24945
25287-48-3: 02121
25288-75-9: 18125
25288-76-0: 11206
25309-64-2: 03417
25312-34-9: 08945
25312-65-6: 09336
25329-35-5: 10488
25332-53-0: 04385
25333-55-5: 00266
25346-31-0: 18191
25346-32-1: 18201
25346-33-2: 18186
25349-12-6: 01104
25349-13-7: 17122
25351-89-7: 00268
25354-97-6: 10965
25354-98-7: 18073
25355-34-4: 20953
25355-59-3: 25456
25393-23-1: 22750
25395-22-6: 00460
25395-31-7: 21036
25412-75-3: 04400
25414-22-6: 12905
25415-67-2: 18449
25415-70-7: 18451
25415-71-8: 18452
25419-06-1: 18170
25419-36-7: 16515
25419-37-8: 16512
25428-26-6: 22383
25438-23-7: 00781
25445-34-5: 02555
25446-32-6: 14955
25446-35-9: 09941
25455-56-5: 12615
25468-91-1: 24512
25474-85-5: 09187
25475-67-6: 15432
25491-20-7: 15798
25522-33-2: 21861
25525-21-7: 13180
25542-62-5: 14217
25543-96-8: 19310
25544-02-9: 09487
25546-43-4: 13187
25555-13-9: 04098
25555-20-8: 00910
25555-21-9: 00909
25555-22-0: 00904
25555-25-3: 00914
25555-26-4: 00915
25555-27-5: 12507
25558-65-0: 11230
25558-66-1: 11232
25558-90-1: 11233
25561-43-7: 00911
25561-48-2: 00908
25562-42-9: 06563
25562-44-1: 23949
25562-65-6: 11203
25570-02-9: 07551
25574-26-9: 11190
25574-37-2: 11191
25574-69-0: 11247
25574-93-0: 21648
25603-67-2: 12782
25611-78-3: 04811
25624-76-4: 23226
25637-97-2: 13213
25641-98-9: 04551
25653-16-1: 19275
25658-36-0: 22895
25659-22-7: 14508
25680-58-4: 22338
25695-77-6: 21965
25709-42-6: 13361
25711-30-2: 22432
25724-34-9: 22238
25726-28-7: 25518
25742-92-1: 05744
25755-73-1: 04829
25770-53-0: 05360
25772-00-3: 15470

25772-06-9: 07249
25781-92-4: 02113
25782-57-4: 04984
25784-91-2: 07202
25790-55-0: 07776
25799-83-1: 05379
25801-73-4: 06970
25804-87-9: 07163
25804-89-1: 07162
25836-11-7: 23614
25867-19-0: 00946
25871-69-6: 15836
25878-23-3: 14742
25941-29-1: 09344
25974-99-6: 06039
25978-95-4: 00868
25986-40-7: 16591
26009-03-0: 20306
26027-67-8: 23433
26028-46-6: 24987
26072-79-7: 14625
26073-26-7: 21603
26078-24-0: 06836
26116-12-1: 23411
26118-38-7: 14340
26120-52-5: 09676
26140-94-3: 03064
26144-38-7: 20350
26153-88-8: 18164
26158-00-9: 05474
26171-83-5: 08204
26177-44-6: 04885
26179-99-7: 09055
26186-01-6: 17178
26186-02-7: 25178
26197-64-8: 08513
26198-63-0: 20557
26207-06-7: 16989
26226-72-2: 06070
26227-54-3: 09779
26228-96-6: 01541
26249-03-6: 00236
26254-07-9: 25059
26254-09-1: 25062
26254-12-6: 25061
26254-17-1: 25051
26254-20-6: 25060
26254-21-7: 25058
26260-02-6: 01737
26266-63-7: 15325
26271-75-0: 19007
26292-65-9: 09481
26292-67-1: 09480
26292-73-9: 09482
26295-42-1: 25046
26295-43-2: 25048
26295-49-8: 25045
26295-50-1: 25047
26301-79-1: 15577
26302-39-6: 13231
26303-90-2: 11008
26308-46-3: 15953
26311-17-1: 11712
26311-33-1: 14461
26330-51-8: 08339
26340-58-9: 09493
26348-70-9: 15550
26356-11-6: 04068
26370-28-5: 17180
26378-73-4: 04736
26385-07-9: 01839
26386-47-0: 18763
26391-16-2: 21689
26423-79-0: 20345
26429-41-4: 03023
26438-26-6: 16468
26438-27-7: 16316
26438-28-8: 16388
26443-03-8: 11659
26456-59-7: 23224
26459-52-9: 03241
26459-95-0: 01581
26464-32-4: 08348
26465-81-6: 15085
26472-41-3: 09753
26473-48-3: 08442
26473-49-4: 08443
26478-94-4: 23009
26489-01-0: 17830
26489-02-1: 17790
26496-37-7: 25428
26513-20-2: 04454
26533-31-3: 17273
26533-33-5: 17271
26533-34-6: 17737
26536-60-7: 24877
26549-24-6: 14290
26549-25-7: 13665
26549-26-8: 13676
26567-10-2: 07645
26577-41-3: 15429
26577-45-7: 15430
26595-27-7: 15564
26608-06-0: 11130
26644-97-3: 00281
26706-12-7: 08628
26730-14-3: 25152
26757-72-2: 17813
26771-06-2: 10804
26771-11-9: 10803
26782-75-2: 08337
26817-14-1: 09252
26817-17-4: 09253
26820-62-2: 16113
26825-83-2: 13409
26825-94-5: 10959
26830-96-6: 06665
26879-83-4: 01733
26886-05-5: 19087
26891-61-2: 22739
26891-95-2: 00825
26901-97-3: 12179
26918-56-9: 15116
26946-33-8: 02364
26947-66-0: 24450
26963-33-7: 24751
26963-39-3: 24730
26971-94-8: 20739
26979-27-1: 01142
27047-62-7: 25384
27049-45-2: 12510
27052-21-7: 05819
27063-27-0: 23519
27076-16-0: 18979
27080-08-6: 07705
27126-93-8: 06670
27127-58-8: 13335
27129-86-8: 02674
27129-87-9: 05046
27138-31-4: 21296
27139-97-5: 02601
27143-09-5: 00523
27151-54-8: 19424
27173-90-6: 24927
27173-98-4: 19729
27183-43-3: 03848
27189-08-8: 24928
27231-33-0: 05646
27241-93-6: 01623
27243-23-8: 08858
27244-03-7: 15128
27246-81-7: 14596
27248-25-5: 18561
27248-44-8: 15129
27251-77-0: 17226
27260-64-6: 02776

27267-69-2: 06793
27284-46-4: 05702
27284-50-0: 05701
27284-51-1: 09148
27284-89-5: 04288
27297-83-2: 08952
27313-32-2: 20385
27326-27-8: 21282
27353-29-3: 19842
27354-06-9: 01343
27356-52-1: 23651
27357-00-2: 24748
27383-32-0: 22385
27383-34-2: 22398
27385-97-3: 24461
27387-79-7: 21103
27402-47-7: 23222
27406-36-6: 23210
27429-76-1: 22998
27429-78-3: 24185
27430-18-8: 23158
27439-08-3: 06809
27464-82-0: 24458
27485-15-0: 01183
27489-12-9: 19387
27505-78-8: 15175
27550-72-7: 21803
27558-25-4: 20133
27559-45-1: 01915
27563-65-1: 04547
27566-37-6: 18289
27567-20-0: 15114
27581-29-9: 13471
27583-37-5: 10624
27598-85-2: 18984
27601-77-0: 16577
27601-78-1: 16578
27601-93-0: 20231
27602-05-7: 23410
27608-01-1: 17757
27610-88-4: 17659
27644-49-1: 17728
27652-35-3: 02460
27654-54-2: 00717
27655-41-0: 15472
27684-84-0: 19125
27693-73-8: 06456
27701-01-5: 09073
27725-15-1: 19600
27750-10-3: 18603
27758-60-7: 14626
27779-01-7: 09506
27816-36-0: 20409
27816-53-1: 15135
27828-56-4: 16767
27831-13-6: 03346
27835-06-9: 21862
27841-33-4: 04430
27854-40-6: 03202
27864-13-7: 04030
27871-49-4: 21190
27875-89-4: 18907
27893-41-0: 01688
27894-50-4: 15552
27895-67-6: 02180
27897-08-1: 05409
27898-66-4: 18306
27912-97-6: 00832
27917-14-2: 09054
27918-14-5: 00979
27926-72-3: 07699
27939-29-3: 13240
27942-64-9: 06059
27944-79-2: 18138
27963-57-1: 09468
27983-18-2: 03190
27988-97-2: 24424
27993-42-6: 18568
27993-66-4: 19912
27993-82-4: 23878
27993-89-1: 23904
27993-90-4: 23905
27993-91-5: 23903
28022-43-7: 04896
28028-86-6: 20881
28029-41-6: 01892
28034-99-3: 07680
28056-06-6: 15425
28056-54-4: 10683
28059-58-7: 24943
28059-64-5: 02469
28066-42-4: 18370
28084-45-9: 05442
28096-37-9: 07248
28096-39-1: 09040
28096-46-0: 09143
28096-87-9: 02252
28097-60-1: 19926
28122-24-9: 03431
28122-25-0: 03430
28123-70-8: 17204
28129-15-9: 06472
28141-44-8: 15473
28145-10-0: 25442
28148-04-1: 20555
28148-08-5: 25065
28165-52-8: 19211
28165-67-5: 19659
28169-46-2: 06484
28178-96-3: 16950
28181-33-1: 23480
28181-34-2: 23479
28188-41-2: 06675
28195-45-1: 06543
28217-87-0: 22335
28217-95-0: 22354
28221-20-7: 08461
28261-97-4: 11450
28292-42-4: 13467
28314-02-5: 12755
28315-93-7: 17010
28320-43-6: 23917
28320-89-0: 13490
28332-44-7: 14518
28340-76-3: 23071
28354-33-8: 20974
28358-98-7: 22070
28443-50-7: 18993
28445-09-2: 16900
28446-68-6: 21763
28446-72-2: 21758
28449-61-8: 06913
28452-93-9: 24708
28455-14-3: 14585
28466-26-4: 22375
28467-92-7: 07823
28469-92-3: 03065
28472-97-1: 17225
28484-70-0: 24418
28490-23-5: 25467
28490-25-7: 25458
28490-26-8: 25403
28490-27-9: 25406
28490-29-1: 25463
28490-30-4: 25457
28520-00-5: 04734
28522-33-0: 11476
28522-80-7: 25404
28522-82-9: 25405
28547-28-6: 06009
28547-84-4: 14713
28549-12-4: 04328
28563-75-9: 25063
28567-90-0: 23486
28567-91-1: 23485
28567-92-2: 23476

28567-93-3: 23481
28567-94-4: 23482
28567-95-5: 23467
28567-96-6: 00916
28567-98-8: 23475
28567-99-9: 23477
28598-81-4: 17243
28613-15-2: 25496
28613-56-1: 00867
28616-74-2: 24872
28616-77-5: 23950
28616-82-2: 23951
28629-57-4: 23911
28660-67-5: 24222
28664-03-1: 09543
28664-08-6: 07330
28712-49-4: 24692
28715-26-6: 05793
28717-12-6: 24610
28732-18-5: 24965
28732-19-6: 00376
28732-20-9: 24973
28732-21-0: 00214
28732-65-2: 09511
28744-45-8: 04009
28745-07-5: 00802
28752-22-9: 11414
28752-82-1: 01775
28785-06-0: 01777
28788-62-7: 07197
28822-58-4: 22205
28832-55-5: 07810
28833-54-7: 08957
28836-20-6: 12312
28899-11-8: 13357
28918-16-3: 21503
28918-17-4: 21504
28918-25-4: 17078
28918-27-6: 17077
28918-29-8: 17072
28918-52-7: 22090
28921-35-9: 13005
28926-31-0: 17071
28974-17-6: 07319
28989-47-1: 24577
28994-41-4: 19576
29003-73-4: 00572
29005-18-3: 12719
29006-25-5: 11003
29021-91-8: 11143
29022-31-9: 17318
29030-57-7: 00193
29030-84-0: 13941
29040-46-8: 05788
29044-06-2: 14169
29048-54-2: 21471
29052-00-4: 18981
29072-93-3: 20666
29074-38-2: 05682
29095-44-1: 15179
29096-75-1: 05605
29097-31-2: 19599
29106-32-9: 10742
29106-33-0: 10745
29108-90-5: 25505
29110-74-5: 14966
29113-63-1: 13001
29117-45-1: 18251
29117-48-4: 17622
29117-56-4: 14161
29125-24-4: 04148
29136-19-4: 03823
29170-08-9: 07517
29172-20-1: 12903
29184-60-9: 01523
29193-34-8: 00849
29193-36-0: 00501
29193-37-1: 00848
29193-41-7: 00478
29204-02-2: 11603
29211-70-9: 22485
29214-57-1: 13598
29228-78-2: 00765
29233-06-5: 23505
29233-09-8: 23509
29235-52-7: 06640
29242-87-3: 16160
29261-44-7: 22984
29263-79-4: 21127
29267-67-2: 04706
29276-40-2: 16149
29277-60-9: 24833
29281-39-8: 07989
29285-59-4: 09064
29289-13-2: 02301
29304-65-2: 16462
29311-53-3: 14449
29325-49-3: 00092
29334-16-5: 04997
29342-44-7: 14005
29342-53-8: 07269
29344-95-4: 01779
29366-44-7: 16804
29394-58-9: 08005
29400-42-8: 17340
29420-78-8: 18500
29440-90-2: 19051
29472-95-5: 19494
29488-27-5: 24754
29504-81-2: 24679
29512-84-3: 11202
29512-95-6: 14214
29538-77-0: 10115
29547-08-8: 07531
29548-14-9: 10358
29548-30-9: 11554
29553-26-2: 08598
29554-49-2: 10464
29555-02-0: 10815
29556-72-7: 11201
29558-77-8: 07679
29559-27-1: 03714
29559-52-2: 20722
29568-69-2: 20526
29577-53-5: 01741
29584-99-4: 14339
29606-79-9: 10261
29608-75-1: 07568
29624-17-7: 09830
29636-75-7: 21446
29653-38-1: 09029
29666-53-3: 09742
29682-12-0: 12440
29682-39-1: 02603
29682-41-5: 03081
29682-46-0: 03101
29686-12-2: 18157
29703-52-4: 11596
29723-00-0: 13575
29745-36-6: 22448
29813-49-8: 13358
29815-94-9: 00479
29816-01-1: 13316
29823-16-3: 18418
29833-69-0: 18081
29834-28-4: 06790
29834-29-5: 06791
29834-32-0: 00217
29834-33-1: 00218
29843-58-1: 08778
29865-90-5: 01705
29881-14-9: 02988
29882-57-3: 18194
29882-59-5: 18252
29883-15-6: 04322
29883-16-7: 04323

29886-19-9: 00988
29897-82-3: 23366
29901-85-7: 13768
29903-32-0: 21350
29916-45-8: 07952
29921-38-8: 11317
29921-50-4: 17074
29927-08-0: 07038
29939-31-9: 22581
29943-42-8: 22295
29949-19-7: 15740
29953-71-7: 21873
29964-84-9: 21904
29965-78-4: 10378
29969-57-1: 23598
29972-79-0: 11563
29976-53-2: 20242
30038-31-4: 05152
30042-98-9: 19870
30070-63-4: 06055
30084-90-3: 12754
30090-17-6: 02413
30095-63-7: 18012
30122-12-4: 18214
30168-50-4: 10551
30203-84-0: 02606
30226-37-0: 19014
30273-11-1: 02412
30293-62-0: 02056
30297-94-0: 20097
30316-17-7: 16536
30355-60-3: 25018
30389-18-5: 09778
30433-57-9: 22412
30436-87-4: 16740
30453-91-9: 23197
30453-92-0: 23215
30462-54-5: 07268
30467-02-8: 22224
30468-02-1: 02011
30478-88-7: 16967
30502-73-9: 17233
30538-80-8: 20194
30542-21-3: 09313
30544-34-4: 12874
30564-13-7: 00856
30567-26-1: 13733
30587-18-9: 12398
30600-19-2: 21796
30614-73-4: 13021
30614-77-8: 13018
30649-59-3: 00405
30654-45-6: 16280
30668-92-9: 00809
30673-60-0: 10980
30685-95-1: 08976
30689-41-9: 10737
30692-16-1: 25171
30720-99-1: 00583
30752-19-3: 02654
30767-14-7: 00829
30823-85-9: 22447
30826-83-6: 14880
30834-75-4: 02250
30842-07-0: 04086
30842-90-1: 15495
30854-15-0: 22446
30877-08-8: 16382
30877-30-6: 15143
30893-21-1: 20463
30903-07-2: 11678
30926-02-4: 11046
30954-98-4: 05872
30956-28-6: 00511
30964-01-3: 17326
31005-02-4: 06837
31053-99-3: 02731
31059-19-5: 16481
31116-84-4: 21563
31160-10-8: 19947
31161-71-4: 24379
31167-04-1: 02205
31173-90-7: 02565
31178-68-4: 01373
31178-69-5: 01374
31183-23-0: 11012
31183-24-1: 11011
31187-53-8: 07403
31221-68-8: 23034
31222-02-3: 09746
31252-42-3: 20206
31283-43-9: 17795
31285-35-5: 13170
31294-92-5: 13535
31294-93-6: 13536
31295-56-4: 11501
31297-82-2: 01291
31330-59-3: 16102
31340-44-0: 21149
31366-00-4: 03454
31367-46-1: 13661
31384-83-5: 04402
31401-59-9: 19582
31409-15-1: 23953
31424-45-0: 00289
31431-13-7: 15905
31431-19-3: 15826
31438-22-9: 09171
31482-07-2: 12131
31501-11-8: 14237
31508-08-4: 13823
31508-12-0: 17311
31528-86-6: 00808
31528-89-9: 00794
31543-75-6: 03021
31563-52-7: 22234
31563-54-9: 22233
31574-87-5: 11211
31576-51-9: 11728
31579-70-1: 09398
31599-60-7: 03529
31599-61-8: 03534
31603-77-7: 07556
31604-39-4: 15397
31638-89-8: 19765
31642-67-8: 17305
31649-47-5: 21064
31656-92-5: 02538
31658-04-5: 10343
31673-59-3: 24272
31699-35-1: 17190
31700-39-7: 22519
31702-33-7: 21744
31704-77-5: 17288
31718-05-5: 22865
31720-69-1: 10205
31721-94-5: 06898
31766-07-1: 06763
31772-05-1: 25146
31772-06-2: 17377
31776-66-6: 21065
31786-68-2: 19979
31793-29-0: 22574
31793-83-6: 24254
31797-64-5: 09846
31798-12-6: 08975
31805-48-8: 24731
31841-77-7: 14296
31844-92-5: 07842
31844-95-8: 18622
31844-96-9: 08712
31849-78-2: 08707
31883-01-9: 24524
31883-98-4: 10435
31907-43-4: 16763
31910-10-8: 19728

31915-94-3: 02740
31949-21-0: 12299
31952-00-8: 23070
32018-86-3: 16614
32018-87-4: 16615
32018-88-5: 16617
32018-89-6: 16618
32064-78-1: 17839
32077-77-3: 15291
32079-28-0: 15487
32084-28-9: 23257
32092-27-6: 23015
32098-78-5: 21068
32113-77-2: 11345
32116-10-2: 13399
32116-65-7: 10676
32117-90-1: 19015
32121-06-5: 21691
32121-07-6: 21690
32136-81-5: 12439
32161-06-1: 20295
32174-26-8: 12720
32184-56-8: 22340
32202-61-2: 14988
32226-54-3: 10392
32231-50-8: 08446
32246-34-7: 01704
32253-74-0: 00649
32263-00-6: 22328
32272-48-3: 24523
32315-10-9: 15812
32318-42-6: 06390
32322-24-0: 00128
32322-33-1: 00129
32322-34-2: 00118
32322-39-7: 00145
32322-48-8: 00183
32322-70-6: 00165
32322-71-7: 00132
32322-73-9: 00166
32322-78-4: 00171
32322-79-5: 00157
32328-03-3: 18311
32337-74-9: 18091
32337-96-5: 02660
32360-05-7: 21911
32367-54-7: 16406
32375-81-8: 16784
32387-96-5: 23894
32395-96-3: 03499
32396-89-7: 05826
32400-86-5: 15242
32400-87-6: 15241
32401-31-3: 15240
32401-33-5: 15252
32402-97-4: 17130
32418-95-4: 14845
32433-28-6: 00369
32444-63-6: 21760
32449-92-6: 13243
32454-25-4: 04872
32460-00-7: 12875
32460-20-1: 13142
32470-70-5: 05829
32511-34-5: 07262
32529-58-1: 08087
32541-33-6: 18523
32566-59-9: 01540
32566-63-5: 01547
32566-64-6: 01546
32566-65-7: 01545
32566-66-8: 01507
32566-70-4: 01506
32566-71-5: 01505
32566-72-6: 01504
32566-73-7: 01503
32566-75-9: 01500
32566-77-1: 01499
32566-86-2: 16170
32580-41-9: 12262
32586-14-4: 07896
32588-46-8: 03650
32601-86-8: 23865
32607-54-8: 18409
32617-67-7: 21110
32620-08-9: 08706
32654-60-7: 19017
32666-56-1: 16162
32666-78-7: 01122
32669-06-0: 02823
32676-82-7: 21063
32680-30-1: 07895
32692-19-6: 15130
32704-10-2: 04046
32712-78-0: 06991
32712-80-4: 19255
32718-56-2: 12189
32723-67-4: 01748
32730-85-1: 09633
32736-94-0: 22330
32741-05-2: 24757
32746-66-0: 09058
32750-01-9: 07642
32750-14-4: 07640
32752-09-3: 18809
32768-54-0: 03088
32777-26-7: 09635
32784-07-9: 24657
32810-20-1: 25453
32811-75-9: 10604
32818-79-4: 18351
32830-97-0: 18576
32832-01-2: 09382
32838-55-4: 20205
32853-24-0: 01843
32854-75-4: 00951
32857-63-9: 04170
32899-41-5: 13887
32917-57-0: 03011
32937-55-6: 12341
32940-15-1: 17026
32950-57-5: 00417
32969-09-8: 09656
32985-23-2: 12681
33018-91-6: 13583
33021-02-2: 20584
33021-68-0: 25398
33021-99-7: 25394
33022-01-4: 25391
33022-02-5: 25392
33022-06-9: 25393
33024-34-9: 05350
33036-62-3: 08532
33037-07-9: 20507
33064-59-4: 20232
33073-60-8: 10010
33082-80-3: 25386
33083-83-9: 25313
33091-15-5: 05138
33092-82-9: 10106
33102-63-5: 25431
33119-65-2: 17928
33125-57-4: 03270
33130-04-0: 05260
33184-48-4: 02850
33184-56-4: 05976
33186-02-6: 25025
33190-16-8: 10011
33227-76-8: 04500
33227-90-6: 11061
33228-44-3: 02449
33229-89-9: 13307
33238-56-1: 23271
33238-57-2: 23273
33238-59-4: 23261
33238-60-7: 23264

33243-33-3: 12340
33251-42-2: 23947
33278-12-5: 20431
33345-17-4: 04394
33350-73-1: 07529
33356-01-3: 21358
33357-44-7: 23652
33401-05-7: 20972
33404-57-8: 17057
33407-02-2: 03076
33417-02-6: 16858
33417-03-7: 04944
33417-05-9: 16175
33417-08-2: 16709
33417-27-5: 12780
33420-37-0: 22372
33426-53-8: 22160
33429-70-8: 04008
33445-07-7: 00679
33467-76-4: 13782
33471-69-1: 18156
33486-21-4: 08391
33486-23-6: 08480
33491-40-6: 08322
33491-46-2: 08319
33491-48-4: 08389
33491-51-9: 08320
33491-53-1: 08315
33491-68-8: 08318
33494-88-1: 08321
33499-44-4: 06212
33515-82-1: 09646
33524-31-1: 05035
33553-93-4: 18223
33560-15-5: 00773
33560-17-7: 00769
33560-47-3: 00138
33560-51-9: 00433
33560-53-1: 00385
33560-54-2: 00386
33560-57-5: 00424
33567-59-8: 04088
33599-07-4: 17396
33602-21-0: 08316
33602-29-8: 08317
33603-90-6: 21615
33622-26-3: 11016
33662-96-3: 21226
33663-37-5: 02069
33666-64-7: 10948
33672-51-4: 11427
33672-52-5: 20494
33672-95-6: 06231
33683-44-2: 14357
33683-96-4: 21049
33683-98-6: 21048
33684-00-3: 21054
33684-01-4: 21055
33685-60-8: 01143
33693-48-0: 05102
33695-59-9: 20989
33697-81-3: 04156
33709-18-1: 01357
33709-23-8: 01358
33715-53-6: 24124
33715-54-7: 24031
33715-64-9: 00387
33717-41-8: 01353
33719-74-3: 03084
33721-54-9: 00296
33722-46-2: 14768
33733-31-2: 19828
33738-48-6: 16599
33742-81-3: 09523
33744-35-3: 12074
33753-12-7: 16799
33781-24-7: 01427
33786-89-9: 04452
33794-66-0: 12983
33798-41-3: 12508
33816-56-7: 21052
33829-48-0: 15627
33840-38-9: 10659
33840-43-6: 08050
33843-55-9: 10404
33868-50-7: 12277
33877-11-1: 04963
33880-83-0: 09855
33890-03-8: 04586
33892-75-0: 17011
33912-24-2: 22752
33917-34-9: 14912
33917-36-1: 14913
33917-38-3: 22187
33933-77-6: 17750
33933-78-7: 17278
33933-79-8: 17742
33952-38-4: 04233
33953-73-0: 06952
33966-50-6: 07886
33973-59-0: 20302
33978-71-1: 25308
33982-85-3: 12356
33985-75-0: 19890
34004-41-6: 00134
34008-71-4: 21478
34038-80-7: 24142
34039-11-7: 04764
34039-83-3: 09607
34052-37-4: 15886
34066-96-1: 11924
34069-94-8: 00927
34086-51-6: 06924
34101-86-5: 03303
34103-97-4: 10658
34105-38-9: 08472
34113-69-4: 06088
34145-05-6: 05023
34155-34-5: 20750
34158-76-4: 01601
34164-06-2: 00935
34180-11-5: 20531
34191-51-0: 19325
34202-45-4: 10829
34234-10-1: 01306
34238-52-3: 18661
34238-81-8: 11068
34240-76-1: 12197
34244-14-9: 22499
34246-54-3: 01681
34246-96-3: 22506
34253-18-4: 19338
34263-68-8: 13991
34272-40-7: 20168
34314-06-2: 07614
34341-28-1: 01795
34341-29-2: 01796
34375-89-8: 23359
34376-24-4: 06248
34386-42-0: 05054
34386-71-5: 23120
34386-73-7: 23119
34386-74-8: 23125
34386-76-0: 23122
34386-78-2: 23121
34386-79-3: 23126
34451-18-8: 21182
34486-71-0: 23221
34561-05-2: 11026
34562-14-6: 10288
34566-04-6: 17573
34570-59-7: 20691
34581-76-5: 20692
34593-74-3: 19239
34604-00-7: 13325
34619-03-9: 09189

34636-79-8: 03686
34636-80-1: 03689
34661-33-1: 01909
34662-32-3: 06690
34662-36-7: 06109
34698-41-4: 14985
34701-14-9: 15021
34703-67-8: 14800
34709-44-9: 24574
34713-94-5: 08548
34732-09-7: 05520
34743-88-9: 12108
34757-14-7: 21125
34761-82-5: 02212
34764-02-8: 10916
34784-07-1: 15442
34786-24-8: 03447
34801-09-7: 00076
34817-14-6: 01331
34828-64-3: 15574
34840-27-2: 09042
34840-79-4: 03057
34844-03-6: 12557
34846-44-1: 24682
34846-64-5: 23725
34859-98-8: 18513
34874-30-1: 23287
34883-39-1: 07470
34883-41-5: 07473
34883-43-7: 07469
34883-46-0: 03552
34885-03-5: 10093
34887-14-4: 18217
34893-50-0: 20148
34897-84-2: 04979
34942-91-1: 13787
34946-84-4: 09763
34970-18-8: 09552
34993-66-3: 04714
35000-22-7: 21664
35002-00-7: 15162
35008-86-7: 03518
35011-47-3: 21414
35018-15-6: 23354
35018-28-1: 23370
35019-06-8: 15958
35021-70-6: 03506
35031-72-2: 21477
35042-48-9: 23474
35060-38-9: 16854
35066-45-6: 18786
35071-04-6: 23229
35087-29-7: 22484
35099-89-9: 09631
35113-90-7: 02145
35120-10-6: 00895
35122-96-4: 02172
35128-44-0: 12877
35160-15-7: 24488
35161-67-2: 12228
35161-71-8: 22098
35182-51-5: 22662
35186-69-7: 25084
35187-28-1: 01982
35187-62-3: 18919
35194-37-7: 13769
35203-93-1: 15070
35216-11-6: 24397
35223-80-4: 00490
35242-05-8: 10113
35246-18-5: 15150
35274-05-6: 21187
35275-62-8: 14999
35280-53-6: 00512
35293-36-8: 03972
35305-13-6: 20190
35306-33-3: 13175
35322-84-0: 15101
35355-36-3: 05796
35364-90-0: 03138
35365-59-4: 17515
35383-51-8: 07785
35413-38-8: 10285
35432-36-1: 21008
35434-73-2: 21590
35446-98-1: 03415
35447-68-8: 08676
35449-34-4: 00770
35472-56-1: 06476
35478-71-8: 19243
35480-26-3: 05094
35486-42-1: 22540
35490-78-9: 05914
35495-63-7: 15722
35502-06-8: 20520
35509-27-4: 23728
35509-93-4: 03545
35513-87-2: 08092
35520-81-1: 07294
35532-35-5: 24565
35533-29-0: 24573
35546-33-9: 25038
35554-98-4: 01013
35569-54-1: 24245
35573-93-4: 20518
35578-47-3: 12011
35608-64-1: 08430
35610-74-3: 24564
35615-98-6: 16645
35635-88-2: 24614
35656-49-6: 15233
35657-97-7: 18407
35673-03-1: 02933
35674-21-6: 06624
35674-27-2: 06423
35674-29-4: 05910
35696-87-8: 01785
35718-03-7: 07260
35763-45-2: 18435
35794-11-7: 20152
35799-08-7: 13376
35801-81-1: 12260
35813-38-8: 00302
35826-57-4: 05835
35911-17-2: 21693
35933-63-2: 15337
35944-64-0: 02303
35963-20-3: 07302
35966-97-3: 22146
35975-12-3: 04433
35998-95-9: 02195
36010-89-6: 10617
36010-90-9: 10616
36020-53-8: 23668
36029-52-4: 09129
36049-78-2: 17621
36065-14-2: 10423
36065-15-3: 07355
36069-05-3: 25542
36076-25-2: 04416
36078-59-8: 19492
36092-42-9: 08199
36107-02-5: 23541
36116-84-4: 19817
36122-35-7: 13035
36130-57-1: 15840
36137-96-9: 14816
36138-54-2: 24371
36138-76-8: 19115
36149-75-4: 15271
36157-39-8: 24811
36157-40-1: 12350
36164-42-8: 00372
36166-75-3: 22929
36171-50-3: 03891
36172-06-2: 12083

36190-77-9: 12216
36192-63-9: 15828
36193-52-9: 10647
36215-07-3: 20529
36216-01-0: 01056
36239-09-5: 21116
36297-22-0: 08859
36301-29-8: 03704
36323-28-1: 02558
36357-38-7: 12478
36357-49-0: 02483
36374-95-5: 14795
36378-49-1: 01057
36383-15-0: 03766
36386-49-9: 07312
36386-52-4: 07315
36400-98-3: 17718
36402-31-0: 08577
36432-46-9: 19822
36433-45-1: 19644
36433-46-2: 19647
36436-65-4: 12406
36439-99-3: 16899
36441-32-4: 16949
36441-61-9: 03976
36448-60-9: 21517
36547-35-0: 03334
36556-06-6: 15466
36566-80-0: 14550
36567-72-3: 05228
36586-31-9: 06061
36653-82-4: 13904
36660-46-5: 16569
36667-05-7: 06597
36668-18-5: 12990
36678-06-5: 07778
36678-08-7: 08995
36685-69-5: 00683
36685-72-0: 00684
36691-76-6: 00643
36691-77-7: 25167
36691-78-8: 00698
36710-53-9: 23522
36768-62-4: 20131
36804-45-2: 15827
36804-63-4: 12788
36823-84-4: 07234
36825-36-2: 23524
36827-65-3: 23150
36847-51-5: 08369
36854-57-6: 04354
36892-19-0: 05211
36897-00-4: 16046
36903-89-6: 17588
36903-92-1: 17829
36903-95-4: 17536
36917-36-9: 22654
36944-58-8: 12467
36944-82-8: 16009
36946-70-0: 15111
36994-69-1: 06317
36994-79-3: 04047
37004-69-6: 19689
37018-52-3: 09130
37025-57-3: 05063
37050-06-9: 17851
37083-27-5: 20284
37103-36-9: 22434
37105-12-7: 24705
37137-00-1: 03535
37140-08-2: 25090
37143-54-7: 20479
37343-48-9: 12680
37418-78-3: 04464
37434-59-6: 20873
37437-17-5: 25365
37438-28-1: 07482
37484-40-5: 22215
37493-70-2: 25300
37496-75-6: 22526
37515-51-8: 09416
37517-81-0: 21117
37521-76-9: 16216
37529-67-2: 01938
37549-74-9: 10897
37549-83-0: 14462
37549-89-6: 14440
37557-97-4: 15691
37593-11-6: 00352
37612-52-5: 12319
37642-55-0: 14810
37657-99-1: 14424
37660-22-3: 23242
37674-63-8: 18706
37674-67-2: 14497
37674-72-9: 06875
37675-33-5: 21983
37676-92-9: 09590
37688-96-3: 24355
37710-49-9: 18095
37721-75-8: 19607
37746-78-4: 08864
37763-43-2: 16252
37777-63-2: 04173
37796-79-5: 16217
37806-29-4: 04913
37832-42-1: 02482
37832-46-5: 20226
37845-14-0: 17058
37866-06-1: 08761
37887-99-3: 21463
37899-61-9: 08621
37921-02-1: 23884
37927-97-2: 14836
37935-39-0: 08360
37951-49-8: 21505
37973-84-5: 25335
37985-18-5: 24634
37994-62-0: 12408
37995-08-7: 20414
38024-62-3: 23207
38031-78-6: 22633
38064-88-9: 06072
38064-90-3: 03175
38093-51-5: 04338
38093-52-6: 00186
38093-53-7: 15748
38093-55-9: 00187
38093-58-2: 01852
38093-61-7: 04816
38093-62-8: 00070
38093-63-9: 01812
38093-65-1: 00081
38093-70-8: 04114
38093-76-4: 24501
38093-77-5: 24500
38119-03-8: 06513
38119-08-3: 16661
38119-11-8: 16620
38127-08-1: 10993
38136-29-7: 18597
38160-63-3: 05916
38175-96-1: 06730
38191-34-3: 18989
38204-89-6: 03054
38232-04-1: 24417
38232-05-2: 18608
38235-58-4: 10393
38235-71-1: 06295
38237-74-0: 18536
38240-21-0: 22949
38243-40-2: 21773
38256-94-9: 11714
38256-95-0: 11713
38257-04-4: 01656
38270-15-4: 22002

38300-67-3: 08064
38326-18-0: 20864
38353-98-9: 04458
38384-05-3: 18203
38396-08-6: 16994
38396-21-3: 16299
38396-26-8: 00341
38396-29-1: 16303
38397-06-7: 16944
38397-07-8: 16996
38397-08-9: 16997
38423-42-6: 00264
38432-57-4: 08497
38435-37-9: 12975
38447-22-2: 14121
38460-95-6: 25340
38470-24-5: 00999
38484-94-5: 05435
38487-94-4: 05166
38489-67-7: 21833
38489-76-8: 21778
38511-07-8: 10383
38514-02-2: 17713
38514-03-3: 17714
38514-04-4: 17715
38514-05-5: 17716
38533-19-6: 06994
38552-72-6: 13815
38562-54-8: 08455
38568-41-1: 04545
38570-92-2: 12761
38573-62-5: 02134
38611-89-1: 21089
38624-62-3: 08877
38628-39-6: 25140
38651-65-9: 07360
38661-81-3: 02639
38661-82-4: 02824
38734-05-3: 09598
38738-60-2: 10331
38744-78-4: 10643
38762-41-3: 02026
38765-78-5: 10380
38775-38-1: 13873
38786-78-6: 22253
38802-18-5: 00233
38857-76-0: 16278
38862-78-1: 03117
38870-89-2: 00937
38903-97-8: 12662
38923-57-8: 08395
38936-60-6: 14729
38939-88-7: 02901
38985-72-7: 14668
38985-79-4: 05847
38990-48-6: 16015
38998-17-3: 01745
39003-07-1: 14325
39052-12-5: 02151
39052-57-8: 06965
39060-20-3: 21092
39075-90-6: 19127
39076-02-3: 09095
39086-61-8: 00154
39097-15-9: 06658
39098-85-6: 08688
39101-64-9: 06246
39101-77-4: 06247
39118-35-9: 17842
39121-37-4: 13714
39137-77-4: 03335
39137-79-6: 03333
39141-45-2: 03414
39141-46-3: 03413
39141-47-4: 03411
39141-49-6: 03412
39141-50-9: 03410
39145-47-6: 02923
39149-34-3: 00190
39161-19-8: 18716
39161-85-8: 04241
39178-11-5: 11417
39186-58-8: 04310
39191-07-6: 04895
39228-29-0: 21942
39234-24-7: 07375
39251-86-0: 12970
39251-88-2: 12986
39252-69-2: 12186
39255-32-8: 18440
39257-08-4: 10284
39266-56-3: 04146
39273-81-9: 13808
39473-62-6: 07363
39478-78-9: 02046
39489-79-7: 19008
39511-08-5: 21626
39512-49-7: 20282
39513-19-4: 04435
39513-75-2: 06892
39515-51-0: 01770
39516-65-9: 17166
39525-69-4: 24392
39538-93-7: 14695
39546-32-2: 20234
39563-50-3: 04499
39565-00-9: 12494
39565-07-6: 17126
39568-70-2: 03936
39581-63-0: 23709
39608-47-4: 06108
39620-04-7: 22537
39622-79-2: 04584
39629-88-4: 23378
39669-95-9: 03509
39677-96-8: 24797
39696-92-9: 23178
39713-71-8: 23356
39719-87-4: 00328
39729-13-0: 21434
39731-48-1: 25558
39742-60-4: 20300
39750-95-3: 13851
39755-03-8: 08617
39755-04-9: 18540
39761-61-0: 14422
39773-32-5: 21784
39781-21-0: 16983
39782-38-2: 07988
39782-43-9: 14421
39782-45-1: 08520
39856-57-0: 22563
39857-38-0: 06044
39881-29-3: 21283
39897-10-4: 21088
39905-50-5: 02451
39910-60-6: 00605
39910-98-0: 12483
39924-47-5: 11470
39928-72-8: 13068
39966-95-5: 08051
39989-39-4: 15452
40048-04-2: 13107
40058-87-5: 21097
40061-45-8: 12509
40061-55-0: 04262
40067-41-2: 12810
40067-44-5: 22221
40070-59-5: 19059
40100-94-5: 09708
40108-12-1: 22861
40136-66-1: 11425
40143-47-3: 04401
40143-48-4: 04395
40172-65-4: 17099
40180-96-9: 04838

40203-74-5: 00560
40225-75-0: 17739
40239-35-8: 25173
40239-49-4: 13721
40242-15-7: 16779
40248-63-3: 00681
40248-84-8: 19400
40276-93-5: 14545
40291-39-2: 04260
40296-15-9: 12538
40307-15-1: 11212
40323-88-4: 24732
40336-13-8: 00803
40338-24-7: 14684
40339-27-3: 16314
40348-66-1: 08205
40358-51-8: 16305
40359-34-0: 00631
40367-32-6: 23212
40385-54-4: 02698
40398-32-1: 24878
40398-34-3: 24856
40441-35-8: 10294
40444-43-7: 15935
40452-16-2: 18935
40452-20-8: 18908
40492-93-1: 16972
40514-70-3: 09861
40517-74-6: 05700
40522-79-0: 08301
40527-16-0: 17938
40560-29-0: 07938
40571-45-7: 10497
40575-41-5: 17748
40583-17-3: 00786
40587-63-1: 06160
40587-64-2: 06144
40589-38-6: 08982
40604-49-7: 16971
40605-42-3: 08933
40614-52-6: 08232
40619-38-3: 22228
40620-59-5: 05227
40641-71-2: 07569
40646-07-9: 13587
40646-08-0: 14163
40648-96-2: 14806
40649-36-3: 10282
40650-41-7: 15029
40650-73-5: 16349
40663-68-1: 01776
40682-56-2: 19717
40709-87-3: 11888
40734-75-6: 02588
40738-26-9: 11518
40745-44-6: 00005
40775-09-5: 03492
40786-99-0: 04591
40787-96-0: 02050
40790-29-2: 22311
40807-61-2: 20285
40812-76-8: 07218
40848-33-7: 23232
40856-44-8: 05204
40856-73-3: 14864
40866-07-7: 15582
40870-59-5: 05124
40914-13-4: 00155
40941-54-6: 23639
40960-69-8: 10194
40960-73-4: 10193
41002-50-0: 18745
41011-01-2: 12293
41017-81-6: 00816
41037-13-2: 16449
41037-29-0: 23592
41046-68-8: 04704
41055-92-9: 13699
41065-91-2: 14230
41085-43-2: 02696
41139-98-4: 12791
41150-48-5: 12325
41165-12-2: 02221
41168-91-6: 12031
41173-75-5: 11186
41187-40-0: 23385
41223-14-7: 10498
41252-83-9: 04826
41265-63-8: 13428
41284-12-2: 11864
41298-85-5: 12474
41342-88-5: 04630
41363-39-7: 05399
41372-08-1: 25257
41377-79-1: 22017
41387-98-8: 17954
41412-11-7: 03647
41419-12-9: 20290
41433-81-2: 20841
41438-18-0: 01722
41458-65-5: 19009
41459-42-1: 21083
41465-21-8: 20039
41492-02-8: 24216
41513-04-6: 02604
41532-81-4: 03588
41575-21-7: 03095
41589-42-8: 20812
41635-77-2: 03507
41653-96-7: 14463
41658-69-9: 20968
41683-62-9: 11820
41692-47-1: 14249
41712-14-5: 15068
41727-70-2: 02135
41738-64-1: 11226
41748-71-4: 14961
41753-99-5: 17400
41755-60-6: 13899
41761-11-9: 14027
41761-12-0: 14026
41780-22-7: 14691
41781-17-3: 07873
41804-94-8: 04302
41822-67-7: 04865
41851-34-7: 09898
41851-35-8: 09922
41858-47-3: 05233
41867-20-3: 08855
41888-62-4: 17969
41901-72-8: 03137
41927-01-9: 04431
41927-50-8: 22480
41973-42-6: 18122
41998-16-7: 06053
42007-73-8: 17726
42016-95-5: 02171
42019-78-3: 15899
42020-21-3: 01871
42027-23-6: 18327
42044-26-8: 16368
42048-11-3: 02848
42056-82-6: 13120
42061-72-3: 10275
42065-60-1: 17168
42067-48-1: 18677
42067-49-2: 18676
42074-40-8: 16847
42087-80-9: 06111
42091-29-2: 05971
42101-38-2: 08731
42123-33-1: 01729
42131-85-1: 18625
42131-89-5: 14024
42135-22-8: 12796
42135-37-5: 12759

42151-56-4: 05050
42180-82-5: 05777
42185-47-7: 10289
42195-92-6: 09771
42204-63-7: 15995
42245-37-4: 18168
42286-84-0: 06859
42287-90-1: 05209
42288-38-0: 00611
42293-35-6: 05447
42297-40-5: 22208
42313-79-1: 19104
42328-76-7: 14294
42345-82-4: 12569
42389-30-0: 04426
42398-73-2: 15456
42412-84-0: 21273
42433-61-4: 19485
42435-56-3: 20144
42437-48-9: 04230
42452-48-2: 13434
42454-06-8: 01734
42456-75-7: 12324
42464-80-2: 00370
42467-39-0: 03328
42467-40-3: 03327
42468-15-5: 09457
42474-20-4: 18207
42474-21-5: 13497
42521-82-4: 16986
42525-60-0: 08053
42533-63-1: 02673
42548-38-9: 02758
42548-78-7: 11726
42551-55-3: 08569
42552-70-5: 12668
42563-83-7: 14178
42569-15-3: 18728
42569-16-4: 18727
42576-47-6: 23349
42587-83-7: 12876
42597-24-0: 19860
42597-26-2: 05201
42607-21-6: 24542
42712-64-1: 23783
42775-75-7: 16407
42791-51-5: 12493
42806-50-8: 00733
42831-64-1: 00514
42842-08-0: 20609
42852-95-9: 13201
42860-10-6: 06001
42881-80-1: 02225
42908-64-5: 24729
42964-73-8: 25043
42965-13-9: 00191
42966-64-3: 18174
42969-85-7: 05782
42974-40-3: 06171
42978-66-5: 21924
43014-68-2: 00342
43020-38-8: 06739
43036-06-2: 23345
43038-45-5: 16633
43057-77-8: 01675
43101-48-0: 01465
43120-22-5: 14971
43145-50-2: 23986
43155-07-3: 22413
43192-03-6: 02296
43195-94-4: 04147
43197-78-0: 14010
43201-07-6: 14240
44652-67-7: 13989
44657-76-3: 20659
44815-64-7: 14195
44829-76-7: 13799
44855-57-4: 17572
44914-03-6: 21922
44917-51-3: 08912
44981-94-4: 13300
45019-28-1: 17255
45043-12-7: 08790
45096-22-8: 21330
45163-99-3: 00660
45173-37-3: 11964
45223-18-5: 13862
45277-60-9: 11589
45582-61-4: 20170
45629-38-7: 11240
45892-47-5: 03042
46005-10-1: 22392
46040-83-9: 07690
46231-21-4: 08671
46255-71-4: 16795
46292-93-7: 08197
46329-68-4: 06410
46439-83-2: 03371
46506-88-1: 05863
46733-28-2: 04725
46742-14-7: 14705
46814-61-3: 11942
46878-66-4: 07589
47041-72-5: 18817
47077-29-2: 23688
47172-89-4: 18305
47195-07-3: 01484
47204-22-8: 09415
47458-32-2: 13031
47616-64-8: 07381
48055-66-9: 19924
48115-38-4: 04972
48144-15-6: 03243
48145-04-6: 21955
48213-06-5: 07651
49540-32-1: 25419
49542-66-7: 12096
49550-23-4: 12738
49563-87-3: 11031
49616-71-9: 23710
49619-44-5: 05987
49623-54-3: 07953
49637-59-4: 19849
49639-26-1: 05146
49642-51-5: 14247
49660-56-2: 06893
49662-15-9: 04438
49711-14-0: 21937
49763-65-7: 07233
49763-66-8: 01764
49763-69-1: 01689
49769-78-0: 20877
49805-32-5: 10594
49816-33-3: 08314
49816-51-5: 21258
49816-54-8: 21066
49816-55-9: 21067
49816-56-0: 21112
49816-57-1: 21062
49816-58-2: 08380
49816-59-3: 08390
49816-61-7: 08388
49816-62-8: 08313
49816-63-9: 08356
49826-98-4: 17333
49845-33-2: 23062
49852-35-9: 13801
49855-19-8: 14891
49855-20-1: 14870
49855-22-3: 14799
49865-78-3: 02170
50274-64-1: 15885
50274-95-8: 02784
50283-86-8: 00597
50290-51-2: 23276
50341-35-0: 07948

50342-01-3: 23730
50353-66-7: 14692
50358-39-9: 23530
50358-40-2: 23533
50358-62-8: 23534
50361-05-2: 01445
50375-10-5: 04016
50390-71-1: 09648
50392-39-7: 19005
50397-95-0: 10666
50402-72-7: 20224
50405-39-5: 08990
50407-18-6: 14981
50410-66-7: 20948
50433-64-2: 00980
50446-43-0: 24317
50461-74-0: 18461
50468-22-9: 08207
50530-12-6: 10958
50548-40-8: 11137
50548-43-1: 11140
50551-88-7: 14503
50552-30-2: 14523
50560-96-8: 22241
50573-74-5: 05941
50583-27-2: 24914
50592-87-5: 14001
50599-73-0: 14370
50606-95-6: 07216
50606-96-7: 07215
50606-97-8: 07230
50609-15-9: 00069
50610-32-7: 00179
50619-00-6: 05410
50619-01-7: 05406
50623-57-9: 17244
50639-02-6: 25316
50651-39-3: 00298
50654-82-5: 21675
50677-27-5: 06060
50688-77-2: 15927
50689-17-3: 12871
50690-12-5: 12331
50709-35-8: 14607
50709-36-9: 14608
50710-40-2: 07929
50729-10-7: 16603
50729-11-8: 16604
50773-21-2: 21831
50779-65-2: 05815
50796-71-9: 05988
50817-44-2: 00455
50817-70-4: 00025
50840-23-8: 23029
50853-52-6: 12027
50868-72-9: 02325
50869-79-9: 19943
50870-10-5: 16177
50870-59-2: 15082
50876-32-9: 09962
50876-33-0: 10575
50877-87-7: 22474
50884-83-8: 23208
50887-69-9: 23103
50888-63-6: 22310
50896-19-0: 07342
50902-82-4: 14233
50906-77-9: 21172
50915-63-4: 16666
50975-84-3: 21923
51004-21-8: 08636
51030-24-1: 01319
51034-46-9: 07786
51036-79-4: 14825
51037-03-7: 16164
51042-51-4: 03802
51054-41-2: 00883
51072-66-3: 01570
51079-52-8: 13658
51104-87-1: 07807
51108-30-6: 15372
51114-68-2: 02089
51115-02-7: 10717
51116-73-5: 18187
51118-27-5: 25050
51122-94-2: 10317
51136-52-8: 04457
51149-87-2: 16932
51152-12-6: 07305
51169-05-2: 22986
51175-71-4: 24824
51181-40-9: 10020
51220-00-9: 21757
51261-14-4: 17583
51264-69-8: 00633
51282-49-6: 06116
51284-68-5: 11041
51290-78-9: 22728
51290-79-0: 22727
51299-55-9: 10291
51306-09-3: 17600
51308-99-7: 17262
51309-10-5: 10983
51309-55-8: 19566
51333-70-1: 17184
51339-41-4: 15917
51373-56-9: 09293
51437-67-3: 11042
51451-26-4: 21491
51472-78-7: 11743
51496-26-5: 02944
51500-32-4: 22394
51524-84-6: 06478
51528-17-7: 19493
51541-42-5: 20218
51546-31-7: 14918
51556-10-6: 13829
51575-85-0: 13834
51584-21-5: 24704
51592-06-4: 14750
51594-55-9: 17997
51616-56-9: 09216
51643-46-0: 22782
51656-58-7: 05822
51674-11-4: 15376
51686-60-3: 10667
51686-78-3: 03028
51698-77-2: 00778
51703-69-6: 17778
51707-35-8: 06372
51717-58-9: 22921
51721-39-2: 11024
51727-42-5: 15869
51741-72-1: 00282
51750-65-3: 14080
51756-08-2: 08413
51772-30-6: 21514
51773-77-4: 07359
51786-10-8: 06080
51786-70-0: 07928
51794-07-1: 04234
51801-69-5: 03700
51806-20-3: 11828
51806-90-7: 25087
51807-53-5: 14761
51818-99-6: 01579
51822-55-0: 15377
51832-31-6: 04530
51834-90-3: 04901
51863-60-6: 12364
51877-58-8: 16912
51884-03-8: 01806
51892-04-7: 07167
51892-26-3: 03114
51903-34-5: 14351
51907-25-6: 06046

51934-45-3: 16644
51952-42-2: 18658
51953-03-8: 22148
51971-64-3: 15313
52008-15-8: 24683
52019-78-0: 14291
52033-97-3: 10675
52045-21-3: 19217
52077-96-0: 05366
52078-56-5: 25106
52092-47-4: 22609
52093-45-5: 04983
52096-49-8: 16938
52097-68-4: 24349
52099-72-6: 05649
52111-19-0: 19800
52111-97-4: 08725
52126-90-6: 13076
52147-97-4: 12612
52152-71-3: 08003
52152-72-4: 18250
52161-54-3: 03201
52182-15-7: 04238
52184-13-1: 19309
52195-70-7: 00805
52196-46-0: 15947
52208-62-5: 03812
52217-02-4: 08653
52220-64-1: 12297
52222-87-4: 15984
52244-70-9: 04376
52251-71-5: 01149
52301-88-9: 05014
52305-66-5: 20146
52313-87-8: 18690
52324-03-5: 18801
52340-78-0: 11999
52358-73-3: 16323
52386-09-1: 07069
52387-29-8: 13192
52387-50-5: 14478
52390-72-4: 13650
52402-00-3: 04632
52405-73-9: 01869
52414-97-8: 02692
52414-98-9: 02700
52418-50-5: 18785
52418-81-2: 14306
52421-81-5: 04590
52427-05-1: 19119
52430-83-8: 10499
52431-74-0: 01320
52460-86-3: 04160
52460-99-8: 04305
52488-28-5: 02693
52494-33-4: 24460
52497-07-1: 08728
52499-06-6: 21435
52525-94-7: 12773
52553-11-4: 24970
52562-19-3: 02365
52562-50-2: 15192
52569-87-6: 05321
52591-10-3: 15996
52605-49-9: 13328
52610-17-0: 02227
52632-06-1: 08441
52662-65-4: 14679
52687-81-7: 20401
52688-89-8: 10574
52704-04-8: 06578
52708-03-9: 17272
52709-42-9: 00712
52710-27-7: 07236
52718-65-7: 10058
52722-79-9: 14657
52744-85-1: 21874
52780-23-1: 04689
52780-67-3: 19238
52802-07-0: 14292
52805-92-2: 03612
52819-37-1: 09122
52896-87-4: 13546
52896-88-5: 13526
52896-89-6: 13527
52896-90-9: 13525
52896-91-0: 13524
52896-92-1: 13558
52896-93-2: 13562
52896-95-4: 13563
52897-00-4: 14044
52897-01-5: 14048
52897-03-7: 14049
52897-04-8: 14046
52897-05-9: 14050
52897-06-0: 14047
52897-08-2: 14078
52897-09-3: 14079
52897-10-6: 14083
52897-11-7: 14084
52897-12-8: 14085
52897-15-1: 14086
52897-16-2: 18226
52897-17-3: 18245
52897-19-5: 18246
52900-79-5: 13014
52914-66-6: 10480
52927-23-8: 16907
52938-79-1: 19242
52944-34-0: 02946
52945-73-0: 08273
53018-27-2: 10092
53022-14-3: 08469
53034-80-3: 21057
53045-71-9: 18623
53046-20-1: 03702
53046-84-7: 05576
53055-05-3: 01750
53066-19-6: 05028
53078-85-6: 02040
53097-56-6: 05194
53109-18-5: 10741
53112-33-7: 15469
53117-17-2: 00718
53119-51-0: 24733
53138-32-2: 06166
53142-19-1: 02030
53152-98-0: 20180
53172-84-2: 03630
53175-28-3: 08691
53178-20-4: 07926
53188-23-1: 12860
53188-82-2: 08702
53209-33-9: 00326
53213-47-1: 17213
53220-82-9: 16254
53222-92-7: 19024
53252-08-7: 13807
53252-19-0: 14519
53252-20-3: 17309
53252-21-4: 14517
53288-40-7: 10233
53299-53-9: 02940
53324-38-2: 02052
53346-71-7: 18072
53348-04-2: 18921
53363-87-4: 15534
53398-86-0: 14236
53398-87-1: 14458
53399-29-4: 23863
53440-12-3: 16692
53457-67-3: 23862
53491-32-0: 05778
53497-32-8: 23543
53503-63-2: 16759
53526-29-7: 16292

53535-68-5: 21534
53554-45-3: 00115
53566-37-3: 14548
53584-29-5: 20905
53584-56-8: 08654
53588-92-4: 11768
53600-19-4: 16205
53606-41-0: 19415
53608-85-8: 09808
53608-93-8: 20811
53611-17-9: 23770
53622-33-6: 18926
53631-70-2: 04974
53657-59-3: 24956
53660-21-2: 13654
53663-16-4: 04367
53663-23-3: 04048
53663-24-4: 05279
53663-26-6: 17245
53663-33-5: 09509
53689-25-1: 00153
53696-14-3: 10972
53702-18-4: 07356
53711-77-6: 15173
53716-51-1: 08081
53716-52-2: 08082
53716-53-3: 08079
53716-54-4: 08083
53716-55-5: 08084
53716-60-2: 07898
53716-61-3: 07899
53716-62-4: 11744
53723-87-8: 11919
53751-40-9: 05037
53760-37-5: 01178
53767-59-2: 07264
53776-69-5: 03855
53778-42-0: 20735
53778-43-1: 10776
53778-70-4: 00854
53778-72-6: 08587
53778-73-7: 08586
53783-86-1: 09545
53783-88-3: 14376
53800-01-4: 00844
53848-17-2: 02041
53854-17-4: 14915
53859-09-9: 23149
53859-78-2: 09614
53867-94-0: 23536
53889-94-4: 19518
53894-19-2: 21379
53897-29-3: 21725
53897-33-9: 08602
53897-34-0: 08601
53897-35-1: 00855
53897-36-2: 07901
53897-38-4: 00853
53897-47-5: 08249
53897-50-0: 18377
53897-51-1: 08025
53907-46-3: 21212
53907-59-8: 18674
53907-70-3: 18734
53907-73-6: 14491
53907-74-7: 08028
53907-80-5: 10696
53907-81-6: 13472
53907-86-1: 17575
53907-87-2: 07902
53907-95-2: 21412
53917-01-4: 08662
53918-78-8: 16725
53933-94-1: 16741
53941-19-8: 14431
53951-40-9: 20646
53951-42-1: 14531
53951-43-2: 11324
53951-44-3: 11327
53956-11-9: 10698
53956-43-7: 15585
53957-22-5: 21421
53957-32-7: 04873
53957-33-8: 05044
53966-36-2: 11423
53966-37-3: 08086
53966-42-0: 12251
53966-44-2: 24684
53966-52-2: 17806
53966-53-3: 17299
53973-14-1: 08954
53977-99-4: 14412
53993-41-2: 08870
54004-27-2: 14474
54004-38-5: 15457
54004-40-9: 18496
54004-42-1: 08386
54004-43-2: 18519
54004-44-3: 08345
54004-46-5: 22265
54013-40-0: 07155
54020-53-0: 15144
54063-11-5: 23309
54063-12-6: 13548
54063-14-8: 11288
54063-15-9: 11286
54063-16-0: 11285
54063-17-1: 25186
54063-18-2: 12579
54069-00-0: 23498
54069-01-1: 23463
54069-04-4: 23081
54069-09-9: 23507
54069-10-2: 23465
54069-11-3: 23466
54069-12-4: 23459
54069-15-7: 23265
54069-21-5: 23250
54069-24-8: 23246
54069-25-9: 23245
54069-33-9: 23260
54069-40-8: 13318
54069-44-2: 23460
54069-48-6: 14919
54069-49-7: 23249
54069-51-1: 23270
54069-60-2: 23235
54069-62-4: 23237
54069-85-1: 23084
54069-88-4: 23247
54088-61-8: 15444
54105-76-9: 09894
54106-40-0: 03715
54120-69-3: 11279
54131-44-1: 07299
54135-80-7: 04015
54147-90-9: 07564
54147-91-0: 07560
54147-94-3: 07565
54151-35-8: 21550
54160-32-6: 20270
54166-53-9: 11424
54198-74-2: 23098
54237-36-4: 07077
54244-74-5: 24738
54244-90-5: 13804
54245-33-9: 16202
54246-65-0: 23076
54256-43-8: 07204
54268-02-9: 20693
54277-60-0: 25353
54295-18-0: 23248
54295-23-7: 14914
54295-24-8: 23506
54298-97-4: 17840
54299-08-0: 25349

54305-87-2: 14025
54305-90-7: 14022
54306-00-2: 14416
54340-87-3: 15031
54340-88-4: 15032
54340-90-8: 14062
54340-92-0: 17746
54340-93-1: 21238
54340-94-2: 08630
54340-96-4: 14443
54340-99-7: 15176
54353-11-6: 20343
54357-08-3: 16456
54365-75-2: 05807
54365-81-0: 20062
54383-22-1: 10060
54388-93-1: 07710
54389-64-9: 07625
54393-71-4: 16248
54408-50-3: 23532
54410-62-7: 20493
54410-69-4: 03726
54410-74-1: 03676
54410-75-2: 03143
54410-83-2: 22329
54410-84-3: 08750
54410-89-8: 10991
54410-90-1: 10191
54410-91-2: 20060
54410-92-3: 20061
54410-93-4: 20040
54410-94-5: 08458
54410-95-6: 02941
54410-96-7: 05776
54410-98-9: 17297
54410-99-0: 09811
54411-00-6: 09903
54411-01-7: 09920
54411-02-8: 09921
54411-03-9: 11015
54411-05-1: 24736
54411-06-2: 24686
54411-07-3: 24720
54411-10-8: 12173
54411-13-1: 20778
54411-16-4: 14453
54429-56-0: 12044
54454-10-3: 22396
54460-96-7: 20676
54460-98-9: 09221
54460-99-0: 13689
54461-00-6: 14279
54461-05-1: 19178
54462-68-9: 14018
54462-70-3: 07954
54467-95-7: 02536
54470-37-0: 11141
54484-61-6: 03074
54484-63-8: 03077
54484-66-1: 14423
54484-72-9: 20648
54484-73-0: 14041
54495-62-4: 03732
54506-04-6: 17947
54515-77-4: 00741
54518-00-2: 03260
54518-01-3: 03904
54518-03-5: 21337
54518-04-6: 15807
54527-68-3: 08487
54530-94-8: 05434
54532-15-9: 13539
54532-97-7: 07456
54547-03-4: 19656
54549-71-2: 21176
54549-80-3: 10529
54553-14-9: 23387
54573-13-6: 13830
54576-32-8: 07907
54576-44-2: 13019
54587-48-3: 21141
54595-34-5: 12678
54619-29-3: 21304
54625-53-5: 02958
54630-50-1: 13662
54638-10-7: 18501
54644-17-6: 21398
54644-21-2: 02942
54644-22-3: 02943
54644-23-4: 02938
54653-26-8: 18210
54657-08-8: 00693
54676-39-0: 09814
54682-16-5: 09458
54696-05-8: 12506
54700-36-6: 01716
54712-18-4: 17724
54715-57-0: 19165
54716-29-9: 21253
54724-99-1: 22125
54737-45-0: 02713
54750-05-9: 06589
54755-20-3: 16458
54755-21-4: 16459
54771-88-9: 15856
54774-82-2: 02971
54774-83-3: 13688
54774-85-5: 07876
54774-91-3: 21926
54774-99-1: 21147
54775-00-7: 18171
54775-01-8: 13687
54784-08-6: 16595
54789-15-0: 03442
54789-21-8: 15098
54789-22-9: 15099
54789-23-0: 15100
54789-24-1: 04274
54789-26-3: 17625
54789-29-6: 03096
54789-30-9: 02831
54789-37-6: 15166
54798-91-3: 05026
54798-95-7: 07048
54811-53-9: 05771
54812-56-5: 14600
54815-12-2: 14066
54815-13-3: 17210
54815-14-4: 05780
54815-15-5: 05779
54815-20-2: 02756
54815-21-3: 05760
54815-22-4: 18179
54815-25-7: 18274
54815-28-0: 13581
54819-86-2: 21093
54824-00-9: 16367
54824-08-7: 21466
54827-17-7: 07615
54828-00-1: 19157
54829-48-0: 12902
54833-04-4: 07923
54833-05-5: 20702
54833-19-1: 05804
54833-23-7: 11582
54833-30-6: 09949
54833-31-7: 09947
54833-34-0: 09870
54837-90-0: 02414
54842-27-2: 22932
54845-24-8: 08376
54845-26-0: 13758
54845-27-1: 14455
54845-28-2: 14457
54845-33-9: 24674
54845-34-0: 24752

54845-35-1: 24688
54845-36-2: 02775
54845-41-9: 04639
54845-45-3: 13027
54852-64-1: 03831
54852-66-3: 12132
54852-68-5: 19100
54852-70-9: 20929
54852-73-2: 07504
54852-77-6: 10000
54862-92-9: 13722
54877-00-8: 14167
54881-18-4: 22863
54884-99-0: 23701
54885-00-6: 16038
54889-48-4: 17599
54889-49-5: 17615
54889-53-1: 14991
54889-54-2: 14992
54889-55-3: 15045
54889-56-4: 16404
54889-59-7: 14951
54889-78-0: 03218
54889-79-1: 03405
54889-80-4: 03749
54889-82-6: 01997
54889-83-7: 03909
54889-97-3: 03227
54889-98-4: 15808
54890-00-5: 09813
54890-01-6: 09857
54890-02-7: 09930
54890-04-9: 08071
54890-05-0: 12138
54898-73-6: 10002
54904-39-1: 23838
54922-86-0: 20590
54932-67-1: 19241
54932-72-8: 02602
54932-74-0: 18192
54932-77-3: 19089
54932-78-4: 19620
54932-79-5: 03728
54932-83-1: 20941
54932-84-2: 03053
54932-85-3: 05781
54932-87-5: 03237
54932-91-1: 03776
54934-49-5: 21138
54934-51-9: 04867
54934-66-6: 05668
54934-69-9: 16390
54934-71-3: 10505
54934-87-1: 12148
54934-90-6: 09900
54934-91-7: 09954
54935-00-1: 03041
54947-74-9: 17685
54964-79-3: 24675
54964-84-0: 16318
54965-00-3: 03044
54965-01-4: 03043
54965-02-5: 03045
54965-04-7: 05775
54965-07-0: 05786
54965-08-1: 05787
54965-39-8: 16237
54965-40-1: 02791
54965-52-5: 21451
54965-53-6: 21450
54965-70-7: 04027
54984-07-5: 17700
54986-31-1: 25324
54986-38-8: 18578
54986-39-9: 10978
54986-44-6: 04065
54986-45-7: 03205
54986-54-8: 14993
54986-62-8: 09401
54986-63-9: 06747
55000-40-3: 07049
55000-44-7: 06203
55000-45-8: 06204
55000-46-9: 06214
55000-47-0: 06215
55000-48-1: 06207
55000-49-2: 06208
55000-53-8: 16370
55000-54-9: 16371
55000-55-0: 16272
55000-56-1: 16271
55012-69-6: 21452
55012-73-2: 24673
55030-16-5: 24737
55030-20-1: 09887
55030-21-2: 09955
55030-45-0: 14995
55030-46-1: 02530
55030-56-3: 08724
55030-57-4: 08726
55030-60-9: 14954
55030-62-1: 25147
55030-63-2: 07525
55038-01-2: 05817
55044-09-2: 03854
55044-10-5: 13413
55044-31-0: 14996
55044-32-1: 09944
55044-33-2: 14997
55044-34-3: 15004
55044-35-4: 15005
55044-36-5: 14998
55044-73-0: 01566
55044-74-1: 01565
55044-77-4: 10544
55044-96-7: 15952
55044-97-8: 03847
55045-05-1: 01512
55045-09-5: 25153
55045-12-0: 24357
55045-13-1: 24358
55057-45-9: 21596
55059-22-8: 07454
55059-23-9: 20049
55066-53-0: 17506
55073-86-4: 14328
55124-77-1: 09939
55124-79-3: 13408
55124-86-2: 10955
55125-02-5: 16441
55125-03-6: 18830
55125-22-9: 14115
55125-25-2: 03679
55130-16-0: 17492
55133-84-1: 25334
55133-89-6: 01126
55133-98-7: 09858
55133-99-8: 06462
55134-05-9: 13580
55134-06-0: 04069
55134-07-1: 03873
55134-08-2: 03203
55134-09-3: 03204
55153-02-1: 07495
55156-16-6: 18563
55160-26-4: 16211
55162-33-9: 05805
55162-34-0: 02593
55162-35-1: 20547
55162-37-3: 11759
55162-38-4: 17200
55162-40-8: 06117
55182-38-2: 10213
55182-83-7: 25350
55186-89-5: 11089
55191-11-2: 21461

55191-12-3: 15126
55191-16-7: 03058
55191-17-8: 03122
55191-18-9: 14131
55191-19-0: 17643
55191-20-3: 15882
55191-25-8: 03258
55191-36-1: 03497
55191-38-3: 03127
55191-39-4: 03259
55191-40-7: 16142
55191-41-8: 22501
55191-43-0: 11509
55191-44-1: 11512
55191-59-8: 12232
55191-60-1: 03298
55191-62-3: 02987
55191-63-4: 16477
55255-45-3: 10942
55255-59-9: 16520
55255-64-6: 06539
55255-70-4: 01131
55255-71-5: 11124
55255-72-6: 24316
55255-73-7: 25175
55255-74-8: 09968
55255-85-1: 10506
55255-91-9: 09943
55256-07-0: 24416
55256-08-1: 25221
55256-09-2: 24951
55256-23-0: 15039
55256-24-1: 11125
55256-26-3: 01513
55268-62-7: 03297
55268-63-8: 03955
55268-64-9: 16315
55268-69-4: 13876
55268-70-7: 13878
55268-72-9: 18062
55270-71-8: 15898
55281-91-9: 09884
55282-03-6: 03887
55282-10-5: 13389
55282-12-7: 17379
55282-14-9: 11449
55282-15-0: 11448
55282-16-1: 11447
55282-17-2: 24332
55282-27-4: 17380
55282-28-5: 13407
55282-29-6: 13868
55282-30-9: 24333
55282-31-0: 11451
55282-32-1: 11581
55282-33-2: 11577
55282-34-3: 09983
55282-64-9: 03886
55282-67-2: 04066
55282-69-4: 00020
55282-70-7: 15000
55289-36-6: 02042
55301-53-6: 06118
55314-57-3: 18787
55319-78-3: 16307
55319-81-8: 16320
55319-83-0: 11453
55320-00-8: 16321
55320-01-9: 17469
55320-02-0: 17470
55320-03-1: 17360
55320-04-2: 17361
55320-06-4: 13388
55320-07-5: 13622
55320-13-3: 24331
55321-99-8: 22361
55334-01-5: 18866
55334-02-6: 09798
55334-08-2: 09959
55334-09-3: 09850
55334-11-7: 07422
55334-13-9: 00019
55334-18-4: 07444
55334-19-5: 09837
55334-22-0: 18865
55334-29-7: 15041
55334-30-0: 02976
55334-31-1: 02985
55334-56-0: 00021
55334-57-1: 12136
55334-71-9: 15001
55334-72-0: 02992
55334-73-1: 09817
55334-75-3: 11460
55334-78-6: 11592
55341-62-3: 12753
55373-86-9: 11452
55373-87-0: 08400
55373-89-2: 18051
55373-90-5: 02993
55373-91-6: 02994
55373-96-1: 16567
55401-54-2: 24335
55401-62-2: 17394
55401-63-3: 17484
55401-64-4: 17485
55401-72-4: 09819
55401-73-5: 15043
55401-76-8: 10542
55401-77-9: 18864
55419-66-4: 00833
55424-74-3: 18456
55429-26-0: 10511
55429-27-1: 09826
55429-35-1: 10507
55429-67-9: 11588
55429-68-0: 11587
55429-83-9: 13846
55429-84-0: 24330
55441-71-9: 22176
55441-72-0: 22178
55449-46-2: 02947
55454-22-3: 13783
55470-97-8: 17381
55470-98-9: 11580
55474-90-3: 09505
55476-89-6: 21660
55482-22-9: 23244
55488-93-2: 16150
55517-73-2: 13845
55517-74-3: 03505
55517-88-9: 03159
55517-89-0: 18049
55521-12-5: 03261
55521-27-2: 09960
55530-51-3: 14952
55538-59-5: 07409
55538-62-0: 02962
55538-69-7: 03112
55538-73-3: 19655
55538-85-7: 07900
55563-72-9: 21508
55590-50-6: 12438
55605-27-1: 15735
55621-90-4: 18502
55638-55-6: 10826
55665-79-7: 14469
55669-54-0: 11062
55671-32-4: 09469
55676-76-1: 07629
55682-90-1: 22500
55682-91-2: 13398
55696-02-1: 09074
55712-59-9: 07114
55712-60-2: 07115
55712-68-0: 00097

55719-85-2: 08905
55751-54-7: 02411
55764-37-9: 18750
55775-41-2: 07814
55776-52-8: 19646
55776-56-2: 19645
55784-69-5: 07338
55789-02-1: 05065
55836-31-2: 14953
55836-69-6: 01881
55836-71-0: 01882
55877-01-5: 14145
55882-21-8: 20707
55899-01-9: 00325
55912-20-4: 05017
55933-85-2: 12382
55937-94-5: 18539
55939-13-4: 00006
55944-70-2: 13640
55955-94-7: 08701
55955-97-0: 07168
56008-61-8: 01578
56020-58-7: 16642
56042-78-5: 01929
56043-01-7: 06666
56046-20-9: 15926
56049-76-4: 22289
56053-84-0: 24804
56059-65-5: 12957
56072-27-6: 13701
56075-50-4: 25359
56079-16-4: 22442
56095-81-9: 00894
56097-31-5: 15392
56097-32-6: 15402
56103-67-4: 19081
56106-37-7: 03292
56107-02-9: 15887
56128-66-6: 06612
56135-03-6: 04253
56148-65-3: 06040
56180-61-1: 18135
56183-17-6: 02008
56196-53-3: 17686
56201-37-7: 19980
56216-12-7: 08871
56241-09-9: 15483
56242-69-4: 24489
56245-02-4: 03765
56246-88-9: 04129
56247-75-7: 14994
56247-76-8: 16566
56247-94-0: 22502
56247-97-3: 03956
56248-65-8: 15042
56272-29-8: 08173
56272-53-8: 10135
56363-84-9: 06081
56375-83-8: 05405
56387-60-1: 14710
56404-36-5: 11001
56406-50-9: 04399
56456-42-9: 03285
56456-50-9: 05010
56475-20-8: 21824
56511-20-7: 20413
56523-26-3: 17531
56527-83-4: 07655
56541-64-1: 16970
56598-72-2: 16418
56600-58-9: 01289
56630-30-9: 17564
56630-31-0: 17324
56630-32-1: 17325
56630-33-2: 17321
56630-34-3: 17323
56661-21-3: 19491
56667-12-0: 03322
56680-65-0: 19614
56688-75-6: 10325
56694-98-5: 25137
56739-95-8: 03547
56745-61-0: 14343
56759-60-5: 19636
56776-55-7: 05383
56800-09-0: 08768
56820-58-7: 16916
56820-70-3: 16915
56862-18-1: 05753
56886-06-7: 05196
56892-32-1: 06951
56908-88-4: 02999
56911-77-4: 03839
56917-14-7: 01679
56920-82-2: 12868
56949-65-6: 12125
56952-31-9: 00522
56961-01-4: 03941
56961-13-8: 19150
56961-14-9: 19155
56961-15-0: 19156
56961-16-1: 02960
56961-17-2: 19229
56961-18-3: 19231
56961-19-4: 19228
56961-20-7: 04680
56961-21-8: 05586
56961-22-9: 05589
56961-23-0: 05594
56961-24-1: 05577
56961-25-2: 05906
56961-26-3: 05998
56961-27-4: 06000
56961-30-9: 06082
56961-31-0: 06083
56961-32-1: 06085
56961-33-2: 06099
56961-34-3: 09745
56961-36-5: 16295
56961-37-6: 16300
56961-38-7: 16301
56961-39-8: 16302
56961-40-1: 17006
56961-75-2: 01783
56961-76-3: 01787
56961-77-4: 02608
56961-79-6: 23587
56961-80-9: 23588
56961-81-0: 23596
56961-82-1: 04007
56961-83-2: 03052
56961-84-3: 03050
56961-85-4: 03051
56961-86-5: 04024
56961-87-6: 00160
56961-88-7: 00164
56961-89-8: 16667
56961-90-1: 16669
56961-91-2: 16775
56961-92-3: 16780
56961-93-4: 16785
56961-95-6: 16822
56961-96-7: 18126
56961-97-8: 18127
56961-98-9: 18928
56961-99-0: 18929
56962-00-6: 18991
56962-01-7: 18994
56962-02-8: 19000
56962-03-9: 19003
56962-04-0: 19098
56962-05-1: 19163
56962-06-2: 04550
56962-07-3: 15312
56962-08-4: 04553
56962-10-8: 01606

56962-11-9: 01607
56962-12-0: 01612
56962-13-1: 01633
56962-14-2: 01635
56962-15-3: 01634
56962-16-4: 01778
56964-65-9: 05252
56970-11-7: 07570
56978-43-9: 00163
56978-45-1: 00140
56978-46-2: 00141
56978-48-4: 02028
56978-49-5: 15881
56978-50-8: 07475
56987-35-0: 11750
57044-25-4: 18037
57049-35-1: 04969
57051-32-8: 15793
57052-99-0: 04608
57053-02-8: 04585
57074-96-1: 17853
57129-02-9: 09360
57129-69-8: 13084
57150-66-0: 19879
57165-81-8: 23527
57179-35-8: 01713
57193-29-0: 21136
57223-18-4: 17310
57296-00-1: 21789
57322-42-6: 16836
57323-93-0: 14369
57338-87-1: 01833
57342-21-9: 23504
57356-27-1: 02860
57383-80-9: 19208
57383-81-0: 19632
57386-83-1: 03124
57396-89-1: 16979
57396-91-5: 14592
57399-97-0: 09043
57403-35-7: 02783
57404-54-3: 18932
57413-33-9: 20038
57415-35-7: 01747
57433-02-0: 21801
57445-42-8: 14813
57450-55-2: 24903
57455-06-8: 05091
57456-98-1: 07818
57477-69-7: 07126
57479-70-6: 06093
57486-68-7: 04151
57502-49-5: 20164
57536-57-9: 11197
57536-88-6: 20079
57558-57-3: 08948
57567-84-7: 10036
57568-95-3: 07561
57576-09-7: 10168
57597-99-6: 19324
57605-95-5: 16849
57616-17-8: 12036
57650-84-7: 03589
57652-76-3: 01999
57671-18-8: 24401
57673-92-4: 19887
57697-64-0: 08888
57706-88-4: 17734
57728-60-6: 13291
57772-50-6: 04977
57772-57-3: 06357
57785-85-0: 22868
57791-59-0: 11275
57794-08-8: 10061
57822-22-7: 09554
57856-10-7: 18546
57883-04-2: 19143
57885-43-5: 21076
57912-35-3: 10876
57933-83-2: 09230
57968-71-5: 08142
57990-92-8: 01262
58010-91-6: 02385
58049-91-5: 10707
58100-26-8: 10865
58119-66-7: 24780
58123-77-6: 06356
58138-81-1: 08618
58153-92-7: 04998
58153-94-9: 05006
58153-97-2: 04992
58160-28-4: 09725
58167-01-4: 08141
58169-99-6: 19606
58175-67-0: 08123
58210-03-0: 10212
58235-81-7: 12966
58243-21-3: 09748
58288-30-5: 15729
58291-77-3: 19175
58313-23-8: 06416
58319-69-0: 09466
58359-53-8: 05519
58371-98-5: 18547
58373-59-4: 02150
58380-11-3: 06010
58401-43-7: 23520
58416-32-3: 23601
58417-88-2: 24556
58469-70-8: 20446
58474-80-9: 10996
58479-61-1: 23989
58502-85-5: 03629
58556-75-5: 16330
58566-66-8: 02146
58588-28-6: 14307
58593-78-5: 04786
58616-94-7: 09561
58616-95-8: 10828
58616-97-0: 09564
58618-39-6: 16723
58618-91-0: 18176
58625-77-7: 08642
58654-67-4: 17758
58665-28-4: 16434
58728-64-6: 16583
58746-36-4: 17229
58766-17-9: 14009
58780-82-8: 04954
58793-63-8: 24994
58795-34-9: 10438
58832-36-3: 12996
58873-44-2: 18882
58901-13-6: 21404
58917-25-2: 13086
58920-49-3: 11213
58979-27-4: 06049
59020-74-5: 05790
59022-71-8: 15276
59039-83-7: 01000
59110-36-0: 07253
59113-36-9: 20891
59189-52-5: 15916
59204-02-3: 17706
59219-75-9: 05828
59228-00-1: 21424
59229-75-3: 04469
59255-95-7: 02049
59289-98-4: 23374
59290-34-5: 22801
59336-59-3: 21635
59357-07-2: 14359
59368-15-9: 09916
59376-54-4: 00619
59424-96-3: 08866
59433-08-8: 20169

59434-05-8: 24718
59443-94-6: 05190
59471-80-6: 10262
59485-34-6: 03301
59551-02-9: 23727
59609-05-1: 11295
59646-36-5: 06468
59654-13-6: 14060
59672-19-4: 11977
59697-51-7: 14393
59697-55-1: 14394
59708-36-0: 09174
59736-00-4: 00078
59755-25-8: 00253
59765-98-9: 01974
59777-92-3: 09645
59840-70-9: 09100
59869-24-8: 19914
59889-45-1: 18196
59893-28-6: 01725
59905-55-4: 07400
59956-48-8: 00488
59969-91-4: 20656
60012-63-7: 05165
60019-49-0: 22817
60021-28-5: 07707
60033-03-6: 13197
60060-44-8: 10071
60096-09-5: 13588
60111-14-0: 14501
60182-96-9: 17345
60185-80-0: 23091
60211-57-6: 05025
60212-29-5: 25344
60212-30-8: 25345
60212-31-9: 25346
60218-42-0: 17687
60229-71-2: 15734
60251-57-2: 17578
60285-29-2: 25481
60285-31-6: 25480
60291-32-9: 09880
60308-82-9: 10969
60420-41-9: 14917
60433-00-3: 09637
60456-23-7: 18038
60468-23-7: 18621
60468-54-4: 02889
60487-86-7: 16045
60490-68-8: 18895
60491-51-2: 12679
60491-60-3: 12655
60512-85-8: 21962
60513-12-4: 12673
60513-39-5: 17874
60553-71-1: 12661
60565-54-0: 11851
60577-30-2: 19398
60607-55-8: 12671
60610-10-8: 14134
60655-16-5: 21779
60658-38-0: 00451
60662-82-0: 20277
60669-40-1: 25323
60671-36-5: 25307
60698-31-9: 12918
60698-32-0: 12988
60703-31-3: 14498
60706-07-2: 15208
60710-39-6: 19111
60772-80-7: 03231
60814-27-9: 23869
60814-29-1: 23875
60836-07-9: 13694
60850-22-8: 00126
60899-39-0: 16389
60901-60-2: 00180
60930-56-5: 11033
60956-23-2: 03018
60956-24-3: 03001
60956-26-5: 02699
60956-27-6: 00229
60965-26-6: 12295
60996-53-4: 13985
60999-92-0: 22028
61019-03-2: 03581
61033-71-4: 02167
61050-68-8: 17660
61098-93-9: 22516
61131-68-8: 19192
61133-37-7: 15356
61133-38-8: 15357
61149-61-9: 02194
61151-88-0: 17946
61157-85-5: 01176
61186-07-0: 19444
61193-04-2: 20251
61203-46-1: 06003
61228-10-2: 13837
61231-45-6: 16169
61296-22-8: 24486
61345-77-5: 09288
61345-78-6: 09280
61365-93-3: 20510
61365-94-4: 17970
61417-93-4: 11976
61424-26-8: 02470
61441-51-8: 19878
61445-49-6: 15854
61466-87-3: 15920
61484-71-7: 11930
61514-89-4: 24499
61531-45-1: 17259
61539-75-1: 17596
61551-49-3: 16893
61563-25-5: 03017
61585-35-1: 10341
61589-68-2: 18593
61592-89-0: 03648
61599-49-3: 06727
61645-80-5: 15445
61675-14-7: 14521
61676-86-6: 06405
61709-37-3: 15919
61715-75-1: 14598
61736-34-3: 23083
61748-86-5: 23407
61755-85-9: 24813
61764-94-1: 13496
61792-07-2: 00788
61813-58-9: 05061
61815-22-3: 14895
61827-71-2: 07037
61827-85-8: 03268
61827-86-9: 03449
61827-87-0: 03452
61827-93-8: 10577
61827-98-3: 10569
61828-04-4: 04033
61828-05-5: 03953
61828-06-6: 03862
61828-07-7: 09847
61828-08-8: 09966
61828-09-9: 09957
61828-10-2: 09945
61828-11-3: 09881
61828-13-5: 09935
61828-15-7: 09965
61828-17-9: 09851
61828-19-1: 09964
61828-21-5: 09885
61828-25-9: 03493
61828-26-0: 03825
61837-66-9: 14857
61838-78-6: 21511
61847-78-7: 14439

61847-80-1: 18640
61847-97-0: 17514
61886-67-7: 16469
61892-85-1: 14181
61893-09-2: 14856
61893-14-9: 10751
61914-03-2: 07337
61915-27-3: 18036
61925-49-3: 17270
61931-68-8: 07180
61949-83-5: 14986
61971-91-3: 08619
61977-46-6: 02503
61994-44-3: 02012
61995-51-5: 15119
61996-35-8: 00324
62016-13-1: 14070
62016-48-2: 08014
62016-53-9: 10565
62016-55-1: 10523
62016-58-4: 10540
62016-61-9: 10530
62016-62-0: 10528
62016-63-1: 10527
62016-75-5: 13405
62032-12-6: 11429
62058-64-4: 02461
62059-56-7: 10040
62067-44-1: 10141
62076-23-7: 21078
62088-36-2: 10321
62108-06-9: 17825
62108-66-1: 21109
62108-68-3: 21108
62108-69-4: 21096
62108-70-7: 21106
62108-71-8: 08349
62119-81-7: 24816
62127-46-2: 08065
62127-47-3: 08066
62127-49-5: 20689
62127-50-8: 20690
62127-51-9: 17165
62127-61-1: 20710
62154-77-2: 05551
62155-41-3: 03883
62155-43-5: 03777
62155-44-6: 02769
62155-45-7: 03778
62157-91-9: 24726
62162-97-4: 18831
62168-25-6: 07944
62184-64-9: 09544
62184-82-1: 10704
62184-83-2: 10724
62199-62-6: 13553
62224-31-1: 07638
62231-08-7: 05824
62246-96-2: 15834
62252-06-6: 06885
62261-05-6: 09092
62261-36-3: 15829
62264-34-0: 18948
62317-12-8: 19804
62351-79-5: 04577
62353-69-9: 18134
62353-80-4: 16606
62356-27-8: 02597
62398-10-1: 12255
62456-32-0: 16910
62496-53-1: 18222
62498-21-9: 14199
62507-76-0: 07847
62521-66-8: 16936
62521-69-1: 08073
62568-81-4: 13026
62572-86-5: 12384
62581-18-4: 10014
62595-52-2: 25321
62624-30-0: 01452
62625-23-4: 17443
62686-69-5: 16892
62690-62-4: 16240
62692-40-4: 02587
62706-16-5: 14434
62707-65-7: 20033
62721-35-1: 08682
62784-48-9: 16773
62808-71-3: 25161
62849-19-8: 21782
62882-00-2: 15458
62985-56-2: 24902
62986-54-3: 11209
63012-94-2: 00842
63017-87-8: 16448
63018-70-2: 02003
63019-16-9: 16745
63019-23-8: 09399
63019-98-7: 07577
63020-49-5: 01996
63020-61-1: 01986
63021-45-4: 01533
63041-71-4: 06945
63051-97-8: 00394
63055-14-1: 16768
63055-15-2: 16769
63060-52-6: 17429
63075-06-9: 09109
63076-62-0: 03029
63098-95-3: 25445
63113-36-0: 19635
63126-08-9: 14277
63136-61-8: 23590
63157-81-3: 21042
63163-09-7: 08119
63180-94-9: 04864
63193-53-3: 24961
63238-41-5: 04044
63370-69-4: 10590
63449-81-0: 19950
63452-61-9: 19300
63468-05-3: 18730
63480-12-6: 18594
63512-34-5: 21230
63526-65-8: 11043
63548-92-5: 05156
63572-76-9: 01211
63746-12-3: 04587
63758-87-2: 13863
63779-92-0: 11657
63783-77-7: 13211
63789-83-3: 13481
63790-14-7: 02724
63805-63-0: 07290
63822-42-4: 15367
63837-76-3: 16797
63867-33-4: 11978
63870-44-0: 02245
63896-86-6: 07866
63917-86-2: 11691
63927-44-6: 14212
63944-52-5: 16072
63980-79-0: 24879
63980-94-9: 12072
63981-12-4: 12071
63986-03-8: 11410
63990-68-1: 12357
63990-84-1: 12437
63991-76-4: 20918
63999-74-6: 20383
64022-00-0: 19974
64057-79-0: 21988
64063-37-2: 02147
64080-14-4: 19096
64085-53-6: 02169
64160-38-9: 01842

64161-50-8: 07334
64165-34-0: 23800
64168-11-2: 20209
64207-03-0: 21493
64226-32-0: 16127
64248-64-2: 06695
64255-63-6: 24822
64287-26-9: 04471
64289-81-2: 17106
64297-60-5: 06425
64334-96-9: 23538
64339-43-1: 04705
64346-45-8: 08741
64357-46-6: 15967
64379-30-2: 18173
64398-84-1: 01951
64400-49-3: 06718
64407-07-4: 06686
64462-51-7: 02164
64470-42-4: 19846
64484-85-1: 18331
64501-77-5: 17814
64502-89-2: 18555
64512-96-5: 18342
64531-26-6: 17847
64632-72-0: 25557
64653-59-4: 02272
64811-81-0: 12456
64821-61-0: 22487
64823-22-9: 02211
64823-23-0: 02210
64831-28-3: 06142
64855-18-1: 21148
64884-47-5: 06866
64887-75-8: 15063
64930-16-1: 08765
64965-46-4: 23789
64965-91-9: 05795
64968-99-6: 06066
64976-62-1: 23686
64985-86-0: 16619
65010-65-3: 22779
65079-19-8: 23535
65137-78-2: 01049
65144-02-7: 20422
65148-70-1: 04229
65151-57-7: 19316
65162-83-6: 15400
65226-98-4: 18952
65248-43-3: 14364
65249-46-9: 10187
65253-32-9: 16978
65384-99-8: 24535
65423-25-8: 11558
65442-07-1: 01937
65469-88-7: 08214
65520-08-3: 22586
65530-93-0: 13990
65536-40-5: 25388
65567-34-2: 14860
65611-31-6: 00768
65611-32-7: 00774
65628-35-5: 01095
65662-88-6: 02718
65685-04-3: 07656
65732-90-3: 24183
65786-09-6: 13505
65786-11-0: 13737
65822-93-7: 13655
65833-01-4: 23342
65927-60-8: 14330
65941-22-2: 21571
65976-60-5: 16089
66050-98-4: 14282
66056-06-2: 00050
66072-67-1: 11927
66072-68-2: 20743
66072-69-3: 20742
66073-40-3: 00978
66142-58-3: 13597
66146-33-6: 04095
66178-47-0: 00196
66205-02-5: 16277
66214-05-9: 16848
66225-16-9: 13752
66225-17-0: 13751
66225-32-9: 18341
66225-35-2: 14166
66226-75-3: 24402
66235-62-9: 08576
66259-12-9: 01133
66271-46-3: 18897
66271-87-2: 18845
66283-22-5: 13677
66291-32-5: 18856
66291-34-7: 18858
66291-35-8: 01148
66292-05-5: 15002
66294-51-7: 03925
66294-58-4: 03924
66295-45-2: 19867
66309-91-9: 07849
66325-15-3: 16527
66325-19-7: 07533
66325-42-6: 16279
66326-11-2: 16342
66326-23-6: 16530
66326-27-0: 16313
66328-47-0: 20744
66359-06-6: 16532
66359-62-4: 15003
66359-65-7: 16528
66359-68-0: 16419
66359-79-3: 16143
66359-81-7: 15044
66374-10-5: 06554
66374-91-2: 16531
66392-83-4: 19811
66395-23-1: 10234
66398-52-5: 08311
66399-07-3: 15406
66424-92-8: 05125
66455-54-7: 16274
66455-55-8: 16273
66455-66-1: 16275
66455-69-4: 16319
66496-82-0: 20143
66538-96-3: 16276
66553-12-6: 16529
66553-32-0: 25273
66553-50-2: 10550
66564-01-0: 16415
66576-31-6: 14317
66577-03-5: 24734
66577-17-1: 16447
66577-22-8: 16539
66591-62-6: 09438
66610-38-6: 00694
66610-39-7: 08012
66671-83-8: 00260
66694-00-6: 11890
66719-34-4: 17709
66719-35-5: 17710
66719-37-7: 13672
66719-40-2: 13673
66719-53-7: 17743
66719-54-8: 17744
66719-55-9: 17732
66731-94-0: 13674
66778-18-5: 07434
66778-24-3: 16461
66793-72-4: 14332
66793-74-6: 14333
66793-83-7: 17729
66793-92-8: 14331
66793-99-5: 13681

66794-02-3: 14297
66794-03-4: 14319
66794-04-5: 14278
66794-29-4: 08994
66799-44-8: 08604
66810-87-5: 14334
66827-00-7: 14130
66827-01-8: 17642
66832-97-1: 12287
66857-17-8: 07856
66898-39-3: 06839
66904-34-5: 07051
66908-07-4: 06222
66910-71-2: 00292
66920-75-0: 08967
66968-75-0: 23153
66970-44-3: 00731
66997-45-3: 20699
67050-82-2: 23080
67083-22-1: 05362
67083-28-7: 02208
67116-33-0: 16266
67191-37-1: 03389
67241-13-8: 15272
67282-69-3: 00307
67329-10-6: 17460
67337-69-3: 14863
67451-76-7: 09994
67464-41-9: 04823
67466-23-3: 23088
67535-07-3: 21163
67535-55-1: 12745
67590-77-6: 03517
67632-66-0: 18804
67682-09-1: 18616
67700-22-5: 05277
67700-26-9: 17823
67747-70-0: 18660
67770-59-6: 05721
67785-72-2: 03179
67828-68-6: 12441
67878-75-5: 16262
67878-77-7: 16267
67885-08-9: 07777
67885-46-5: 12589
67915-26-8: 00272
67916-68-1: 25550
67952-28-7: 15555
67952-57-2: 13664
67952-96-9: 20764
67953-05-3: 13477
67990-20-9: 20163
67990-66-3: 05900
67992-12-5: 07528
68047-33-6: 20370
68057-83-0: 08426
68077-28-1: 21771
68084-17-3: 04172
68101-31-5: 17927
68107-05-1: 13992
68142-19-8: 17354
68144-45-6: 20138
68161-67-1: 24463
68165-06-0: 23430
68189-19-5: 19295
68384-70-3: 08470
68413-94-5: 08788
68420-92-8: 23685
68420-93-9: 23545
68474-12-4: 20139
68555-41-9: 11965
68676-77-7: 10896
68680-98-8: 17708
68723-25-1: 16014
68723-48-8: 09412
68750-16-3: 12095
68776-15-8: 16757
68791-95-7: 21183
68833-53-4: 01283
68844-87-1: 01310
68844-88-2: 01311
68922-17-8: 20185
68925-98-4: 11219
69018-87-7: 06129
69112-21-6: 14387
69127-80-6: 06930
69169-56-8: 08861
69190-56-3: 02600
69245-80-3: 19282
69248-35-7: 08188
69295-21-2: 08719
69305-42-6: 19666
69321-60-4: 03019
69371-59-1: 14524
69429-57-8: 21501
69442-23-5: 22942
69470-50-4: 19083
69473-96-7: 24631
69498-28-8: 07958
69542-91-2: 10125
69564-68-7: 07182
69668-88-8: 13803
69676-59-1: 13262
69704-44-5: 08098
69737-54-8: 20428
69745-49-9: 10357
69770-98-5: 10678
69778-83-2: 23439
69813-63-4: 03662
69833-18-7: 01457
69855-49-8: 08624
69947-01-9: 20444
69986-22-7: 00082
70071-22-6: 23867
70110-24-6: 21599
70110-25-7: 21598
70116-68-6: 08590
70127-93-4: 14969
70202-92-5: 10736
70223-10-8: 09368
70261-49-3: 19314
70267-26-4: 14241
70288-96-9: 15897
70288-97-0: 02789
70289-01-9: 13626
70289-05-3: 19859
70301-29-0: 01826
70354-71-1: 23436
70382-04-6: 04111
70387-28-9: 10274
70396-13-3: 13833
70419-06-6: 17267
70453-50-8: 00251
70466-89-6: 04207
70489-30-4: 16387
70492-67-0: 13993
70644-45-0: 22736
70644-46-1: 22613
70690-24-3: 17591
70706-60-4: 10874
70715-19-4: 22682
70732-43-3: 17319
70732-44-4: 17549
70732-45-5: 11023
70766-54-0: 19073
70786-82-2: 18410
70805-94-6: 07307
70862-15-6: 08097
70958-03-1: 04188
71029-37-3: 03675
71057-15-3: 08414
71119-13-6: 00329
71127-64-5: 16774
71151-84-3: 00824
71159-31-4: 21394
71162-62-4: 25385

71172-48-0: 12824
71172-57-1: 18554
71172-59-3: 20228
71172-71-9: 07291
71173-15-4: 07310
71200-79-8: 08793
71261-05-7: 23187
71277-14-0: 21627
71284-96-3: 16919
71297-97-7: 11044
71319-20-5: 16016
71319-21-6: 16017
71424-71-0: 07320
71436-66-3: 16253
71563-86-5: 21210
71568-87-1: 06006
71605-85-1: 17251
71608-10-1: 19322
71633-43-7: 23716
71672-75-8: 05059
71691-56-0: 20836
71720-52-0: 15077
71745-63-6: 19329
71747-37-0: 05269
71815-31-1: 12470
71838-16-9: 02659
71959-13-2: 05679
72078-30-9: 21637
72083-19-3: 15893
72093-13-1: 22600
72101-07-6: 11350
72115-06-1: 06667
72137-22-5: 10407
72277-16-8: 05258
72278-08-1: 12453
72286-34-1: 24481
72312-48-2: 10352
72365-58-3: 23803
72436-99-8: 25085
72469-59-1: 11218
72506-83-3: 21374
72517-15-8: 06004
72622-74-3: 21301
72690-56-3: 09341
72701-63-4: 22862
72707-66-5: 21786
72851-41-3: 05767
72927-80-1: 04092
72928-38-2: 00641
72934-37-3: 10801
72948-52-8: 01330
72955-79-4: 20089
73033-58-6: 05018
73095-40-6: 16955
73120-52-2: 11493
73178-43-5: 18687
73206-57-2: 21494
73210-25-0: 10210
73267-76-2: 19166
73282-23-2: 23790
73334-01-7: 00232
73383-24-1: 08723
73397-68-9: 14342
73398-15-9: 13943
73404-00-9: 04836
73441-51-7: 04937
73453-68-6: 01275
73453-71-1: 01271
73454-83-8: 13948
73464-47-8: 10835
73506-86-2: 13419
73506-87-3: 17172
73513-50-5: 14466
73522-17-5: 08206
73522-42-6: 07300
73548-71-7: 14420
73548-72-8: 13754
73557-60-5: 03989
73622-68-1: 07854
73635-75-3: 01651
73642-91-8: 17206
73673-27-5: 14762
73680-98-5: 23202
73713-56-1: 05992
73741-61-4: 10405
73747-53-2: 15165
73758-54-0: 08357
73758-57-3: 04242
73790-84-8: 20227
73791-26-1: 06594
73791-38-5: 01387
73813-61-3: 21549
73816-58-7: 15190
73840-20-7: 18155
73859-73-1: 06524
73868-16-3: 23600
73882-13-0: 17787
73931-44-9: 03991
73953-80-7: 25433
73987-32-3: 19858
74037-50-6: 22647
74039-86-4: 06944
74051-47-1: 25251
74084-58-5: 21386
74111-27-6: 07309
74112-36-0: 17722
74113-34-1: 16837
74204-00-5: 19112
74221-86-6: 09917
74244-64-7: 09622
74282-28-3: 20559
74283-25-3: 18760
74294-73-8: 14819
74299-52-8: 17594
74319-86-1: 05408
74319-93-0: 05407
74328-37-3: 13723
74421-05-9: 13444
74421-26-4: 04860
74438-98-5: 14851
74448-32-1: 08364
74478-95-8: 14817
74502-96-8: 23092
74513-58-9: 21969
74581-94-5: 14232
74586-54-2: 02304
74614-53-2: 15374
74663-99-3: 01763
74683-66-2: 17266
74685-42-0: 03734
74685-84-0: 08149
74708-34-2: 08177
74709-95-8: 03210
74742-08-8: 17269
74742-10-2: 10985
74744-28-8: 25341
74744-36-8: 11564
74764-56-0: 13819
74778-22-6: 17740
74845-46-8: 16141
74873-98-6: 15403
74924-94-0: 16203
74925-48-7: 15007
74985-88-9: 09433
74990-49-1: 22347
75024-21-4: 06533
75032-63-2: 08968
75055-73-1: 08234
75122-96-2: 04163
75238-15-2: 00337
75312-36-6: 01272
75313-07-4: 01307
75343-53-2: 22927
75457-13-5: 23817
75579-88-3: 13461
75597-94-3: 25191

75597-95-4: 25190
75634-10-5: 03286
75731-44-1: 16001
75793-58-7: 23531
75821-44-2: 06355
75854-68-1: 17256
75854-70-5: 10971
75884-39-8: 21481
75885-75-5: 06897
75908-52-0: 23569
75908-56-4: 22363
75908-63-3: 24505
75908-68-8: 06568
75908-74-6: 17958
75908-79-1: 15950
75908-80-4: 18283
75917-07-6: 16977
75919-82-3: 08157
75919-83-4: 23082
75919-84-5: 23089
75919-87-8: 23085
75919-88-9: 23087
75919-90-3: 23086
75919-91-4: 12488
75919-92-5: 12489
75919-94-7: 15948
75919-95-8: 23013
76063-06-4: 11562
76078-85-8: 17228
76118-81-5: 12766
76129-28-7: 07572
76149-14-9: 04662
76216-35-8: 19486
76216-36-9: 19425
76343-97-0: 21245
76361-99-4: 19121
76429-68-0: 05785
76469-08-4: 17854
76469-33-5: 12322
76472-83-8: 07566
76553-86-1: 02122
76619-89-1: 19026
76655-66-8: 02341
76684-72-5: 04130
76708-85-5: 07509
76727-11-2: 18044
76792-22-8: 08336
77008-58-3: 15888
77008-62-9: 02786
77008-64-1: 02787
77053-92-0: 14487
77117-15-8: 06822
77145-59-6: 12837
77215-03-3: 13607
77295-79-5: 12563
77304-63-3: 19834
77373-34-3: 16097
77392-54-2: 08562
77437-98-0: 13792
77464-05-2: 25368
77464-06-3: 25371
77502-18-2: 16743
77609-04-2: 23163
77619-93-3: 15348
77753-21-0: 08786
77753-24-3: 08039
77791-64-1: 00204
77825-53-7: 08872
77825-54-8: 08873
77866-58-1: 10201
77928-86-0: 10914
77938-63-7: 13239
77958-21-5: 13820
78073-03-7: 12465
78073-04-8: 12466
78137-76-5: 19124
78142-00-4: 07530
78150-61-5: 12450
78156-03-3: 04437
78176-22-4: 03796
78210-35-2: 19531
78238-14-9: 06397
78270-92-5: 23238
78277-23-3: 21966
78279-13-7: 11374
78281-59-1: 15874
78313-88-9: 17593
78369-81-0: 24960
78546-03-9: 13302
78547-40-7: 19489
78627-20-0: 01023
78666-95-2: 25091
78679-15-9: 05709
78738-36-0: 21385
78738-37-1: 20166
78751-58-3: 22518
78759-31-6: 03311
78769-83-2: 11225
78775-71-0: 20854
78778-91-3: 21387
78881-21-7: 06668
78893-76-2: 07719
78918-91-9: 14064
78927-73-8: 12320
78938-11-1: 24087
78941-93-2: 23594
78987-59-4: 14102
78987-65-2: 00763
78987-68-5: 00776
78987-84-5: 08100
78987-85-6: 08791
78987-86-7: 14103
79025-28-8: 06200
79127-41-6: 12263
79179-56-9: 14850
79300-08-6: 11606
79301-80-7: 04730
79314-59-3: 18586
79325-55-6: 16688
79392-44-2: 11395
79435-04-4: 21098
79482-00-1: 16011
79504-01-1: 07913
79560-27-3: 16954
79605-28-0: 17421
79630-70-9: 08052
79663-99-3: 10372
79735-02-7: 01485
79762-76-8: 21132
79786-43-9: 22007
79802-79-2: 10207
79868-38-5: 07867
79876-13-4: 07563
79912-55-3: 18416
79912-57-5: 18415
79965-62-1: 23682
79979-69-4: 16613
80204-19-9: 18655
80325-37-7: 13746
80396-17-4: 05264
80438-67-1: 02774
80466-34-8: 13928
80468-92-4: 01305
80472-59-9: 05489
80513-04-8: 14314
80638-53-5: 00696
80651-00-9: 16984
80651-03-2: 16926
80656-12-8: 10610
80656-14-0: 10611
80710-39-0: 06916
80710-40-3: 06918
80710-41-4: 06917
80710-76-5: 16348
80777-94-2: 21814
80789-66-8: 09594

80789-69-1: 10596
80789-72-6: 22564
80789-75-9: 10601
80824-85-7: 22022
80832-54-8: 01677
80864-10-4: 18335
80866-80-4: 05015
80867-75-0: 05402
80879-86-3: 02204
80947-25-7: 23525
81012-95-5: 00381
81028-94-6: 19207
81177-02-8: 14143
81280-12-8: 13789
81335-87-7: 04978
81363-76-0: 21003
81525-12-4: 15928
81568-83-4: 13290
81608-88-0: 07811
81686-41-1: 03099
81699-50-5: 12764
81722-06-7: 06192
81726-81-0: 18061
81726-82-1: 13406
81730-48-5: 03701
81764-16-1: 23539
81785-47-9: 21318
81824-67-1: 16163
81873-90-7: 09346
81903-36-8: 14642
81925-81-7: 13800
81957-18-8: 21362
81957-22-4: 06486
82064-14-0: 18936
82064-15-1: 18887
82064-16-2: 18827
82142-72-1: 21159
82191-81-9: 14213
82322-93-8: 10682
82356-36-3: 15386
82364-21-4: 01388
82507-04-8: 14436
82525-64-2: 06071
82584-73-4: 17990
82697-79-8: 24480
82801-46-5: 17334
82807-35-0: 08828
82871-10-1: 16057
82898-85-9: 00310
82906-78-3: 19881
82994-21-6: 03841
82997-81-7: 04775
83004-97-1: 00575
83027-11-6: 08752
83072-25-7: 04251
83074-06-0: 14698
83088-71-5: 00827
83088-72-6: 00799
83320-61-0: 11918
83439-54-7: 11105
83466-88-0: 09011
83669-51-6: 06655
83682-72-8: 21317
83699-51-8: 15875
83704-00-1: 20415
83785-27-7: 08269
83803-88-7: 15974
83846-48-4: 02849
83863-33-6: 02306
83968-50-7: 11662
84098-45-3: 21199
84100-59-4: 05609
84228-43-3: 05943
84228-45-5: 05947
84256-17-7: 19745
84258-49-1: 15279
84297-13-2: 05254
84309-33-1: 04955
84333-02-8: 01568
84355-05-5: 03932
84375-72-4: 06365
84483-25-0: 02515
84501-73-5: 15373
84539-34-4: 22570
84540-59-0: 02890
84583-53-9: 23804
84612-77-1: 14231
84741-02-6: 06345
84902-21-6: 18288
84998-84-5: 00759
85061-22-9: 23907
85144-14-5: 10373
85180-69-4: 10668
85196-33-4: 00610
85290-80-8: 22441
85328-29-6: 16853
85417-71-6: 02571
85600-11-9: 09115
85617-05-6: 25306
85617-06-7: 25305
85642-07-5: 19530
85712-05-6: 13644
85733-97-7: 18510
85775-61-7: 19313
85925-12-8: 00029
85978-93-4: 14702
86051-37-8: 17254
86073-38-3: 17434
86123-20-8: 21242
86199-39-5: 24545
86240-21-3: 23375
86241-62-5: 18419
86453-13-6: 21551
86453-14-7: 21558
86475-37-8: 22455
86571-08-6: 06576
86607-81-0: 11136
86623-80-5: 18707
86636-38-6: 00468
86661-53-2: 18198
86668-33-9: 17809
86704-82-7: 05549
86711-71-9: 21101
86926-11-6: 02242
86984-32-9: 23603
87018-30-2: 08539
87026-45-7: 23050
87145-66-2: 25030
87171-21-9: 23157
87172-81-4: 08427
87259-17-4: 17422
87298-11-1: 20593
87310-12-1: 17257
87324-44-5: 15791
87345-49-1: 24600
87406-33-5: 03139
87473-78-7: 16863
87541-87-5: 21811
87760-48-3: 14160
87812-99-5: 11134
87976-23-6: 06599
88019-89-0: 13631
88122-95-6: 06409
88153-39-3: 08514
88200-01-5: 09330
88331-62-8: 15930
88373-40-4: 01414
88472-61-1: 17442
88497-30-7: 24284
88499-89-2: 23718
88499-98-3: 23708
88521-64-6: 00629
88529-79-7: 22439
88613-84-7: 22345
88613-97-2: 22306
88613-98-3: 22307

88663-39-2: 17322
88722-94-5: 11669
88766-09-0: 16490
88780-84-1: 25088
88783-79-3: 07583
88795-49-7: 19919
88850-40-2: 00312
88893-11-2: 11237
88900-03-2: 19461
88905-86-6: 24454
89015-27-0: 04201
89033-70-5: 09524
89123-28-4: 24669
89124-06-1: 00505
89130-56-3: 04583
89157-36-8: 13772
89166-28-9: 22051
89169-06-2: 00280
89223-57-4: 10549
89282-31-5: 24839
89284-45-7: 12811
89300-69-6: 21158
89346-57-6: 07461
89364-27-2: 13110
89365-48-0: 03998
89415-67-8: 08370
89422-42-4: 08298
89459-38-1: 06421
89465-63-4: 25381
89489-70-3: 25517
89532-61-6: 24503
89533-45-9: 25474
89533-67-5: 14539
89581-05-5: 08374
89581-41-9: 05344
89581-58-8: 23097
89582-17-2: 11088
89583-12-0: 14019
89583-61-9: 20563
89599-01-9: 05343
89599-37-1: 18764
89599-47-3: 10110
89602-56-2: 08164
89660-51-5: 19864
89677-85-0: 24917
89677-86-1: 24918
89677-87-2: 04045
89691-80-5: 19204
89691-87-2: 15518
89693-51-6: 22043
89694-89-3: 24924
89709-22-8: 04228
89761-18-2: 21940
89775-61-1: 00653
89776-60-3: 25516
89793-90-8: 19528
89796-76-9: 13507
89813-56-9: 06985
89827-45-2: 11131
89853-03-2: 25086
89854-42-2: 25466
89854-43-3: 25450
89886-25-9: 10138
89894-97-3: 09549
89895-45-4: 09918
89896-73-1: 13732
89910-62-3: 23094
89937-51-9: 25519
89938-16-9: 24940
89942-36-9: 25540
89979-85-1: 12628
90000-75-2: 23093
90007-18-4: 15424
90007-36-6: 23101
90049-81-3: 07120
90087-94-8: 24596
90091-32-0: 00743
90112-26-8: 10685
90112-75-7: 13773
90112-82-6: 14243
90139-63-2: 21139
90151-04-5: 01731
90154-78-2: 23280
90162-24-6: 17647
90199-77-2: 18705
90200-83-2: 17824
90201-13-1: 14220
90202-08-7: 14215
90202-21-4: 17802
90210-56-3: 15462
90221-54-8: 07243
90223-46-4: 13137
90225-75-5: 11245
90226-23-6: 10439
90227-07-9: 25513
90227-43-3: 25512
90325-30-7: 06642
90329-57-0: 24430
90349-65-8: 06501
90357-58-7: 00642
90365-63-2: 17741
90389-70-1: 04938
90418-41-0: 09068
90434-25-6: 16058
90437-76-6: 02318
90536-04-2: 07259
90536-71-3: 06611
90536-74-6: 06615
90560-40-0: 13353
90560-79-5: 22054
90566-30-6: 13264
90606-97-6: 06982
90610-83-6: 00470
90611-37-3: 00755
90611-43-1: 18520
90612-18-3: 09995
90638-48-5: 21921
90641-79-5: 13301
90643-12-2: 12100
90644-60-3: 17312
90645-54-8: 10292
90645-61-7: 10669
90645-92-4: 17837
90648-30-9: 22362
90648-83-2: 00747
90670-72-7: 14322
90673-26-0: 22943
90676-00-9: 21535
90676-23-6: 10142
90676-55-4: 17828
90716-89-5: 00354
90722-14-8: 17317
90745-27-0: 16453
90760-75-1: 10208
90767-01-4: 17107
90769-70-3: 10713
90771-32-7: 23667
90772-52-4: 14959
90793-41-2: 09097
90798-84-8: 24479
90799-45-4: 17108
90833-18-4: 18972
90833-20-8: 23892
90833-30-0: 23890
90833-41-3: 23895
90833-77-5: 23898
90833-79-7: 23897
90841-14-8: 08755
90902-84-4: 22562
90920-28-8: 12248
90923-80-1: 17070
90925-48-7: 08208
90942-87-3: 24599
90948-04-2: 16263
90970-69-7: 09638
90978-85-1: 24598

90978-91-9: 09542
90991-68-7: 00820
91004-91-0: 00745
91013-44-4: 00722
91013-73-9: 00742
91013-74-0: 00748
91049-43-3: 18329
91057-79-3: 09568
91091-02-0: 22348
91091-63-3: 25465
91108-69-9: 09596
91121-79-8: 14898
91131-86-1: 15117
91132-48-8: 24594
91137-27-8: 16451
91137-28-9: 16457
91137-78-9: 22346
91140-68-0: 22322
91170-70-6: 05288
91180-88-0: 09136
91190-03-3: 23896
91244-98-3: 02869
91248-67-8: 00746
91252-74-3: 01106
91253-07-5: 20116
91266-99-8: 09636
91270-68-7: 16908
91270-69-8: 16909
91270-74-5: 19862
91278-70-5: 07751
91313-63-2: 21583
91330-52-8: 13121
91356-11-5: 24597
91362-36-6: 00658
91391-83-2: 22352
91392-06-2: 22350
91423-84-6: 14211
91558-02-0: 13286
91572-15-5: 11242
91718-52-4: 07447
91719-59-4: 14136
91769-08-3: 25093
91776-68-0: 19921
91799-71-2: 16286
91840-99-2: 08272
91859-33-5: 00526
91890-87-8: 08924
91902-60-2: 17064
91913-99-4: 08359
91965-23-0: 07750
91982-53-5: 24919
91999-14-3: 11679
92017-89-5: 25092
92018-32-1: 21084
92028-83-6: 19053
92040-00-1: 15519
92042-04-1: 01857
92042-21-2: 13269
92107-46-5: 13267
92148-64-6: 09490
92152-01-7: 07624
92153-11-2: 06148
92153-13-4: 06153
92153-14-5: 06759
92153-17-8: 06760
92188-81-3: 17889
92190-68-6: 24935
92191-45-2: 24311
92245-40-4: 00074
92292-40-5: 22464
92293-04-4: 23873
92296-30-5: 12807
92326-91-5: 20095
92340-81-3: 18489
92367-69-6: 24672
92375-23-0: 24520
92378-92-2: 00002
92385-32-5: 10007
92426-93-2: 23871
92498-14-1: 09325
92498-22-1: 09315
92550-15-7: 25361
92552-62-0: 05809
92553-02-1: 12694
92554-11-5: 23929
92565-72-5: 00685
92576-03-9: 22349
92646-32-7: 19047
92687-41-7: 10382
92700-37-3: 14209
92736-02-2: 13774
92737-91-2: 13691
92748-09-9: 05341
92765-71-4: 25192
92795-55-6: 12198
92847-27-3: 22309
92847-41-1: 23377
92855-96-4: 09446
92856-32-1: 09321
92902-61-9: 13123
92962-91-9: 09391
92963-36-5: 05806
92965-40-7: 05825
93011-89-3: 13080
93012-35-2: 11344
93031-24-4: 14326
93048-35-2: 19896
93049-73-1: 14734
93125-13-4: 22959
93142-97-3: 15526
93179-35-2: 12732
93186-50-6: 15663
93257-06-8: 20034
93310-91-9: 06765
93317-20-5: 22334
93321-57-4: 09392
93322-14-6: 12772
93323-03-6: 25514
93323-45-6: 22308
93404-72-9: 25082
93409-03-1: 22320
93432-80-5: 12695
93457-19-3: 24606
93484-34-5: 10885
93484-96-9: 10886
93484-97-0: 10887
93505-96-5: 16056
93532-35-5: 19538
93568-54-8: 00598
93585-23-0: 22490
93655-07-3: 12254
93658-33-4: 13422
93681-82-4: 18714
93713-40-7: 21034
93718-74-2: 24933
93812-81-8: 13266
93905-63-6: 05608
93921-42-7: 10232
93972-93-1: 14198
93999-88-3: 09147
93999-94-1: 24519
94004-88-3: 24595
94045-48-4: 21554
94064-75-2: 22496
94065-57-3: 22355
94084-62-5: 12973
94096-31-8: 15792
94109-57-6: 05325
94127-04-5: 23595
94210-70-5: 23793
94210-71-6: 23794
94211-06-0: 12715
94230-83-8: 22449
94255-53-5: 15486
94258-53-4: 23916
94307-48-9: 12709

94332-82-8: 11347
94385-09-8: 13621
94434-61-4: 17430
94434-62-5: 17431
94434-63-6: 17423
94434-64-7: 17424
94434-65-8: 17425
94487-99-7: 15747
94502-19-9: 23851
94523-45-2: 25097
94574-49-9: 12708
94574-85-3: 13066
94580-37-7: 25565
94602-15-0: 25488
94622-10-3: 06766
94623-28-6: 09396
94632-64-1: 15066
94685-68-4: 15115
94709-11-2: 06769
94709-12-3: 06770
94730-96-8: 15527
94731-83-6: 23253
94826-08-1: 09464
94869-50-8: 16829
94883-14-4: 00469
94908-49-3: 07186
94916-06-0: 01377
94916-08-2: 25571
94916-82-2: 15237
94942-89-9: 05248
94963-68-5: 01851
94996-48-2: 11649
95019-82-2: 18101
95121-04-3: 22163
95133-29-2: 05808
95163-45-4: 23551
95163-46-5: 23556
95163-58-9: 23550
95166-04-4: 20009
95173-82-3: 05432
95173-83-4: 05484
95194-86-8: 14740
95195-48-5: 05282
95225-21-1: 23880
95279-90-6: 09033
95282-75-0: 11656
95288-82-7: 24518
95429-50-8: 21417
95435-96-4: 09028
95493-62-2: 24937
95496-91-6: 15378
95533-81-6: 24265
95541-18-7: 05848
95547-10-7: 15600
95594-16-4: 15361
95642-49-2: 18368
95696-17-6: 16124
95696-45-0: 11655
95746-76-2: 24342
95766-63-5: 01102
95768-98-2: 22298
95769-42-9: 24266
95782-14-2: 18729
95811-57-7: 22515
95811-91-9: 23868
95817-71-3: 20010
95835-77-1: 05802
95844-90-9: 16126
95845-11-7: 19910
95845-72-0: 22297
95873-08-8: 13265
95915-20-1: 09390
95946-09-1: 11665
95957-96-3: 22676
95970-07-3: 03937
95970-10-8: 03990
95970-13-1: 03988
95980-60-2: 06967
95982-07-3: 24553
96014-36-7: 09158
96034-00-3: 08059
96097-67-5: 05485
96097-69-7: 24932
96117-14-5: 00744
96123-64-7: 17563
96187-18-7: 07571
96191-75-2: 01050
96244-13-2: 10206
96260-50-3: 23872
96264-89-0: 23549
96432-16-5: 18296
96432-49-4: 06577
96432-96-1: 09071
96446-96-7: 23931
96458-82-1: 01597
96479-96-8: 17991
96558-72-4: 03996
96591-06-9: 17160
96668-33-6: 14728
96707-44-7: 20334
96784-54-2: 06723
96836-97-4: 10034
96933-86-7: 09484
96977-78-5: 09145
96977-92-3: 13247
96980-43-7: 23114
97021-39-1: 01462
97049-71-3: 07648
97073-23-9: 21525
97153-34-9: 01908
97172-82-2: 21577
97176-61-9: 24264
97329-43-6: 02595
97388-49-3: 00644
97412-23-2: 19431
97414-58-9: 20360
97441-57-1: 09144
97456-81-0: 02658
97472-93-0: 15362
97498-49-2: 21557
97505-87-8: 23289
97511-05-2: 11210
97599-94-5: 19046
97618-71-8: 12688
97620-84-3: 06754
97766-62-6: 15358
97790-61-9: 01380
97809-33-1: 00353
97961-66-5: 14460
97980-65-9: 00715
98020-86-1: 24456
98023-09-7: 19742
98025-28-6: 00137
98070-91-8: 08074
98074-68-1: 09146
98139-72-1: 08078
98140-03-5: 23044
98180-43-9: 11236
98203-04-4: 23571
98272-34-5: 12904
98273-49-5: 17901
98335-68-3: 12812
98431-97-1: 24270
98486-62-5: 09022
98546-20-4: 22067
98564-41-1: 17037
98596-11-3: 17039
98596-13-5: 17038
98693-42-6: 24920
98751-32-7: 03284
98785-33-2: 11777
98879-03-9: 19946
98930-89-3: 25166
98962-99-3: 14141
98991-54-9: 13777
99055-62-6: 13334

99075-33-9: 09021
99076-71-8: 00721
99099-00-0: 01093
99100-02-4: 04051
99109-29-2: 06840
99167-08-5: 14599
99172-57-3: 12873
99277-71-1: 06023
99349-02-7: 25109
99363-12-9: 10723
99363-16-3: 13506
99365-26-1: 06882
99418-51-6: 00142
99421-19-9: 15426
99447-77-5: 05496
99465-09-5: 23829
99465-10-8: 23831
99512-73-9: 23798
99515-47-6: 06757
99586-47-7: 12623
99699-42-0: 21832
99706-68-0: 25155
99767-70-1: 00125
99814-65-0: 18082
99840-88-7: 09045
99852-98-9: 22048
99854-34-9: 00406
99885-51-5: 23406
99903-07-8: 14831
99926-14-4: 15787
99970-57-7: 20001
99971-81-0: 15506
99972-07-3: 17063
99972-22-2: 22065
99973-91-8: 22068
99980-00-4: 09494
99980-28-6: 06971
99983-48-9: 22052
100008-43-3: 16345
100030-11-3: 00210
100046-93-3: 07112
100057-07-6: 09643
100058-19-3: 22165
100060-95-5: 07176
100062-58-6: 20016
100062-67-7: 22057
100103-70-6: 12933
100116-62-9: 06120
100117-83-7: 00446
100120-21-6: 22049
100123-38-4: 25094
100128-07-2: 15427
100129-18-8: 23926
100130-25-4: 00133
100144-29-4: 02760
100144-93-2: 13354
100184-66-5: 16657
100191-60-4: 14603
100224-75-7: 16010
100295-82-7: 14321
100295-83-8: 14324
100296-26-2: 13675
100314-90-7: 12916
100319-40-2: 03425
100319-48-0: 17550
100367-45-1: 13950
100395-94-6: 00146
100396-73-4: 09486
100397-98-6: 12778
100516-88-9: 23765
100521-89-9: 17336
100524-19-4: 25023
100606-77-7: 10013
100614-29-7: 01351
100710-80-3: 17029
100721-49-1: 06121
100723-01-1: 23876
100725-06-2: 01821
100791-92-2: 00136
100791-95-5: 23432
100791-98-8: 00135
100802-80-0: 12691
100859-51-6: 13832
100871-11-2: 13840
100877-88-1: 17675
100883-07-6: 25096
100888-42-4: 11248
100955-29-1: 16876
100958-82-5: 13143
100961-23-7: 22069
100964-96-3: 00228
100965-13-7: 14756
100983-12-8: 21082
100996-43-8: 01510
101014-13-5: 00259
101038-91-9: 22137
101057-63-0: 00192
101080-16-4: 23045
101081-07-6: 20014
101084-14-4: 14111
101096-60-0: 22056
101097-83-0: 20015
101107-08-8: 15562
101144-90-5: 03391
101144-94-9: 03392
101166-96-5: 06119
101201-84-7: 24921
101231-37-2: 15235
101256-97-7: 09034
101257-63-0: 14008
101257-71-0: 20223
101257-79-8: 09539
101258-43-9: 09538
101259-49-8: 12065
101277-90-1: 19534
101279-08-7: 24457
101283-13-0: 12757
101287-05-2: 17030
101292-50-6: 22050
101342-75-0: 13780
101349-91-1: 08046
101420-81-9: 06021
101427-54-7: 24344
101451-33-6: 12707
101488-61-3: 05188
101488-73-7: 01187
101501-17-1: 15786
101541-43-9: 11188
101567-36-6: 10023
101567-63-9: 00800
101588-41-4: 00423
101597-94-8: 23278
101611-98-7: 00208
101654-23-3: 18982
101654-30-2: 14006
101715-60-0: 17537
101729-13-9: 12756
101736-34-9: 24455
101772-33-2: 01899
101774-91-8: 22188
101777-29-1: 23336
101865-85-4: 22060
101870-06-8: 13938
101870-15-9: 13007
101877-12-7: 12758
101885-30-7: 14129
101885-32-9: 17641
101935-28-8: 13446
101935-40-4: 19117
102013-79-6: 17177
102014-43-7: 14205
102023-81-4: 00415
102024-14-6: 09403
102082-85-9: 22902
102127-39-9: 19141
102153-58-2: 16296

102153-88-8: 10552
102158-83-8: 18902
102169-99-3: 24944
102174-12-9: 23006
102177-34-4: 14718
102300-80-1: 24497
102308-83-8: 17207
102312-38-9: 06639
102322-89-4: 13241
102344-58-1: 11176
102370-18-3: 10140
102433-82-9: 11684
102433-83-0: 11685
102439-69-0: 17239
102449-95-6: 14011
102457-65-8: 05979
102464-65-3: 24234
102468-13-3: 20013
102541-44-6: 22047
102547-97-7: 14473
102558-41-8: 01488
102571-53-9: 08281
102599-00-8: 14602
102660-38-8: 19889
102664-66-4: 05568
102701-87-1: 12626
102747-81-9: 01556
102776-00-1: 08751
102871-67-0: 23719
102877-11-2: 00344
102877-13-4: 00345
102878-61-5: 17112
102898-11-3: 14956
102947-74-0: 11428
102955-13-5: 06585
103034-71-5: 20011
103037-67-8: 17080
103039-42-5: 17119
103064-91-1: 09342
103095-51-8: 22376
103100-75-0: 00477
103114-32-5: 07673
103162-08-9: 16351
103165-66-8: 12686
103206-23-1: 23012
103244-32-2: 16641
103260-67-9: 00216
103260-68-0: 20007
103260-74-8: 09454
103269-23-4: 07169
103273-25-2: 17337
103279-05-6: 01557
103279-10-3: 09357
103385-16-6: 09031
103393-12-0: 12698
103404-57-5: 13606
103408-11-3: 16584
103458-90-8: 24925
103623-31-0: 17556
103697-14-9: 17368
103788-60-9: 24548
103854-67-7: 13139
103861-06-9: 07132
103861-22-9: 07246
103862-61-9: 23583
103983-94-4: 23722
103985-40-6: 10389
103987-83-3: 17067
104035-66-7: 13814
104097-71-4: 23900
104115-88-0: 05497
104164-79-6: 08735
104177-72-2: 01103
104209-33-8: 06047
104216-55-9: 05569
104237-19-6: 00835
104246-27-7: 00184
104514-49-0: 03000
104542-38-3: 06479
104693-46-1: 17862
104713-60-2: 10087
104763-81-7: 15415
104870-56-6: 10167
104997-66-2: 25095
105043-65-0: 12634
105065-24-5: 10339
105105-63-3: 09997
105224-04-2: 12476
105401-59-0: 14720
105476-84-4: 17246
105564-86-1: 05761
105688-74-2: 23653
105728-23-2: 23608
105752-67-8: 16452
105836-68-8: 11346
105906-97-6: 18618
105911-04-4: 20035
105972-87-0: 19812
105991-77-3: 11177
106047-04-5: 11051
106047-07-8: 11066
106120-04-1: 04685
106164-79-8: 22179
106166-67-0: 19730
106214-24-8: 06067
106247-35-2: 15529
106378-18-1: 20760
106405-40-7: 17252
106472-11-1: 01834
106591-60-0: 10024
106654-25-5: 20317
106738-34-5: 00124
106738-59-4: 00123
106819-03-8: 17668
106837-44-9: 25566
106864-78-2: 19961
106949-24-0: 17068
106949-26-2: 17076
107167-08-8: 20501
107327-43-5: 15149
107412-71-5: 07597
107422-77-5: 14735
107522-48-5: 09258
107558-84-9: 20012
107559-24-0: 12625
107574-57-2: 21512
107697-82-5: 22353
107770-91-2: 05537
107770-92-3: 24277
107771-41-5: 14263
107775-84-8: 01116
107785-25-1: 20091
107894-44-0: 23000
107916-57-4: 09317
107920-10-5: 00913
107920-11-6: 00912
107922-19-0: 06154
107943-23-7: 09312
107943-75-9: 12692
108040-13-7: 24784
108040-17-1: 24468
108123-07-5: 01114
108200-26-6: 18184
108237-78-1: 12699
108268-28-6: 09639
108299-58-7: 21341
108340-61-0: 18330
108365-83-9: 08760
108368-38-3: 09452
108368-41-8: 09453
108369-20-6: 06643
108438-40-0: 16480
108473-88-7: 00865
108630-85-9: 12985
108673-30-9: 02517
108801-16-7: 23046

108836-86-8: 18074
108846-58-8: 24398
108878-18-8: 22053
108897-96-7: 18567
108952-44-9: 24467
108965-84-0: 10900
108965-86-2: 10901
108972-73-2: 00906
108974-19-2: 02546
108981-92-6: 06514
108981-94-8: 17065
108989-86-2: 22268
109000-42-2: 15434
109000-48-8: 15439
109000-61-5: 24436
109038-67-7: 00404
109044-03-3: 06069
109066-93-5: 06035
109067-05-2: 13248
109105-43-3: 19895
109157-31-5: 06152
109157-64-4: 06758
109220-85-1: 17341
109241-57-8: 05773
109252-38-2: 14224
109311-05-9: 00905
109402-94-0: 06034
109423-22-5: 06311
109444-32-8: 19891
109445-77-4: 24936
109466-84-4: 01577
109503-48-2: 06164
109511-03-7: 24472
109593-91-1: 22055
109643-29-0: 00481
109747-39-9: 12714
109799-62-4: 04353
109806-12-4: 23954
109812-63-7: 17032
109818-37-3: 19898
109843-15-4: 06068
109844-00-0: 19857
109847-57-6: 06033
109940-58-1: 10379
109963-37-3: 15299
109963-38-4: 15298
109967-13-7: 24471
109979-40-0: 12697
110028-87-0: 06734
110048-99-2: 09328
110058-38-3: 06574
110058-39-4: 06427
110081-55-5: 06032
110150-56-6: 17033
110194-18-8: 20008
110244-88-7: 12721
110252-31-8: 19956
110252-33-0: 19959
110273-47-7: 25342
110299-87-1: 14783
110318-09-7: 13956
110333-81-8: 06567
110333-82-9: 06566
110334-95-7: 02009
110335-26-7: 07247
110356-46-2: 16898
110356-50-8: 05950
110356-89-3: 06510
110376-81-3: 09061
110383-31-8: 14493
110393-19-6: 06225
110435-97-7: 23560
110436-44-7: 23559
110684-44-1: 25044
110704-59-1: 24536
110790-31-3: 11480
110795-27-2: 09052
110927-14-5: 24347
110967-33-4: 14970
110978-53-5: 14101
111002-96-1: 13541
111033-28-4: 24276
111325-76-9: 07158
111330-20-2: 11371
111412-63-6: 06028
111443-88-0: 09027
111474-76-1: 11222
111562-65-3: 23483
111618-88-3: 08124
111768-02-6: 18488
111768-04-8: 14281
111768-05-9: 14284
111768-08-2: 14283
111772-90-8: 25144
111790-01-3: 06802
111797-37-6: 12696
111823-35-9: 18233
111897-18-8: 17455
112005-21-7: 10359
112071-38-2: 25564
112112-90-0: 24968
112126-34-8: 09652
112298-82-5: 06586
112345-57-0: 00374
112456-46-9: 07281
112489-85-7: 13032
112534-43-7: 11370
112535-24-7: 12700
112553-57-8: 18937
112614-96-7: 25053
112901-80-1: 25582
113121-55-4: 14784
113222-74-5: 15360
113665-35-3: 05574
113666-64-1: 25304
113802-43-0: 11073
113845-11-7: 09472
113873-76-0: 25363
114133-84-5: 19958
114159-65-8: 12635
114165-21-8: 22066
114180-21-1: 07937
114191-96-7: 01329
114205-74-2: 24473
114223-30-2: 12629
114328-80-2: 19797
114383-24-3: 24470
114489-96-2: 21100
114882-91-6: 20902
114960-52-0: 13377
114998-58-2: 07689
115098-30-1: 09483
115102-47-1: 13136
115113-98-9: 13784
115117-81-2: 16934
115310-98-0: 23814
115374-35-1: 01341
115581-09-4: 23799
115759-71-2: 18180
115823-43-3: 05195
115989-20-3: 10022
116013-82-2: 05440
116295-56-8: 16883
116295-66-0: 18131
116295-76-2: 22191
116295-77-3: 22192
116295-78-4: 10882
116295-79-5: 01383
116295-80-8: 01386
116295-81-9: 01381
116295-93-3: 01382
116323-37-6: 22194
116401-69-5: 00750
116401-71-9: 12041
116530-16-6: 14809